Porous Plastics

Scrivener Publishing
100 Cummings Center, Suite 541J
Beverly, MA 01915-6106

Publishers at Scrivener
Martin Scrivener (martin@scrivenerpublishing.com)
Phillip Carmical (pcarmical@scrivenerpublishing.com)

Porous Plastics

Johannes Karl Fink
Montanuniversität Leoben, Austria

Scrivener Publishing

WILEY

This edition first published 2022 by John Wiley & Sons, Inc., 111 River Street, Hoboken, NJ 07030, USA and Scrivener Publishing LLC, 100 Cummings Center, Suite 541J, Beverly, MA 01915, USA

For more information about Scrivener publications please visit www.scrivenerpublishing.com.

Wiley Global Headquarters
111 River Street, Hoboken, NJ 07030, USA

For details of our global editorial offices, customer services, and more information about Wiley products visit us at www.wiley.com.

Library of Congress Cataloging-in-Publication Data

ISBN 978-1-119-89638-8

Cover image: Pixabay.com
Cover design by Russell Richardson

Set in size of 11pt and Minion Pro by Manila Typesetting Company, Makati, Philippines

Printed in the USA

10 9 8 7 6 5 4 3 2 1

Contents

Preface

Porous polymers are materials that are having pores in their design. Porous polymers are important for various fields of application, as described below. They are used with pores of different sized, i.e. from macropores to micropores.

This book focuses on the issues of porous polymers as well as low molecular compounds that can be introduced in porous polymers.

The book begins with a chapter about polymers that are used for porous materials. Here, among others, microporous polymer networks, hypercrosslinked polymers, and rigid ladder-type porous polymers are detailed. Related issues will also be detailed in the subsequent chapters. In the next chapter, the major synthesis methods for porous polymers are described.

Then, the properties and material testing methods, such as standards, are described in a chapter.

In the following chapters, special fields of applications of porous polymers are described in detail, such as:

Chapter 4: Medical uses,
Chapter 5: Thermal insulation,
Chapter 6: Membranes,
Chapter 7: Separation methods, and
Chapter 8: Other fields of use.

The text focuses on the literature of the past decade. Beyond education, this book will serve the needs of industry engineers and specialists who have only a passing knowledge of the plastics and composites industries but need to know more.

How to Use This Book

Utmost care has been taken to present reliable data. Because of the vast variety of material presented herein, however, the text cannot be complete in all aspects, and it is recommended that the reader study the original literature for more complete information.

The reader should be aware that mostly US patents have been cited where available, but not the corresponding equivalent patents in other countries. For this reason, the author cannot assume responsibility for the completeness, validity or consequences of the use of the material presented herein. Every attempt has been made to identify trademarks; however, there were some that the author was unable to locate.

Index

There are four indices: an index of trademarks, an index of acronyms, an index of chemicals, and a general index.

In the index of chemicals, compounds that occur extensively, e.g., "acetone," are not included at every occurrence, but rather when they appear in an important context.

Acknowledgements

I am indebted to our university librarians, Dr. Christian Hasenhüttl, Margit Keshmiri, Friedrich Scheer, Christian Slamenik, Renate Tschabuschnig, and Elisabeth Groß for support in literature acquisition. I also want to express my gratitude to all the scientists who have carefully published their results concerning the topics dealt with herein. This book could not have been otherwise compiled.

Last, but not least, I want to thank the publisher, Martin Scrivener, for his interest in publishing this book. In addition, my thanks go to Jean Markovic, who made the final copyedit with utmost care.

Johannes Fink
Leoben, May 2022

1

Materials

Porous materials are typically categorized into three classes that have different pore sizes (1):

1. Macroporous with pore diameter larger than 50 *nm*,
2. Mesoporous (pore diameter between 2 *nm* and 50 *nm*), and
3. Microporous materials (pore diameter smaller than 2 *nm*).

While conventional polymer networks undergo pore collapse upon solvent removal as polymer strands can adopt many conformations in order to pack space efficiently, recent research efforts have popularized several classes of polymer networks that possess permanent porosity based on the use of rigid components.

1.1 Styropor

Otis Ray McIntire (1918-1996), a chemical engineer at Dow Chemical, rediscovered a process first patented by Swedish inventor Carl Munters (2).

According to the Science History Institute, "Dow bought the rights to the Munters method and began producing a lightweight, water-resistant, and buoyant material that seemed perfectly suited for building docks and watercraft and for insulating homes, offices, and chicken sheds (3). In 1944, Styrofoam was patented.

Before 1949, chemical engineer Fritz Stastny (1908-1985) developed pre-expanded poly(styrene) beads by incorporating aliphatic hydrocarbons such as pentane. These beads are the raw material for molding parts or extruding sheets. BASF and Stastny applied for a patent that was issued in 1949. The molding process was demonstrated at the Kunststoff Messe in Düsseldorf in 1952. These products were named Styropor (3).

The crystal structure of isotactic poly(styrene) was reported by Giulio Natta (4). In 1954, the Koppers company in Pittsburgh, Pennsylvania, developed expanded poly(styrene) foam under the trade name Dylite (5).

1.2 Porous Coordination Polymers

The design, analysis and applications of coordination polymers have been descried in a monograph (6).

A coordination polymer is an inorganic or organometallic polymer structure containing metal cation centers linked by ligands. More formally, a coordination polymer is a coordination compound with repeating coordination entities extending in 1, 2, or 3 dimensions (7, 8)

Examples of coordination polymers are lanthanoid coordination polymers, organometallic networks, and organic-inorganic hybrids (6).

1.2.1 *Multifunctional Pillared-Layer Material*

A multifunctional pillared-layer porous coordination polymer, has been constructed based on a flexible viologen derivative, 1,1′-bis(4-carboxybenzyl)-4,4′-bipyridinium dichloride, and an oxalate co-ligand. 1,1′-Bis(4-carboxybenzyl)-4,4′-bipyridinium dichloride is shown in Figure 1.1.

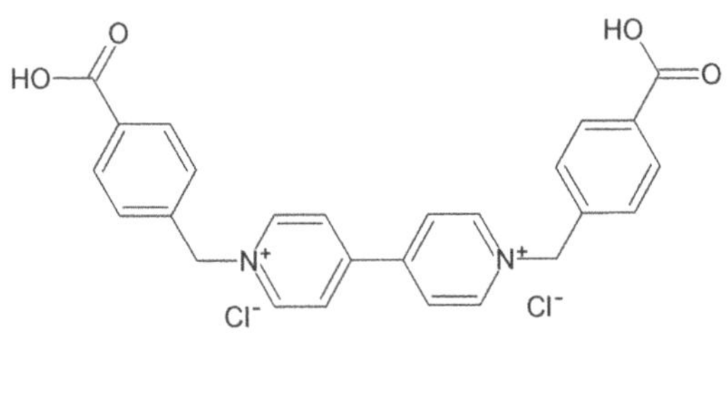

Figure 1.1 1,1′-Bis(4-carboxybenzyl)-4,4′-bipyridinium dichloride.

Single-crystal X-ray analysis showed that the compound possesses multichannels with dimensions of about 6.1×6.6 Å along the [110] and [-110] directions and 4.2×7.6 Å along [100], and a void space of about 41.4%.

Hydrogen adsorption measurements at 77 *K* and 1 *atm* indicated that the compound exhibits a hydrogen uptake of 0.71%. Owing to the incorporation of bipyridinium acceptor units, the compound can selectively accommodate aromatic donors into its nano-sized pores based on charge-transfer interactions in an elastic way, and afford a specific color to different guests.

Furthermore, the effect of perturbation exerted by the guest molecules on its magnetic properties has been investigated. The results indicated that the donor inclusion has little effect on its antiferromagnetic behavior, whereas dehydration of the compound decreases the strength of the magnetic exchange couplings and results in a change of the antiferromagnetic transition temperature from 14.7 *K* to 9.8 *K* (9).

1.2.2 Porous Coordination Polymer-Ionic Liquid Composite

A porous coordination polymer-ionic liquid composite has been described that includes an insulating structure composed of a porous coordination polymer, and an ionic liquid retained inside pores of the porous coordination polymer. The porous coordination polymer preferably has a main chain containing a typical metal element (10).

It has been proposed to apply an ionic liquid owing to high ionic conductivity thereof to an electrochemical device as an electrolyte for a battery or an electrical double-layer capacitor. The ionic liquid has extremely high flame retardance, and hence when used as the electrolyte for the electrochemical device, there is no need for a combustible organic solvent, thus ensuring the electrochemical device with high safety (10).

A schematic diagram that shows that an ionic liquid is filled with particles of the porous coordination polymer to form particles after filling is shown in Figure 1.2.

Here a a plurality of particles 111 composed of a porous coordination polymer are filled with a ionic liquid 12. The composite 131 obtained by a molding process. The structure 11, which is used as an electrolyte for a battery or an electrical double-layer capacitor,

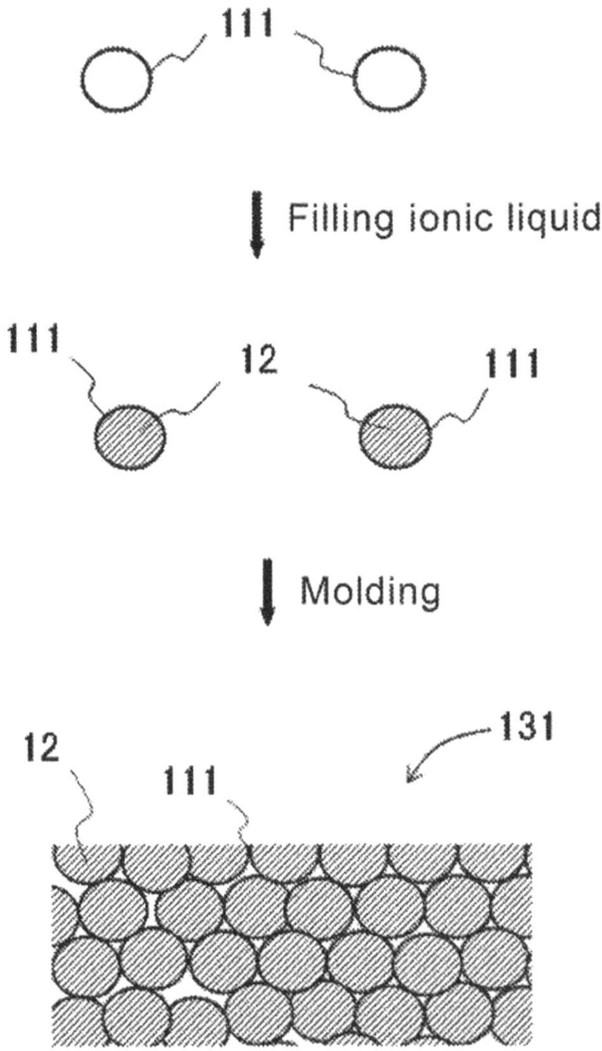

Figure 1.2 Synthesis of a porous coordination polymer-ionic liquid composite (10).

has a dense structure, thus making it easier for ion conduction pathways between the particles to be connected to each other. Hence, the composite 131 is a satisfactory ion conductor. In the case of using the structure 11 obtained by subjecting a plurality of the particles 111 composed of the porous coordination polymer to compression molding, a plurality of voids are respectively formed between the particles 111 of the porous coordination polymer (10).

Examples of the porous coordination polymer are collected in Table 1.1.

Table 1.1 Porous coordination polymers (10).

Compound	Shortcut
Zn(MeIM)2	ZIF-8
Al(OH)[BDC]	MIL-53(A1)
Cr(OH)[BDC]	MIL-53(Cr)
Fe(OH) [BDC]	MIL-53(Fe)
Zn2 (DOBDC)	MOF-74(Zn)
Mg2 (DOBDC)	MOF-74(Mg)
Al(OH)(1,4-NDC)	
Cr3F(H2O)2O(BDC)3	MIL-101(Cr)
Al8(OH)12(OH)3(H2O)3 [BTC]3	MIL-110(Al)
Abbreviation	Compound
HMeIM	2-Methylimidazole
H2BDC	1,4-Benzenedicarboxylic acid
H4DOBDC	2,5-Dihydroxyterephthalic acid
H2NDC	1,4-Naphthalenedicarboxylic acid
H3BTC	1,3,5-Benzenetricarboxylic acid
H2BPDC	4,4′-Biphenyldicarboxylic acid
H2TPDC	4,4″-*p*-Terphenyldicarboxylic acid

The acids in Table 1.1 are shown in Figure 1.3.

Examples of hard acids, hard bases, soft acids, soft bases, intermediate acids, and intermediate bases are described in a monograph (11).

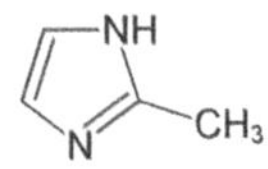

2-Methylimidazole

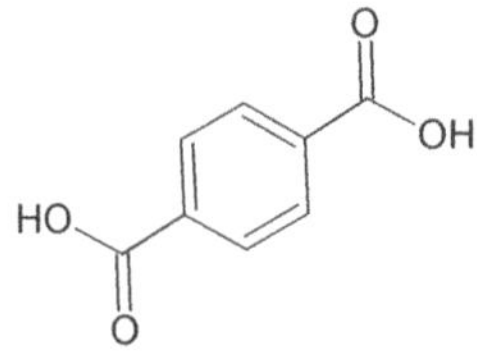

1,4-Benzenedicarboxylic acid

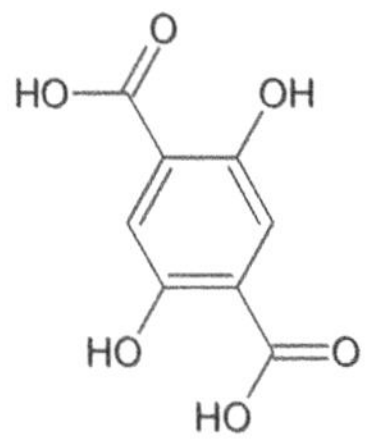

2,5-Dihydroxyterephthalic acid

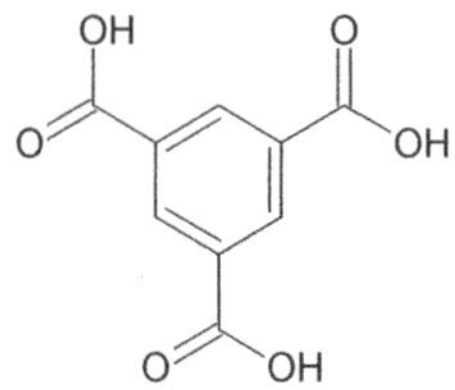

1,4-Naphthalenedicarboxylic acid

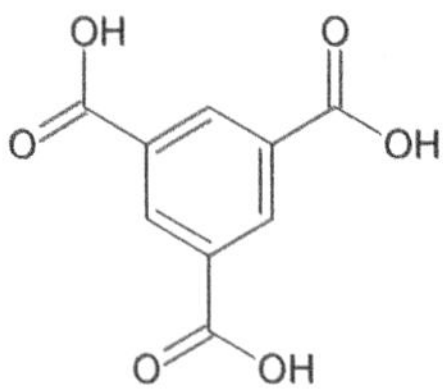

1,3,5-Benzenetricarboxylic acid

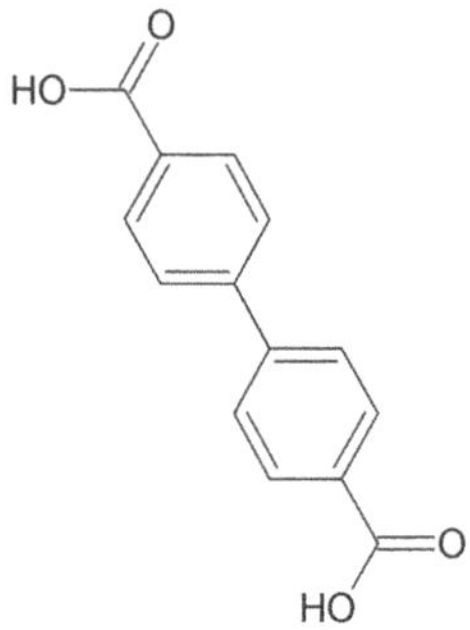

4,4′-Biphenyldicarboxylic acid

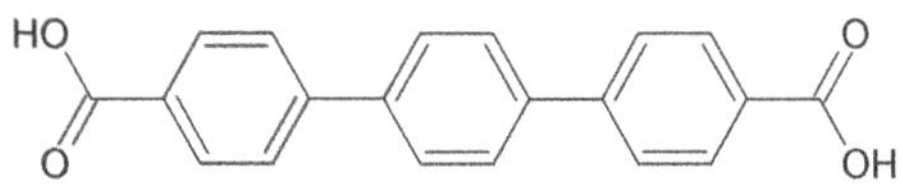

4,4″-*p*-Terphenyldicarboxylic acid

Figure 1.3 Acids in Table 1.1.

1.3 Networks

1.3.1 Microporous Polymer Networks

Microporous materials are defined as materials containing interconnected pores of less than 2 *nm* in diameter (12)

Due to their large surface area, many conventional microporous materials, such as zeolites and activated carbons, are widely used as catalysts, sorbents, and separation membranes. Recently, the field has evolved rapidly with the development of several novel types of microporous polymer networks. These materials not only benefit fundamental research by introducing modular approaches to accessing numerous sophisticated structures, but also provide new opportunities for various emerging applications (1).

The central design principle for introducing permanent microporosity into polymer networks involves the use of rigid building blocks. Such rigidity precludes the network strands from behaving effectively as entropic molecular springs and prevents the collapse of microporous structures upon solvent removal; consequently, the mechanical properties of these materials are stiff yet brittle.

Furthermore, the rigidity of the monomers prevents small loop formation and allows for establishing long-range order in the presence of self-error-correcting mechanisms, e.g., a reversible bond formation (1).

So, microporous polymer networks can be either amorphous or crystalline. Aside from the general use of very rigid components, the basic concepts of microporous polymer network synthesis are similar to those discussed above for either covalent or physical polymer networks.

1.3.2 Amorphous Microporous Polymer Networks

Amorphous microporous polymer networks of different types have been denoted by various names, such as:

1. Polymers with intrinsic microporosity (PIMs) (13),
2. Porous organic polymers (POPs) (14),
3. Conjugated microporous polymers (CMPs) (15), and
4. Hyper-crosslinked polymers (16).

It has been suggested to divide these materials into two categories, based on whether or not the strands are covalently crosslinked (1).

1.3.2.1 *Conjugated Microporous Polymers*

Conjugated microporous polymers (CMPs) are a class of organic porous polymers that combine *p*-conjugated skeletons with permanent nanopores, in sharp contrast to other porous materials that are not *p*-conjugated and with conventional conjugated polymers that are nonporous. As an emerging material platform, CMPs offer a high flexibility for the molecular design of conjugated skeletons and nanopores.

A lot of chemical reactions, building blocks and synthetic methods have been developed and a broad variety of CMPs with different structures and specific properties have been synthesized, driving the rapid growth of the field. CMPs are unique in that they allow the complementary utilization of *p*-conjugated skeletons and nanopores for functional exploration; they have shown great potential for challenging energy and environmental issues, as exemplified by their excellent performance in gas adsorption, heterogeneous catalysis, light emitting, light harvesting and electrical energy storage. This review describes the molecular design principles of CMPs, advancements in synthetic and structural studies and the frontiers of functional exploration and potential applications.

Building blocks with different geometries are listed in Table 1.2.

The structures of some building blocks with different geometries, sizes and reactive groups for the synthesis of CMPs are shown in Figure 1.4.

To construct a conjugated skeleton, the synthetic reaction must covalently link the building blocks with a *p*-conjugated bond.

The chemical reactions utilized for the preparation of linear conjugated polymers can also be employed for the synthesis of CMPs. The special reactions are listed in Table 1.3 and are shown in Figure 1.5.

Because building blocks can have different geometries, reactive groups, and P systems, this structural diversity significantly enhances the flexibility of the design of both skeletons and pores.

Table 1.2 Building blocks with different geometries (15).

C2 Compounds
1,4-Dibromobenzene
1,2-Dibromobenzene
1,4-Dibromo-2-methyl-benzene
1,4-Dibromo-2-trifluoromethyl-benzene
2,5-Dibromofluorobenzene
1,4-Dibromo-2-nitro-benzene
4,7-Dibromo-2,1,3-benzothiadiazole
2,5-Dibromo pyridine
2,5-Dibromobenzoic acid
2,5-Dibromonitrobenzene
1,4-Dibromo-2,5-difluorobenzene
1,4-Dibromo-2,5-dimethylbenzene
1,4-Dibromo-2,5-dihydroxybenzene
1,4-Dibromo-2,5-dimethoxybenzene
1,3-Dibromobenzene
2,6-Dibromophenol
2,4-Dibromoaniline
2,6-Dibromoaniline
3,5-Dibromo pyridine
3,5-Dibromo-*N*,*N*-dimethyl-4-pyridinamine
2,7-Dibrom-9*H*-carbazol
1,6-Dibromo-2-naphthol
2,6-Dibromonaphthalene
9,10-Dibromoanthracene
4,4′-Dibromobiphenyl
5-Bromo-2-(4-bromophenyl)pyridine
5,5′-Dibromo-2,2′-bipyridine
4,4′-Dibromooctafluorobiphenyl
1,4-Diiodobenzene
4,4′-Diiodobiphenyl
1,4-Diaminobenzene
4,4′-Diaminobiphenyl 1,2-Dicyanobenzene
1,3-Dicyanobenzene
1,4-Dicyanobenzene
4,4′-Diacyanobiphenyl
1,1′:4′,1″-Terphenyl-4,4″-dicarbonitrile

Table 1.2 (cont) Building blocks with different geometries (15).

C3 Compounds
1,3,5-Tribromobenzene
1,3,5-Tris(3-bromophenyl)benzene
1,3,5-Tris(4-bromophenyl)benzene
2,4,6-Tris(p-bromophenyl)-s-triazine
1,3,5-Tris(4-bromophenylethynyl)benzene
1,3,5-Tris(4-aminophenyl)benzene
C4 Compounds
1,2,4,5-Tetrabromobenzene
1,3,6,8-Tetrabromopyrene
1,3,6,8-Pyrenetetracarbaldehyde
1,2,4,5-Tetraaminobenzene
2,2′,7,7′-Tetrabromo-9,9′-spirobifluorene
2,2′,7,7′-Tetraamino-9,9′-spirobifluorene
1,1′,1″,1‴-(1,1,2,2-Ethenetetrayl)tetrakis (4-bromobenzene)
C6 Compounds
1,2,3,4,5,6-Hexabromobenzene
1,2,3,4,5,6-Hexakis(4-bromophenyl)benzene

Table 1.3 Reactions for preparation of conjugated polymers.

Reaction name	References
Suzuki cross coupling reaction	(17, 18)
Yamamoto reaction	(19, 20)
Sonogashira-Hagihara reaction	(21, 22)
Oxidative coupling reaction	(23–25)
Schiff base reaction	(26–28)
Friedel-Crafts reaction	(29)
Phenazine ring fusion reaction	(30)
Cyclotrimerization	(31, 32)

1,4-Dibromobenzene

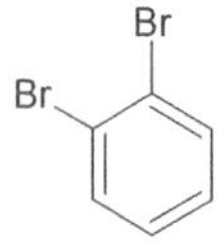

1,2-Dibromobenzene

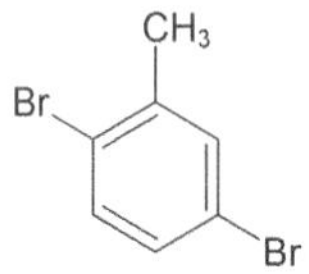

1,4-Dibromo-2-trifluoromethyl-benzene

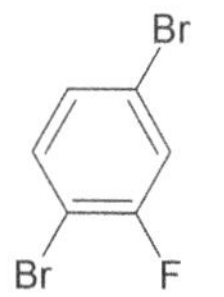

2,5-Dibromofluorobenzene

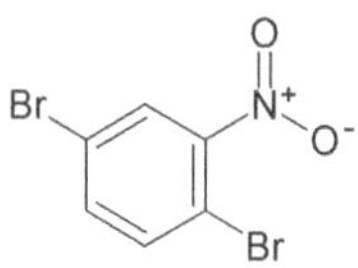

1,4-Dibromo-2-nitro-benzene

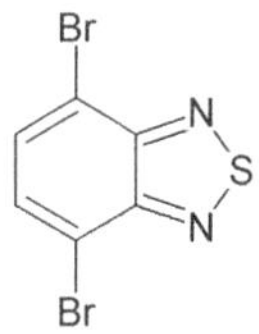

4,7-Dibromo-2,1,3-benzothiadiazole

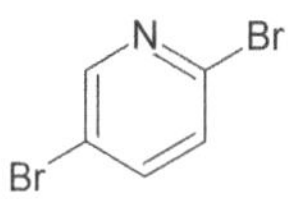

2,5-Dibromo pyridine

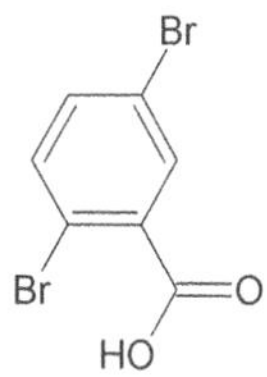

2,5-Dibromobenzoic acid

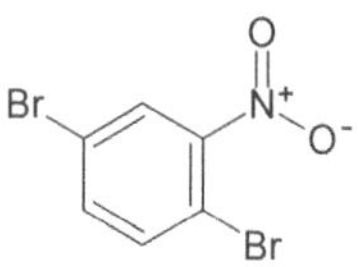

2,5-Dibromonitrobenzene

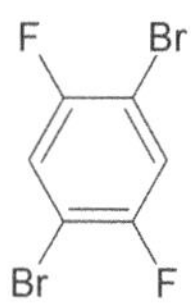

1,4-Dibromo-2,5-difluorobenzene

Figure 1.4 Building blocks (15).

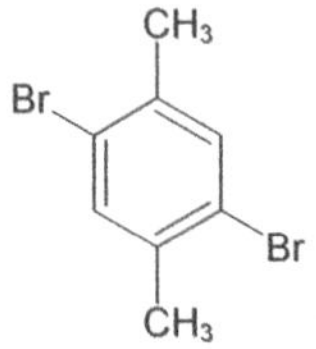

1,4-Dibromo-2,5-dimethylbenzene

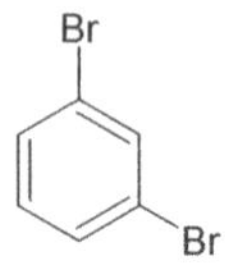

1,3-Dibromobenzene

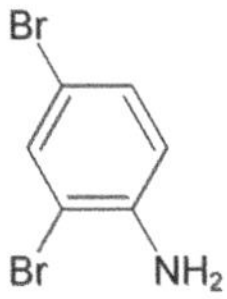

2,4-Dibromoaniline

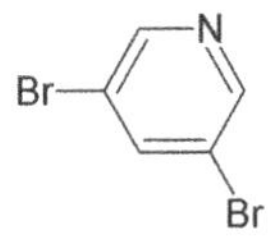

3,5-Dibromo pyridine

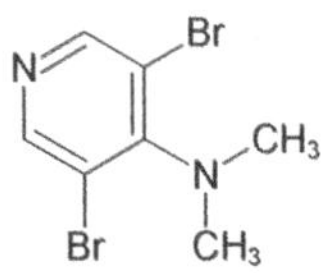

3,5-Dibromo-*N*,*N*-dimethyl-4-pyridinamine

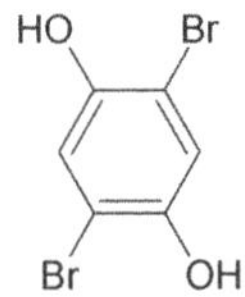

1,4-Dibromo-2,5-dihydroxy benzene

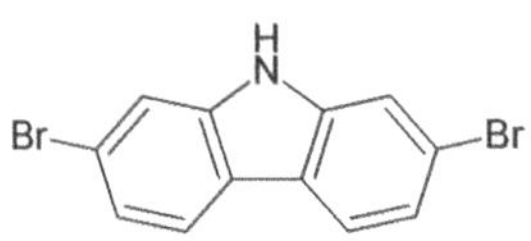

2,7-Dibromo-9*H*-carbazole

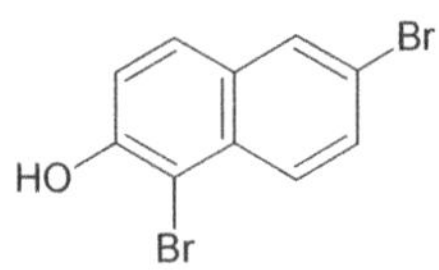

1,6-Dibromo-2-naphthol

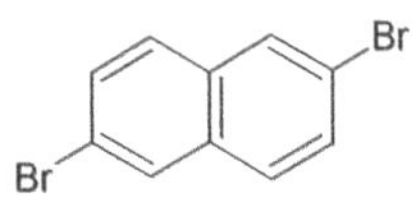

2,6-Dibromonaphthalene

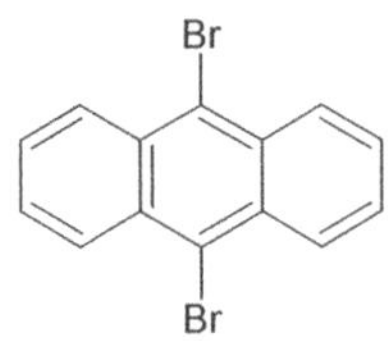

9,10-Dibromoanthracene

Figure 1.4 (cont) Building blocks (15).

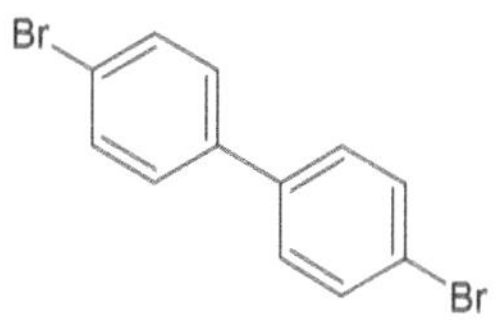

4,4′-Dibromobiphenyl

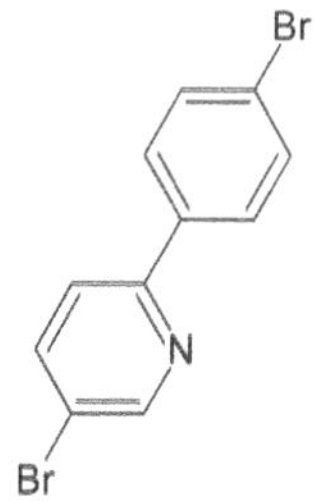

5-Bromo-2-(4-bromophenyl)pyridine

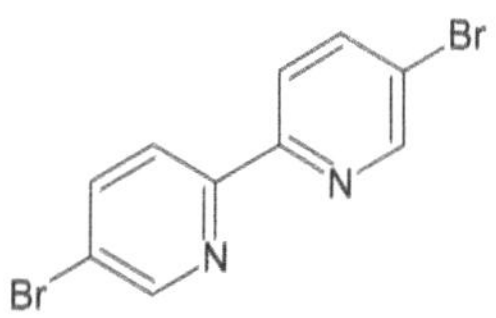

5,5′-Dibromo-2,2′-bipyridine

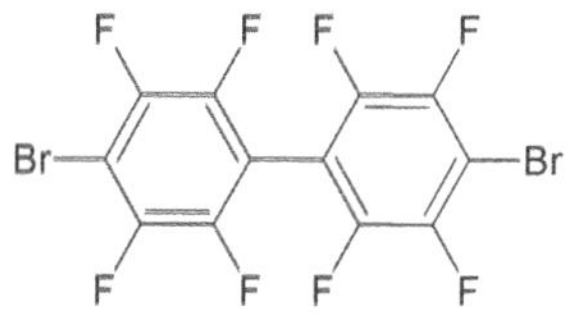

4,4′-Dibromooctafluorobiphenyl

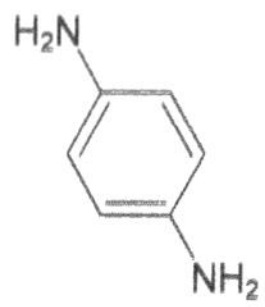

1,4-Diaminobenzene

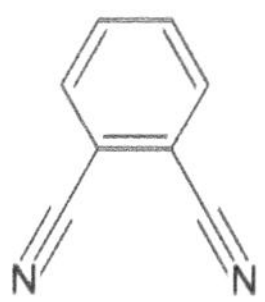

1,2-Dicyanobenzene

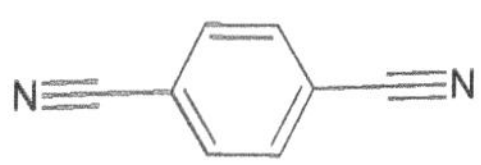

1,4-Dicyanobenzene

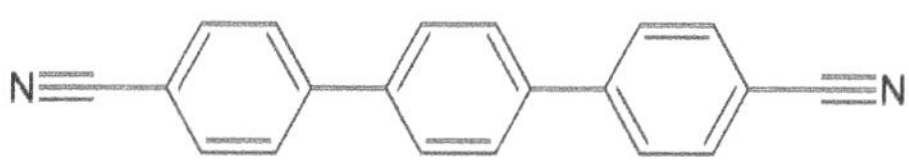

1,1′:4′,1″-Terphenyl-4,4″-dicarbonitrile

Figure 1.4 (cont) Building blocks (15).

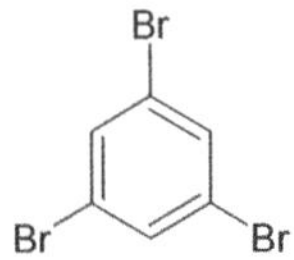

1,3,5-Tribromobenzene

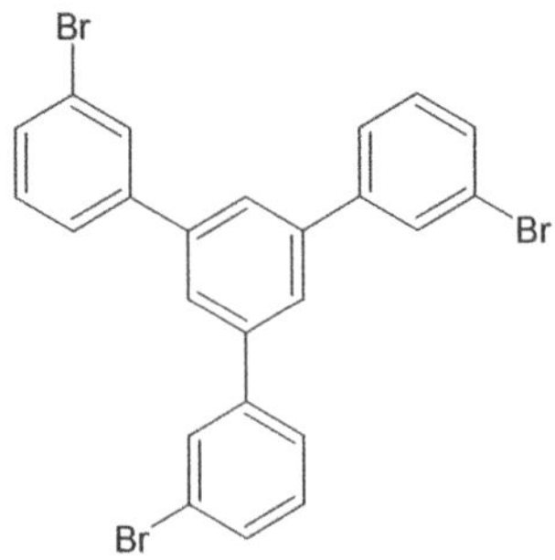

1,3,5-Tris(3-bromophenyl)benzene

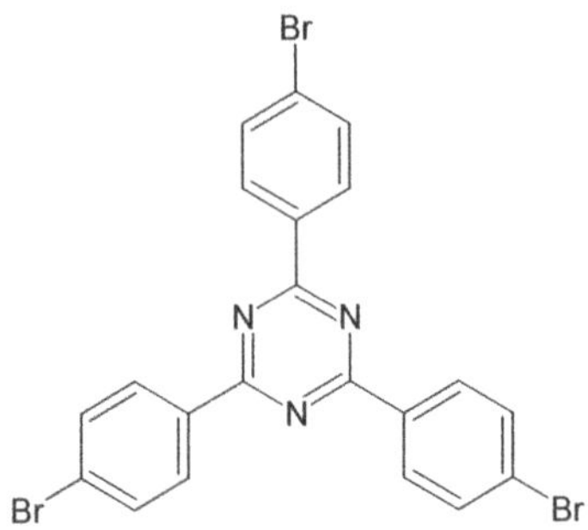

2,4,6-Tris(*p*-bromophenyl)-*s*-triazine

Figure 1.4 (cont) Building blocks (15).

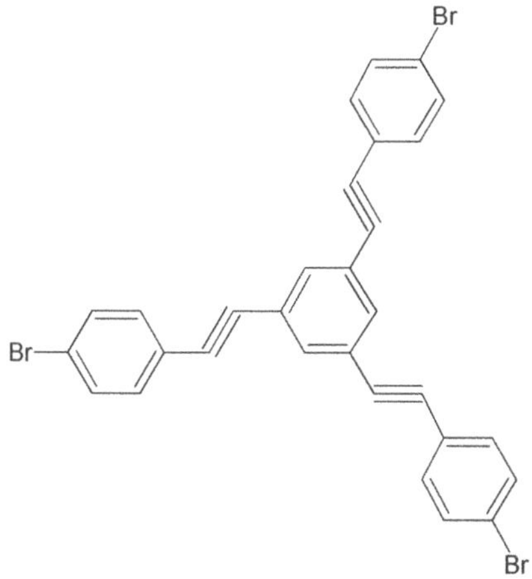

1,3,5-Tris(4-bromophenylethynyl)benzene

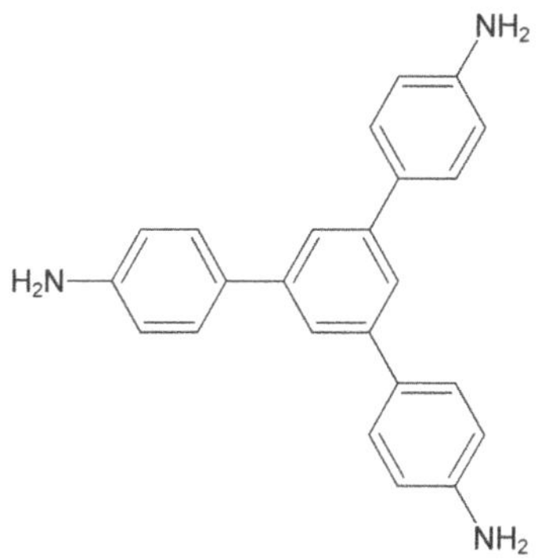

1,3,5-Tris(4-aminophenyl)benzene

Figure 1.4 (cont) Building blocks (15).

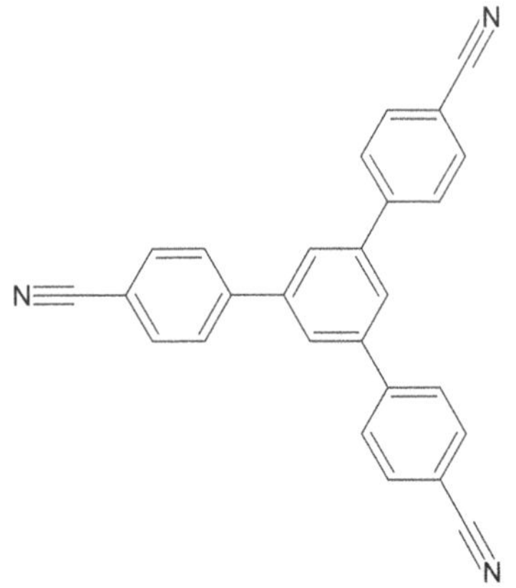

1,3,5-Tris(4-cyanophenyl)benzene

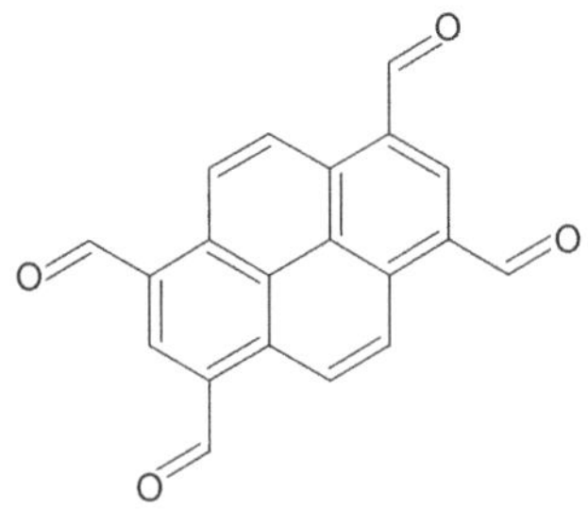

1,3,6,8-Pyrenetetracarbaldehyde

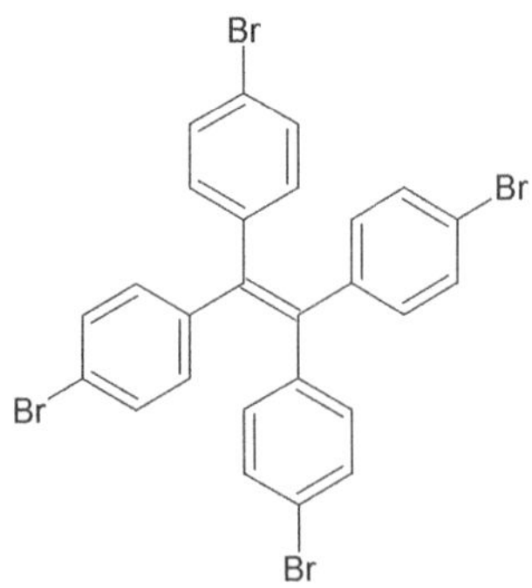

1,1′,1″,1‴-(1,1,2,2-Ethenetetrayl)tetrakis (4-bromobenzene)

Figure 1.4 (cont) Building blocks (15).

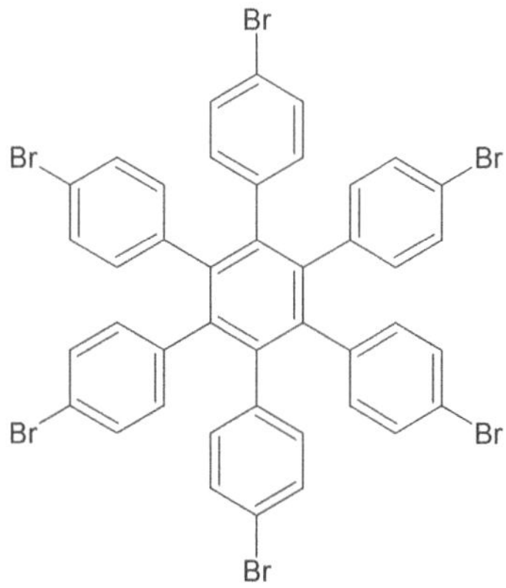

1,2,3,4,5,6-Hexakis(4-bromophenyl)benzene

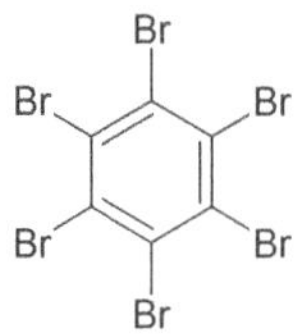

1,2,3,4,5,6-Hexabromobenzene

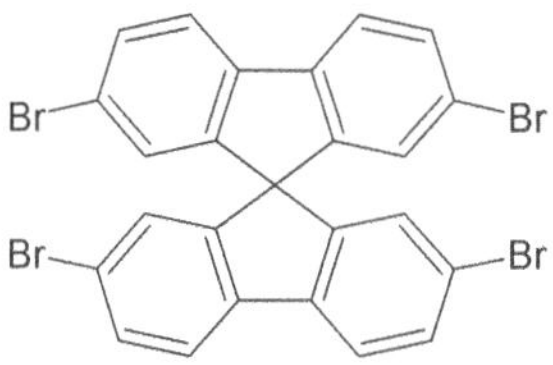

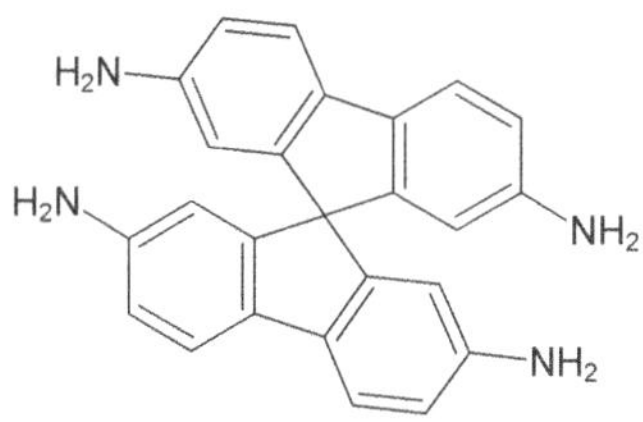

2,2′,7,7′-Tetrabromo-
9,9′-spirobifluorene

2,2′,7,7′-Tetraamino-
9,9′-spirobifluorene

Figure 1.4 (cont) Building blocks (15).

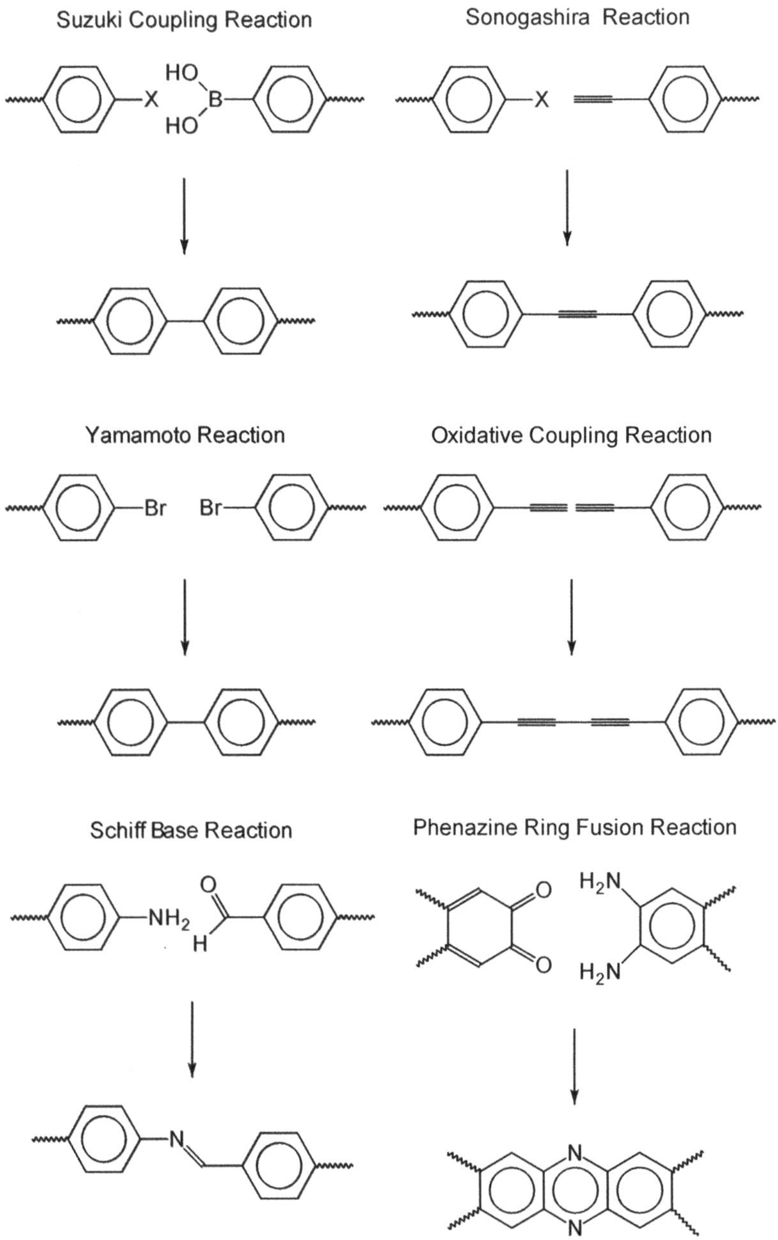

Figure 1.5 Reactions for the synthesis (15).

1.3.2.2 *Hyper-Crosslinked Polymers*

Hyper-crosslinked polymers (HCPs) are a series of permanent microporous polymer materials (16).

They were initially reported by Davankov ad Tsyurupa (33), and have received an increasing level of research interest.

In recent years, HCPs have experienced rapid growth due to their remarkable advantages such as diverse synthetic methods, easy functionalization, high surface area, low cost reagents and mild operating conditions (16).

The judicious selection of monomers, appropriate length crosslinkers and optimized reaction conditions yielded a well-developed polymer framework with an adjusted porous topology. Post fabrication of the as developed network facilitates the incorporation of various chemical functionalities that may lead to interesting properties and enhance the selection toward a specific application. To date, numerous HCPs have been prepared by post-crosslinking poly(styrene)-based precursors, one-step polycondensation or external crosslinking strategies.

The advent of these methodologies has prompted researchers to construct well-defined porous polymer networks with customized micromorphology and functionalities. In this review, we describe not only the basic synthetic principles and strategies of HCPs, but also the advancements in the structural and morphological study as well as the frontiers of potential applications in energy and environmental fields such as gas storage, carbon capture, removal of pollutants, molecular separation, catalysis, drug delivery, sensing, and other fields (16).

1.4 Rigid Ladder-Type Porous Polymers

A porous ladder polymer network was designed and synthesized as the model material via cross coupling polymerization and subsequent ring-closing olefin metathesis, followed by a characterization with solid-state nuclear magnetic resonance spectroscopy (34).

It could be shown that a rigid ladder-type backbone is more entropically favorable for gas adsorption and leads to a high gas uptake per unit surface area (34). The material exhibited a remarkable methane uptake per unit surface area, which outperformed those of most reported porous organic materials. Variable-temperature thermodynamic adsorption measurements corroborated the significantly less negative entropy penalty during high-pressure gas adsorption, compared to its non-ladder-type counterpart.

So, the here described method provides an orthogonal strategy for multiplying volumetric methane uptake capacity of porous materials. In addition, the entropic approach offers the opportunity to increase deliverable gas upon pressure change while mitigating the performance decline in high-temperature applications (34).

1.5 Photocatalysts

Conjugated microporous polymers (CMPs) are an emerging class of promising photocatalysts because of their large specific surface areas and adjustable optical band gaps. To avoid metal contamination, metal-free synthetic procedures for making CMPs for photocatalytic water splitting are highly desired.

Triazine-based conjugated microporous polymers were synthesized through a simple, efficient, metal-free catalyzed approach. The compound is shown in Figure 1.6.

Through linking donor-acceptor type pyrazole-benzothiadiazole-pyrazole light-absorbing units by triazine units, the polymer has an ideal optical band gap of about 2.3 eV and exhibits a high hydrogen evolution rate of 50 $\mu mol\, h^{-1}$ under visible light illumination ($\lambda \geq 420\ nm$) and apparent quantum efficiency as high as 3.58% at 420±20 nm. However, by replacing electron-neutral benzene instead of the electron-withdrawing benzothiadiazole, c.f., Figure 1.7, such a polymer shows an obvious hypochromatic shift in the absorption and large optical band gap of about 2.9 eV, as well as poor photocatalytic property (35).

1.5.1 Compounds for Photocatalytic Aerobic Oxidation

Heptazine-based porous organic polymers (POPs), as new potential photocatalytic materials, have spurred extensive research interest (36).

Four heptazine-based POPs were prepared by using cyameluric chloride as precursor, piperazine (POP-1), *p*-phenylenediamine

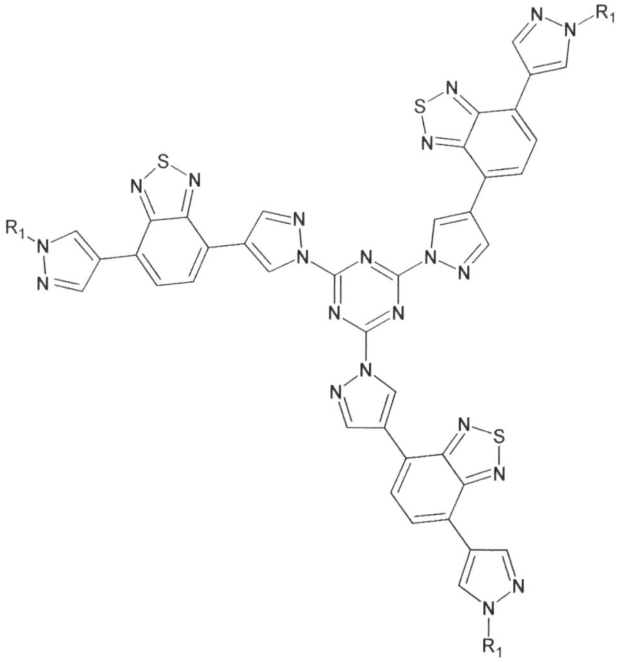

Figure 1.6 Pyrazole-benzothiadiazole-pyrazole polymer (35).

(POP-2), biphenyl diamine (POP-3) and 4,4″-diamino-*p*-terphenyl (POP-4) as basic blocks. Compounds for photocatalytic materials are shown in Figure 1.8.

The effects of the composition and structure of POPs on their spectrum, specific surface area, interface contact in acetonitrile, electronic structure, carrier separation and photocatalytic aerobic oxidation of amine through experiments and density functional theory calculations were systematically studied.

The results show that with the increase of the benzene rings and electron density of diamine linker, the electron-donating ability of these heptazine-based POPs is significantly enhanced, which reduces the band gap of the materials, improves their electronic donor-acceptor interaction with heptazine and promotes the transfer and separation of photogenerated carriers, thus endowing them with significantly higher photocatalytic performance (36).

Notably, POP-4 shows the best photocatalytic aerobic oxidation activity of benzylamine, which is significantly higher than that of g–C3N4. The conversion and selectivity of benzylamine to corre-

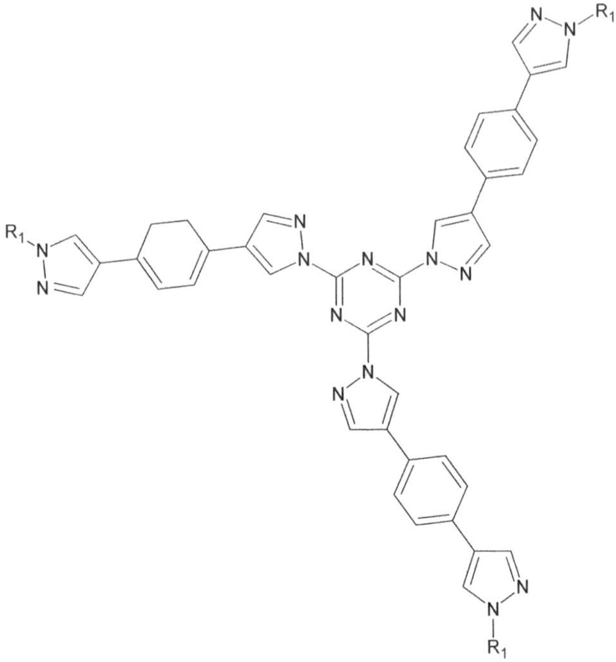

Figure 1.7 Pyrazole-benzene-pyrazole polymer (35).

sponding imine are both >99% within 4 *h* under 10 *W* 420 *nm* light-emitting diode light irradiation (36).

1.5.2 *Floating Photocatalysts*

A floating photocatalysis material has a big advantage over a powder- suspension one for the deep purification of pollutants in colored wastewater. So, the design and synthesis of floating photocatalysts are of great importance (37).

The preparation of metal-free triazine-based porous organic polymer has been presented. Here, a visible-light photocatalyst based on biomass waste poplar catkins is used to construct a reusable floating photocatalysis system via a soft synthesis. A trigonal-symmetrical melamine is coupled with triformylphloroglucinol as monomer to obtain a photoactive conjugated material loaded on soft and light poplar catkin microfibers. Triformylphloroglucinol is shown in Figure 1.9.

When this microfiber material is floated on the polluted water, it

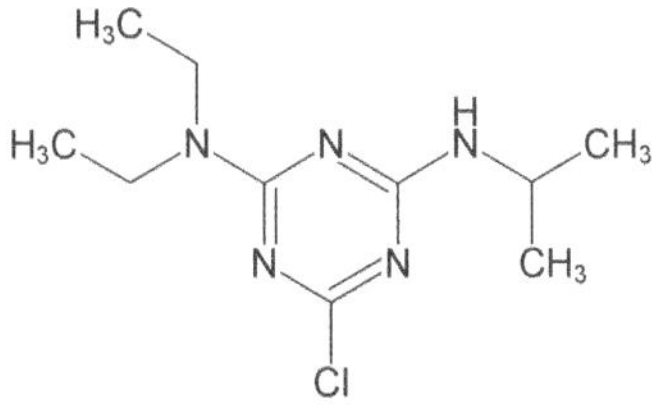

Heptazine

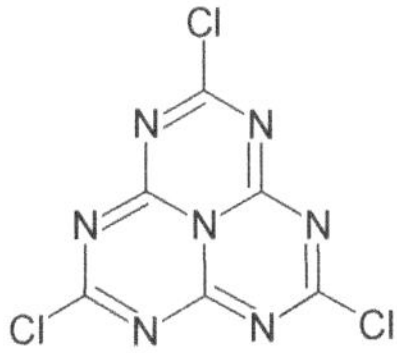

Cyameluric chloride

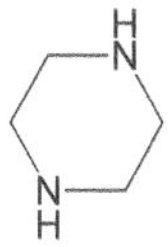

Piperazine

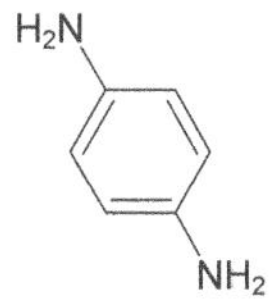

p-Phenylenediamine

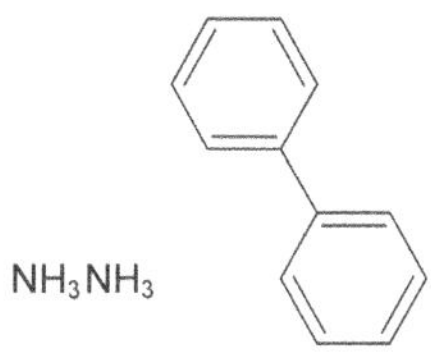

Biphenyl diamine

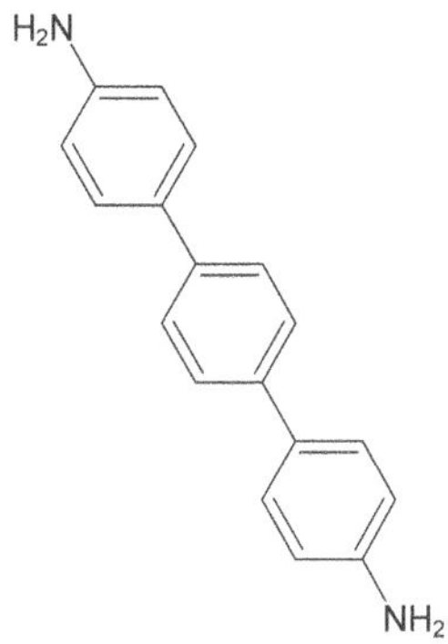

4,4″-Diamino-*p*-terphenyl

Figure 1.8 Compounds for photocatalytic materials.

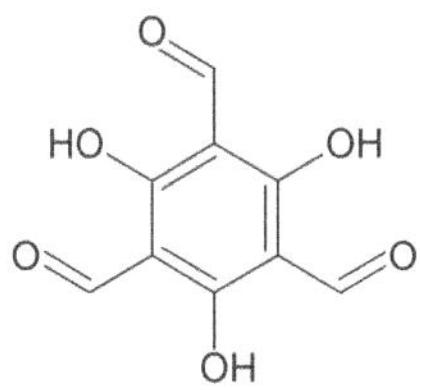

Figure 1.9 Triformylphloroglucinol.

can convert 100% Cr(VI) 80 *min* into Cr(III) species, which are then partly synchronously immobilized (56.9%) to lessen the rerelease of Cr element, or degrade 100% methylene blue after 100 *min* of visible-light irradiation.

Moreover, the film-like flexible self-supporting material also exhibits good catalytic activities and an excellent reusability for multipurpose water purification under real natural conditions.

So, the assembly of green floating photocatalysis systems allows a versatile solid surface activation for establishing a more energy efficient and robust catalysis process under visible light (37).

1.5.3 Photocatalysts with Side Chains

Conjugated polymers have been investigated as photocatalysts for hydrogen production. An efficient design has been demonstrated to enhance the hydrogen evolution rate of conjugated polymers through the modification of surface chemistry by introducing the hydrophilic adenine group onto the side chain (38).

An adenine group with plentiful nitrogen atoms can form multiple hydrogen bonds with water molecules, which improve the interactions between the resulting polymer surface and water molecules, leading to an improved hydrophilicity and dispersity of the polymer photocatalyst in photocatalytic reaction solution.

A density functional theory calculation indicated that the introduction of adenine groups also leads to the enhanced separation of the electrostatic potential on the surface of polymer photocatalyst, which is favorable for the photocatalytic hydrogen evolution reaction. Therefore, a high hydrogen evolution rate of 36.43 $mmol\,h^{-1}g^{-1}$ could be achieved by the adenine-functionalized polymer PF6A-SF, c.f. Figure 1.10, without using a Pt cocatalyst, which is almost 42 times higher than that of the alkyl-functionalized polymer (38).

So, a rational design of side-chain engineering is an effective design for organic photocatalysts with a high photocatalytic activity (38).

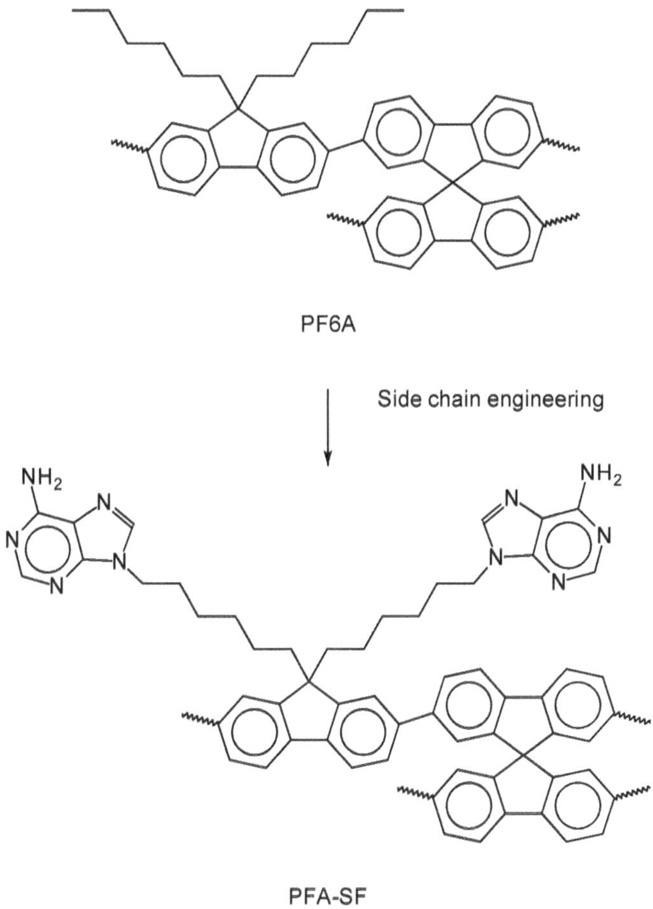

Figure 1.10 Side chain engineering.

References

1. Y. Gu, J. Zhao, and J.A. Johnson, *Angewandte Chemie International Edition*, Vol. 59, p. 5022, 2020.
2. M.O. Ray, Manufacture of cellular thermoplastic products, US Patent 2 450 436, assigned to Dow Chemical Co., October 05, 1948.
3. Wikipedia contributors, Polystyrene — Wikipedia, the free encyclopedia, 2021. [Online; accessed 29-July-2021].
4. G. Natta, P. Corradini, and I.W. Bassi, *Nuovo Cimento*, Vol. 15, p. 52, 1960.
5. T.H. Ferrigno, *Rigid Plastics Foams*, p. 207. Reinhold Pub. Corp., New York, 2nd edition, 1967.
6. S.R. Batten, S.M. Neville, and D.R. Turner, *Coordination Polymers: Design, Analysis And Application*, Royal Society of Chemistry, 2008.
7. Wikipedia contributors, Coordination polymer — Wikipedia, the free encyclopedia, 2021. [Online; accessed 21-January-2022].
8. S.R. Batten, N.R. Champness, X.-M. Chen, J. Garcia-Martinez, S. Kitagawa, L. Öhrström, M. O'Keeffe, M.P. Suh, and J. Reedijk, *Pure and Applied Chemistry*, Vol. 85, p. 1715, 2013.
9. Q.-X. Yao, L. Pan, X.-H. Jin, J. Li, Z.-F. Ju, and J. Zhang, *Chemistry — A European Journal*, Vol. 15, p. 11890, 2009.
10. H. Kitagawa, T. Yamada, and K. Fujie, Porous coordination polymer-ionic liquid composite, US Patent 10 535 474, assigned to Kyoto University, January 14, 2020.
11. G. Thomas, *Medicinal Chemistry: An Introduction*, John Wiley & Sons, 2nd edition, 2011.
12. A.G. Slater and A.I. Cooper, *Science*, Vol. 348, p. aaa8075, May 2015.
13. N.B. McKeown and P.M. Budd, *Chem. Soc. Rev.*, Vol. 35, p. 675, 2006.
14. P. Kaur, J.T. Hupp, and S.T. Nguyen, *ACS Catalysis*, Vol. 1, p. 819, 2011.
15. Y. Xu, S. Jin, H. Xu, A. Nagai, and D. Jiang, *Chem. Soc. Rev.*, Vol. 42, p. 8012, 2013.
16. L. Tan and B. Tan, *Chem. Soc. Rev.*, Vol. 46, p. 3322, 2017.
17. L. Chen, Y. Honsho, S. Seki, and D. Jiang, *Journal of the American Chemical Society*, Vol. 132, p. 6742, 2010.
18. G. Cheng, T. Hasell, A. Trewin, D.J. Adams, and A.I. Cooper, *Angewandte Chemie International Edition*, Vol. 51, p. 12727, 2012.
19. J.-X. Jiang, A. Trewin, D.J. Adams, and A.I. Cooper, *Chemical Science*, Vol. 2, p. 1777, 2011.
20. Y. Xu, L. Chen, Z. Guo, A. Nagai, and D. Jiang, *Journal of the American Chemical Society*, Vol. 133, p. 17622, 2011.
21. J.-X. Jiang, F. Su, A. Trewin, C.D. Wood, N.L. Campbell, H. Niu, C. Dickinson, A.Y. Ganin, M.J. Rosseinsky, Y.Z. Khimyak, et al., *Angewandte Chemie International Edition*, Vol. 46, p. 8574, 2007.

22. J.-X. Jiang, F. Su, A. Trewin, C.D. Wood, H. Niu, J.T.A. Jones, Y.Z. Khimyak, and A.I. Cooper, *Journal of the American Chemical Society*, Vol. 130, p. 7710, 2008.
23. J.-X. Jiang, F. Su, H. Niu, C.D. Wood, N.L. Campbell, Y.Z. Khimyak, and A.I. Cooper, *Chemical Communications*, pp. 486–488, 2008.
24. A. Li, R.-F. Lu, Y. Wang, X. Wang, K.-L. Han, and W.-Q. Deng, *Angewandte Chemie*, Vol. 122, p. 3402, 2010.
25. A. Li, H.-X. Sun, D.-Z. Tan, W.-J. Fan, S.-H. Wen, X.-J. Qing, G.-X. Li, S.-Y. Li, and W.-Q. Deng, *Energy & Environmental Science*, Vol. 4, p. 2062, 2011.
26. C. Xu and N. Hedin, *Journal of Materials Chemistry A*, Vol. 1, p. 3406, 2013.
27. P. Pandey, A.P. Katsoulidis, I. Eryazici, Y. Wu, M.G. Kanatzidis, and S.T. Nguyen, *Chemistry of Materials*, Vol. 22, p. 4974, 2010.
28. M.G. Rabbani, A.K. Sekizkardes, O.M. El-Kadri, B.R. Kaafarani, and H.M. El-Kaderi, *Journal of Materials Chemistry*, Vol. 22, p. 25409, 2012.
29. Q. Chen, J.-X. Wang, F. Yang, D. Zhou, N. Bian, X.-J. Zhang, C.-G. Yan, and B.-H. Han, *Journal of Materials Chemistry*, Vol. 21, p. 13554, 2011.
30. Y. Kou, Y. Xu, Z. Guo, and D. Jiang, *Angewandte Chemie*, Vol. 123, p. 8912, 2011.
31. P. Kuhn, M. Antonietti, and A. Thomas, *Angewandte Chemie International Edition*, Vol. 47, p. 3450, 2008.
32. S. Yuan, B. Dorney, D. White, S. Kirklin, P. Zapol, L. Yu, and D.-J. Liu, *Chemical Communications*, Vol. 46, p. 4547, 2010.
33. V.A. Davankov and M.P. Tsyurupa, *Reactive Polymers*, Vol. 13, p. 27, 1990.
34. S. Che, J. Pang, A.J. Kalin, C. Wang, X. Ji, J. Lee, D. Cole, J.-L. Li, X. Tu, Q. Zhang, H.-C. Zhou, and L. Fang, *ACS Materials Letters*, Vol. 2, p. 49, 2020.
35. J. Yu, X. Sun, X. Xu, C. Zhang, and X. He, *Applied Catalysis B: Environmental*, Vol. 257, p. 117935, 2019.
36. X. Wang, C. Wang, H.-Q. Tan, T.-Y. Qiu, Y.-M. Xing, Q.-K. Shang, Y.-N. Zhao, X.-Y. Zhao, and Y.-G. Li, *Chemical Engineering Journal*, Vol. 431, p. 134051, 2022.
37. M. Xu, J. Wu, J. Wang, Y. Mao, M. Liu, Y. Yang, C. Yang, L. Sun, Y. Du, Y. Li, and H. Li, *Applied Surface Science*, Vol. 581, p. 152409, 2022.
38. R. Li, C. Zhang, C.-X. Cui, Y. Hou, H. Niu, C.-H. Tan, X. Yang, F. Huang, J.-X. Jiang, and Y. Zhang, *Polymer*, Vol. 240, p. 124509, 2022.

2

Synthesis Methods

Microfluidics, membrane/microchannel, suspension, dispersion, precipitation, multistage polymerizations and a few other less known methods were detailed (1). Also, the ability to yield non-spherical particles for these methods and size monodispersity, pore characteristics and chemical functionality of the obtained particles was discussed.

2.1 Porogens

Porogens are substances that can yield a porous nature of certain particles. Reactive porogens and immiscible porogens cannot really dilute a monomer mixture (1).

2.1.1 *Polymers and Organic Solvents*

Several polymers and oligomers can be used for polymerization, generally together with a solvent as the porogen. In this case the pore formation occurs via ξ-induced syneresis.

Examples of polymers and oligomers that can be used as porogens are shown in Table 2.1.

It was found that the amount and the nature of a polymeric porogen may either induce a porous or a nonporous hollow final structure (2, 3).

Also, differences between the use of a good solvent (toluene), a non-solvent (dodecanol) and a polymeric porogen (poly(styrene) (PS) in toluene, 15%) for the synthesis of a copolymer of glycidyl

Table 2.1 Polymers and oligomers for porogens (1).

Compound	Reference
Poly(methyl methacrylate)	(4)
Poly(styrene)	(5)
Poly(ethylene oxide)	(6)
Poly(propylene oxide)	(7)
Poly(dimethyl siloxane)	(7)

methacrylate (GMA) and ethylene glycol dimethacrylate (EGDM) were shown (8). With a good solvent, non-solvent and a polymeric porogen, the specific surface area decreases below 1 m^2g^{-1}, whereas the size of the microglobules and the total pore volume increase.

A bimodal pore size distribution can be obtained in some cases by using a mixture of toluene (good solvent, inducing micropores) and poly(dimethyl siloxane) (PDMS) (polymeric porogen, inducing macropores) for a bead composed of divinylbenzene (DVB) alone (7).

It was not expected that from a single porogen the combination of high surface area and high pore volume could be reached. However, it was found that polymeric DVB particles exhibit a surface area equal to 720 m^2g^{-1}, together with a very high pore volume of 68%, when prepared in the presence of 1-chlorodecane alone, which is a non-solvent for a DVB polymer (9).

An important problem due to the use of a non-solvent as the porogen is the possibility of the formation of a dense and often impermeable polymer layer on the surface of particles, although the internal structure is highly porous (1).

A non-solvent should possess a solubility parameter that significantly differs from that of the polymer. However, when the difference in the solubility parameter is too large, a *skin* formation is promoted (10). Here, the difference in the solubility parameter was increasing by a decreasing polarity of the porogen.

Since the continuous phase was water in their system, highly nonpolar porogens disliked being present in the water/oil interface due to the high interfacial tension. Thus, the interface became rich in monomer and polymer, resulting in a skin layer, while the interior was porous (1).

2.1.2 *Water as Porogen*

A porogen that is immiscible with the initial monomer mixture can be utilized to synthesize porous particles. The most common example of such a strategy is the usage of water as porogen. A water-in-oil-in-water (W/O/W) double emulsion is formed by adding oil soluble surfactants to the discrete monomer (oil) phase. Water is absorbed from the continuous water phase by the monomer droplets as a result of the stabilizing effect of the oil-soluble surfactants (11).

Although porogen water droplets should have been separated initially inside the monomer phase, highly porous polymer beads with pore sizes around 80 *nm* and surface area values reaching up to 200 m^2g^{-1} (proving the interconnectivity of pores) are obtained after polymerization. The same authors also published that a combination of surfactants can produce hollow porous beads (12).

2.1.3 *Solid Porogens*

Immiscible porogens can be a solid instead of water, which results in the realization of solid-in-oil-in-water dispersions. Here, pores are formed after the removal of solid particles embedded on polymer beads via washing or etching. Washing is needed to get the porous structure.

As an example for a solid-in-oil-in-water dispersion, ~0.8 μm $CaCO_3$ particles were dispersed in a monomer mixture of EGDM and GMA and suspension polymerized this solid to oil dispersion in water (13). After removal of $CaCO_3$ via HCl etching, the beads exhibited pores as large as 10 μm and a surface area value of 79 m^2g^{-1}.

In another report (14), a mixture of solid $CaCO_3$, non-solvent (dodecanol) and good solvent (cyclohexanol) porogens are utilized all together for the suspension polymerization of the same EGDM-GMA monomer mixture. Together with a total surface area of 91 m^2g^{-1}, the formation of a bimodal distribution of micropores (10–90 *nm*) and macropores (180–4000 *nm*) is observed.

2.2 Living Radical Polymerization

Porous particles can be made of an organic polymer. They have a substantially spherical shape (15). Each of the porous particles has an interconnected pore structure in which through-holes provided inside the porous particle communicate with each other, and ends of the through-holes are open toward the outside of the porous particle.

The raw material monomer constituting the polymer forming the porous particles may be radical polymerizable or ion polymerizable. When a radical polymerizable monomer is subjected to free radical polymerization, a polymer of a particle agglutination type is liable to be formed.

Thus, in order to cause a spinodal decomposition, it is preferable to use living radical polymerization that proceeds in a consecutive manner. Ion polymerization and condensation polymerization also are consecutive reactions, so that it can be said they also are suitable for causing spinodal decomposition (15).

The ion polymerizable monomer may be an anion polymerizable monomer or a cation polymerizable monomer, for example. Examples of the radical polymerizable monomer include poly(ethylene) derivatives and poly(acrylic acid) or poly(methacrylic acid) derivatives. Examples of the ion polymerizable monomer include epoxy monomers; styrene, 1,3 butadiene and derivatives thereof, vinylpyridine, methacrylic esters; and acrylonitrile.

The monomer used for forming the porous particle of the present invention may be a monomer other than the radical polymerizable monomer, and may be an ion polymerizable monomer (15).

2.3 Emulsion Polymerization

2.3.1 *High Internal Phase Emulsion Polymerization*

High internal phase emulsion (HIPE) polymerization is a relatively new technique that has found a wide variety of applications in tissue engineering scaffolds, enzyme immobilization, gas storage and separation media (16–19).

Recently, it was shown that *p*-conjugated highly porous polymer scaffolds can combine the hierarchically interconnected pore structure of highly porous polymer scaffolds and the *p*-conjugated polymer backbone throughout the network, showing high efficiency as a heterogeneous photosensitizer for singlet oxygen generation under visible light (20).

Also, the micropore engineering and photocatalytic activity of conjugated microporous highly porous polymer scaffolds was reported for highly selective oxidation of organic sulfides to sulfoxides under visible light (21).

Actually, the photopolymerization of methyl methacrylate using *p*-conjugated porous HIPE polymers as a heterogeneous photoinitiator could be efficiently achieved under visible light (22).

A method for producing a HIPE foam has been described (23).

Here, a water bath is used to polymerize a HIPE. The characteristics of the polymerized HIPE foam may be varied depending on the composition and characteristics of the water bath. The HIPEs are produced using a continuous process, for example, by having a HIPE pumped into a water bath while traveling through the bath.

The HIPE foams produced from this method are useful for the absorption of liquid materials. A HIPE consists of two phases. One phase is a continuous oil phase that contains monomers that are polymerized to form a HIPE foam and an emulsifier to help to stabilize the HIPE. The monomer component may be present in an amount of from about 80% to about 99%, and under certain conditions from about 85% to about 95% by weight of the oil phase. The emulsifier component, which is soluble in the oil phase and suitable for forming a stable water-in-oil emulsion, may be present in the oil phase in an amount of from about 1% to about 20% by weight of the oil phase. The emulsion may be formed at an emulsification temperature of from about 20°C to about 130°C and in certain embodiments from about 50°C to about 100°C.

In general, the monomers will include from about 20% to about 97% by weight of the oil phase at least one substantially water-insoluble monofunctional alkyl acrylate or alkyl methacrylate.

For example, monomers of this type may include C_4–C_{18} alkyl acrylates and C_2–C_{18} methacrylates, such as ethylhexyl acrylate, butyl acrylate, hexyl acrylate, octyl acrylate, nonyl acrylate, decyl acrylate, isodecyl acrylate, tetradecyl acrylate, benzyl acrylate, nonyl phenyl acrylate, hexyl methacrylate, 2-ethylhexyl methacrylate, octyl methacrylate, nonyl methacrylate, decyl methacrylate, isodecyl methacrylate, dodecyl methacrylate, tetradecyl methacrylate, and octadecyl methacrylate. Some of these compounds are shown in Figure 2.1.

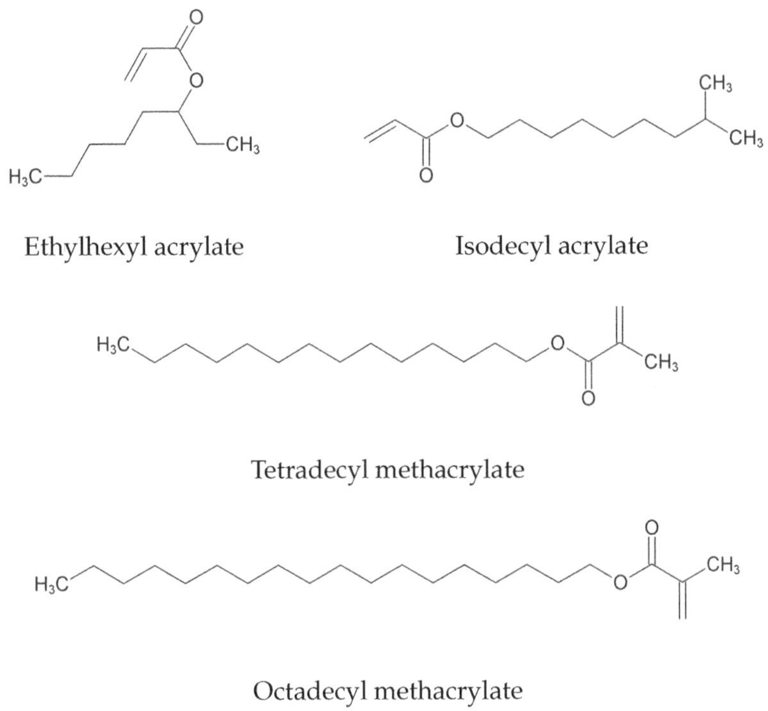

Figure 2.1 (Meth)acrylic monomers.

Blends of these monomers can have the desired glass transition temperature (T_g) of the resulting HIPE foams. The oil phase may also comprise from about 2% to about 40%, and in certain embodiments from about 10% to about 30%, by weight of the oil phase, a substantially water-insoluble, polyfunctional crosslinking alkyl acrylate or methacrylate.

This crosslinking comonomer is added to confer strength and resilience to the resulting HIPE foam. Examples of crosslinking monomers of this type comprise monomers containing two or more activated acrylate, methacrylate groups, or combinations thereof.

Examples of this group include 1,6-hexanediol diacrylate, 1,4-butanediol dimethacrylate, trimethylol propane triacrylate, trimethylol propane trimethacrylate, 1,12-dodecyldimethacrylate, 1,14-tetradecanedioldimethacrylate, ethylene glycol dimethacrylate, neopentyl glycol diacrylate (2,2-dimethylpropanediol diacrylate), hexanediol acrylate methacrylate, glucose pentaacrylate, sorbitol pentaacrylate. Some of these compounds are shown in Figure 2.2.

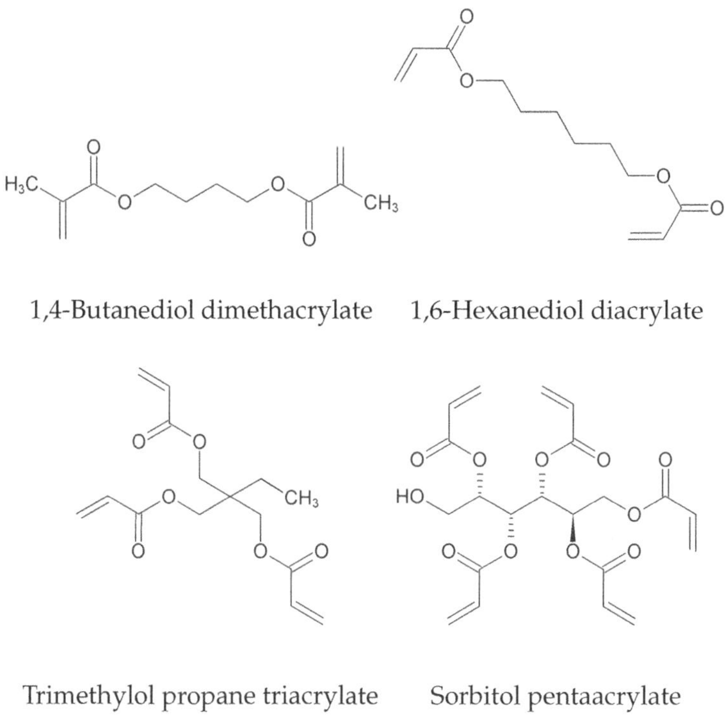

Figure 2.2 Crosslinking monomers.

Other examples of crosslinking agents contain a mixture of acrylate and methacrylate moieties, such as ethylene glycol acrylate-methacrylate and neopentyl glycol acrylate-methacrylate. The ratio of methacrylate:acrylate group in the mixed crosslinking agent may be varied from 50:50 to any other ratio as needed.

Any third substantially water-insoluble comonomer may be added to the oil phase in weight percentages of from about 0%

to about 15% by weight of the oil phase, in certain embodiments from about 2% to about 8%, to modify the properties of the HIPE foams.

In certain cases, *toughening* monomers may be desired which impart toughness to the resulting HIPE foam. These include monomers such as styrene, vinyl chloride, vinylidene chloride, isoprene, and chloroprene. Some of these compounds are shown in Figure 2.3

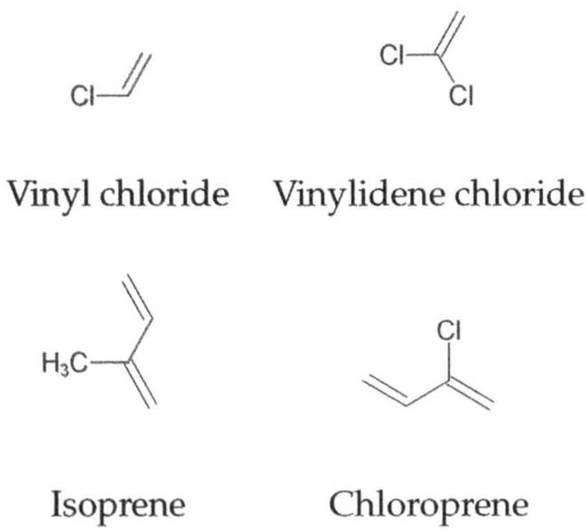

Figure 2.3 Toughening monomers.

It is believed that such monomers aid in stabilizing the HIPE during polymerization to provide a more homogeneous and better formed HIPE foam which results in better toughness, tensile strength, and abrasion resistance.

Monomers may also be added to confer flame retardancy as shown previously (24). The monomers may be added to confer color, e.g., vinyl ferrocene, fluorescent properties, radiation resistance, opacity to radiation, (e.g. lead tetraacrylate), to disperse charge, to reflect incident infrared light, to absorb radio waves, to form a wettable surface on the HIPE foam struts, or for any other desired property in a HIPE foam.

In some cases, these additional monomers may slow the overall process of conversion of HIPE to HIPE foam, the tradeoff being necessary if the desired property is to be conferred. Thus, such monomers can be used to slow down the polymerization rate of a HIPE. Examples of monomers of this type comprise styrene and vinyl chloride. The oil phase may further contain an emulsifier used for stabilizing the HIPE.

Emulsifiers used in a HIPE can include (24):

1. Sorbitan monoesters of branched $C_{16}-C_{24}$ fatty acids,
2. Linear unsaturated $C_{16}-C_{22}$ fatty acids; and linear saturated $C_{12}-C_{14}$ fatty acids, such as sorbitan monooleate, sorbitan monomyristate, and sorbitan monoesters, sorbitan monolaurate diglycerol monooleate (DGMO), polyglycerol monoisostearate (PGMIS), and polyglycerol monomyristate (PGMM),
3. Polyglycerol monoesters of branched $C_{16}-C_{24}$ fatty acids, linear unsaturated $C_{16}-C_{22}$ fatty acids, or linear saturated $C_{12}-C_{14}$ fatty acids, such as diglycerol monooleate, diglycerol monomyristate, diglycerol monoisostearate, and diglycerol monoesters,
4. Diglycerol monoaliphatic ethers of branched $C_{16}-C_{24}$ alcohols, linear unsaturated $C_{16}-C_{22}$ alcohols, and linear saturated $C_{12}-C_{14}$ alcohols, and mixtures of these emulsifiers.

Some of these compounds are shown in Figure 2.4.

Another emulsifier that may be used is polyglycerol succinate, which is formed from an alkyl succinate, glycerol, and triglycerol. Such emulsifiers, and combinations thereof, may be added to the oil phase so that they comprise between about 1% and about 20%, in certain embodiments from about 2% to about 15%, and in certain other embodiments from about 3% to about 12% by weight of the oil phase. In certain embodiments, coemulsifiers may also be used to provide additional control of cell size, cell size distribution, and emulsion stability, particularly at higher temperatures of, for example, greater than about 65°C.

Examples of coemulsifiers include phosphatidyl cholines and phosphatidyl choline-containing compositions, aliphatic betaines, long chain $C_{12}-C_{22}$ dialiphatic quaternary ammonium salts, short chain C_1-C_4 dialiphatic quaternary ammonium salts, long chain $C_{12}-C_{22}$ dialkoyl(alkenoyl)-2-hydroxyethyl, short chain C_1-C_4 dialiphatic quaternary ammonium salts, long chain $C_{12}-C_{22}$ dialiphatic imidazolinium quaternary ammonium salts, short chain C_1-C_4 dialiphatic imidazolinium quaternary ammonium salts, long chain $C_{12}-C_{22}$ monoaliphatic benzyl quaternary ammonium salts, and

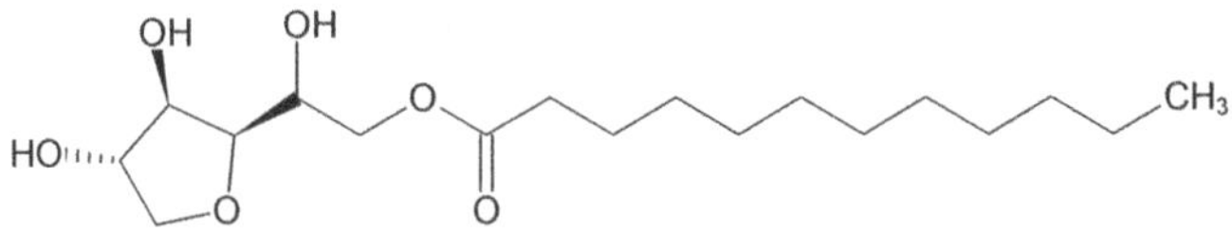

Sorbitan monolaurate

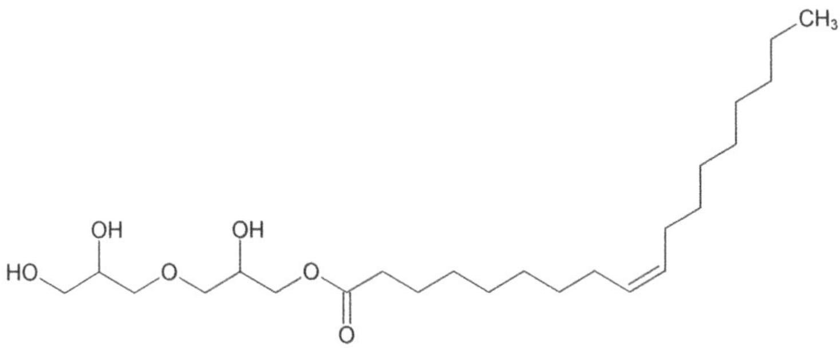

Diglycerol monooleate

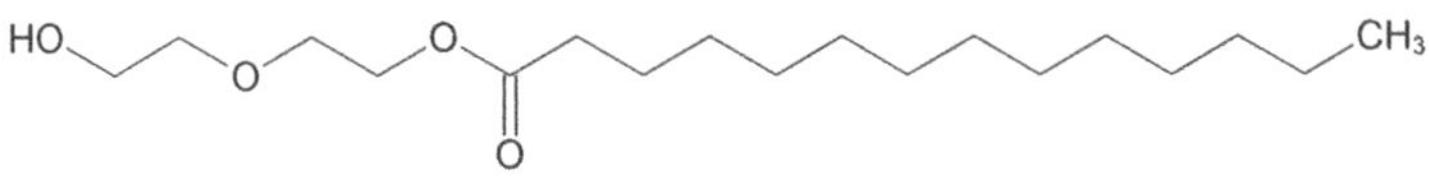

Diglycerol monomyristate

Figure 2.4 Emulsifiers.

long chain $C_{12}-C_{22}$ dialkoyl(alkenoyl)-2-aminoethyl, short chain C_1-C_4 monoaliphatic benzyl quaternary ammonium salts, short chain C_1-C_4 monohydroxyaliphatic quaternary ammonium salts. In certain embodiments, ditallow dimethyl ammonium methyl sulfate (DTDMAMS) may be used as a coemulsifier.

Another component that may be present in the aqueous phase is a water-soluble free radical initiator. The initiator can be present at up to about 20 *mol%* based on the total moles of polymerizable monomers present in the oil phase. In certain embodiments, the initiator is present in an amount of from about 0.01 *mol%* to about 10 *mol%* based on the total moles of polymerizable monomers in the oil phase. Suitable initiators include ammonium persulfate, sodium persulfate, potassium persulfate, 2,2′-azobis(*N*,*N*′-dimethyleneisobutyramidine)dihydrochloride, and other suitable azo initiators. To reduce the potential for premature polymerization which may clog the emulsification system, the addition of the initiator to the monomer phase may be done just after or near the end of the emulsification process.

The HIPEs and their subsequent polymerization into absorbent foams are illustrated in the following example (23):

Preparation 2–1: HIPE foams are prepared from a HIPE (5% oil in 95% water). The oil phase is prepared from a 3.35:1 ratio of EHA: ethylene glycol dimethacrylate and 12% polyglycerol succinate as surfactant. The aqueous phase is 2% NaCl and 0.33% NaPS in water. The HIPE is injected into the water bath in the form of fibers. The water baths can be maintained at 85°C and have varying compositions of deionized water, 2% NaCl, 0.33% NaPS, and both 2% NaCl.

It has been found that placing electrolytes in the water bath produces a thicker layer of monomer at the HIPE-bath interface prior to curing. This allows one to modify the porosity of the HIPE foam. The water bath composition can create a HIPE foam with a wide range of porosity. When the bath contains only NaCl, a shrinkage occurs during polymerization, leading to fewer windows connecting the pores within the cured HIPE. The presence of an initiator in the bath may result in an additional growth of a surface layer of the polymer.

As shown in Table 2.2, the porosity of a HIPE foam can be manipulated by additional components that are added to the deionized water. The porosity of the fibers in Table 2.2 below was determined gravimetrically, by measuring the mass and the dimensions of the fibers.

Table 2.2 Porosity (23).

Water bath	Porosity/[%]	Tensile modulus/[*MPa*]
Deionized Water	99.1	1.2
2% NaCl	78.9	2.5
0.33% NaPs	96.5	2.5
2% NaCl + 0.33% NaPs	80.6	0.5

2.3.2 *Microchannel Emulsification*

A microchannel emulsification technique was developed, for which every hole (channel) for a discrete phase droplet formation was custom made (1).

Silicon (25–28), metal (29, 30) and polymer (31, 32) based highly uniform microchannels have been used for monodisperse particle manufacture with a coefficient of variation of less than 5% (33).

Two types of pristine porous PDMS microparticles were fabricated using a oil-in-water single emulsion template by the needle-based microfluidic multiphase system (34).

By manipulating the flow rate of either the dispersed or continuous phase, microparticles of various sizes were obtained. The capillary number of the continuous phase for all flow conditions applied in this study was less than 0.1, suggesting that the flow regime was dripping.

The coefficient of variation of the sizes under different flow conditions were less than 1.5% which indicated the particles to be highly monodispersed. The surface morphology and particle size were characterized by optical microscope and scanning electron microscopy (SEM).

Pristine PDMS microparticles and other PDMS microparticles that were templated from tetrachloromethane and white granulated sugars, were prepared under the same flow conditions. Subsequently, the microparticles were adopted for a treatment of a synthetic wastewater that contained organic compounds such as toluene under static and dynamic states for comparison (34).

The effects, including the releasing amount, size of particles, porosity of microparticles and initial concentration of pollutants, were investigated based on the toluene concentration variation, which was quantified by a gas chromatograph-headspace sampler. It has been found that 50 *mg* of porous PDMS microparticles are capable of realizing over 65% of toluene removal efficiency of 200 ppm toluene aqueous solution within 2 *h*. The microparticles were collected and reused 30 times with unchanged treatment capacity (34).

2.4 Solvent-Free Polymerization

A methanesulfonic acid-mediated solvent-free synthesis of conjugated porous polymer networks has been described. The method consists of (35):

1. Providing a first reactant with at least one acetyl group, and a second reactant with an alkylsulfonic acid,
2. Creating a solution of the reactants,
3. Casting the solution in a form, and
4. Treating the solution under such conditions to produce a porous polymer network.

Aldol condensation is an effective method to construct carbon-carbon bonds and that widely used in organic synthesis (36). In 1991, a tandem aldol triple condensation (ATC) reaction was reported that constructs a central benzene ring from aromatic acetyl groups (37). In the presence of acid, this reaction involves two aldol condensation steps followed by a [3+3] electrocyclic reaction and aromatization (Figure 2.5).

Due to its highly efficient nature and the intriguing C3 symmetry of the products, aldol triple condensation reaction has been widely used in the synthesis of star-shaped molecules (38) and dendrimers (39). Recently, the preparation of organic microporous materials was also reported by using the aldol triple condensation reaction (40–42).

In these reactions, however, either toxic and irritate acid was used as catalysis, or the acids were easily decomposed at high temperature. Furthermore, all these methods require specialized reactors to protect the reaction from oxygen and moisture (35).

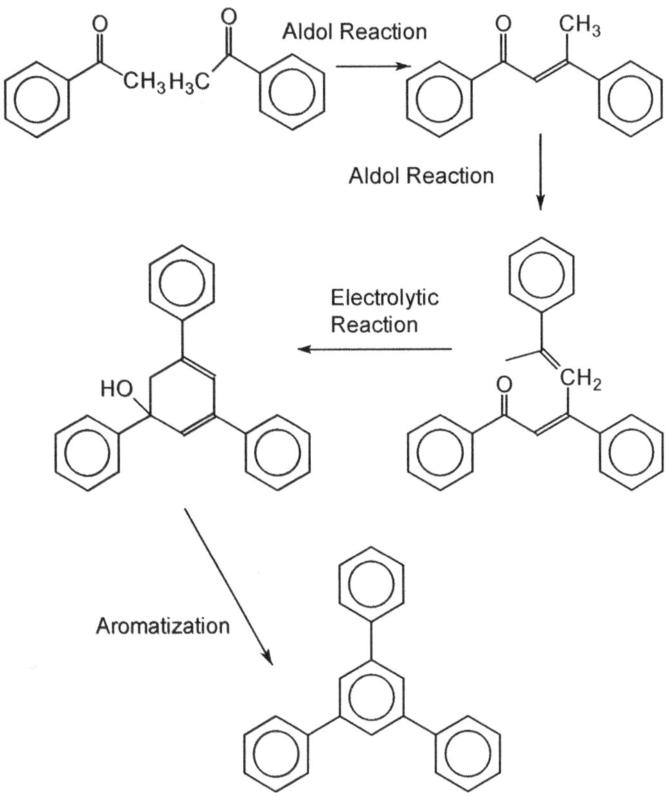

Figure 2.5 Electrocyclic reaction and aromatization (35).

Methanesulfonic acid could be an ideal acid for the aldol triple condensation synthesis of porous polymer networks. Compared to the widely-explored acidic reagents for aldol triple condensation reaction, such as silicon tetrachloride, 4-methylbenzenesulfonic acid, sulfuric acid and thionyl chloride, methanesulfonic acid is nontoxic, environmentally benign and stable at high temperature. Methanesulfonic and 4-methylbenzenesulfonic acids are shown in Figure 2.6. Importantly, methanesulfonic acid is a liquid acid so that

Methanesulfonic acid 4-Methylbenzenesulfonic acid

Figure 2.6 Sulfonic acids.

it can serve not only as the catalyst but also as the solvent in the reaction (35).

The structure of various monomers for preparing porous polymer networks is shown in Figure 2.7.

2.5 Suspension Polymerization

Suspension polymerization is a heterogeneous radical polymerization process that uses mechanical agitation to mix a monomer or mixture of monomers in a liquid phase, such as water, while the monomers polymerize, forming spheres of a polymer (43).

The suspension polymerization method has been used for the preparation of porous polymer microspheres with different surface morphology (44). The morphology results show that the smooth, convex, and wrinkled microspheres could be controlled by adjusting the ratio of the monomer to the porogens.

The micelles forming the framework support the *eggshell*, and its size and shape directly affect the morphology of the eggshell. The addition of GMA as monomer, whose polymer has low glass transition temperature (T_g), is important for the formation of a wrinkled

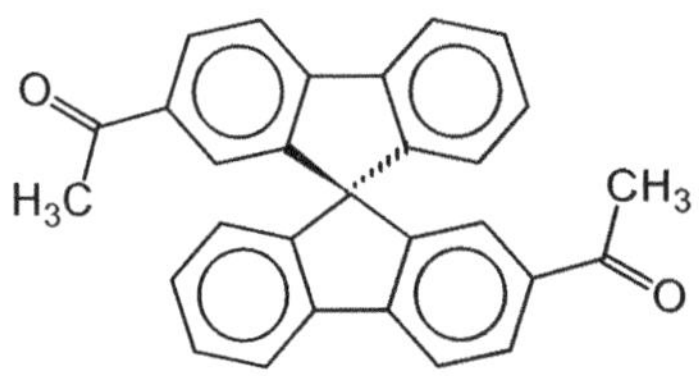

(S)-1,1′-(9,9′-Spirobi[fluorene]-7,7′-diyl)diethanone

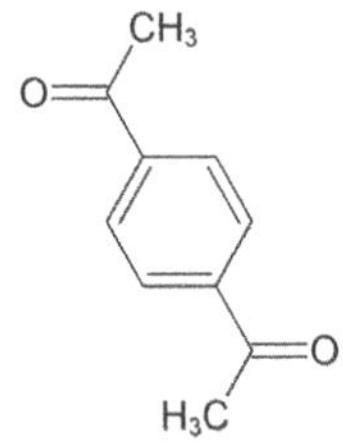

1,1′-(1,4-Phenylene) diethanone

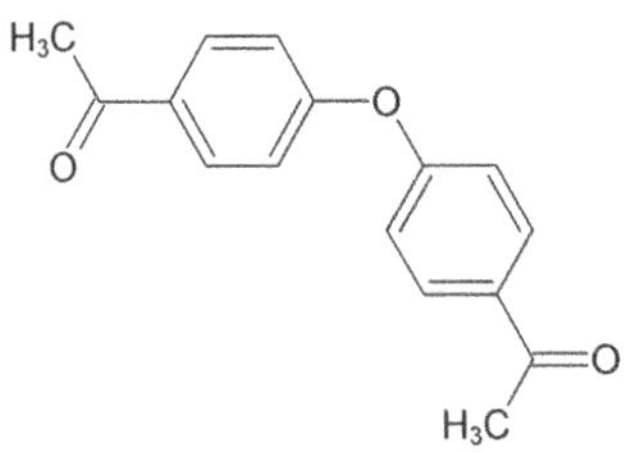

1,1′-(Oxybis(4,1-phenylene)) diethanone

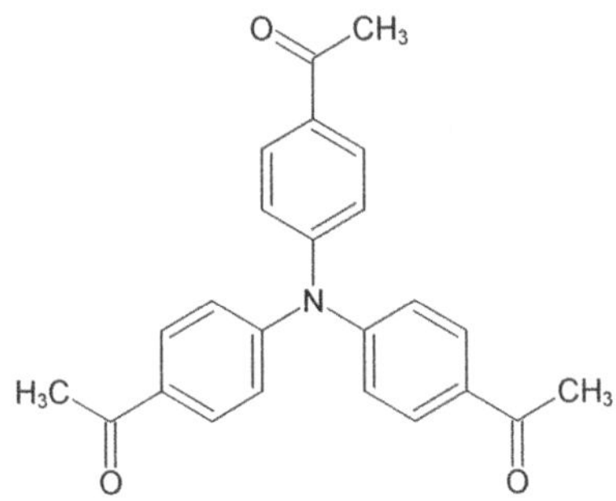

1,1′,1″-(Nitrilotris(benzene-4,1-diyl))triethanone

Figure 2.7 Monomers for porous polymer networks (35).

morphology. The amount of toluene and PDMS also affects the surface morphology of microspheres. In addition, the effect of PDMS is also more significant. The preparation process of the wrinkled microspheres from glycidyl methacrylate, styrene and ethylene glycol dimethacrylate with abundant epoxy groups can be amplified (44).

The morphology of the material prepared in a 100 *l* reactor is well maintained, and the yield in the size range of 80–160 μm is more than 80%. So, the wrinkled-surface porous polymer microspheres may have potential applications in the fields of enzyme carrier, separation and purification, and light scattering (44).

Several porous materials have been prepared using the suspension polymerization technique. These materials are collected in Table 2.3.

Table 2.3 Porous materials prepared by suspension polymerization.

Material	References
Porous poly(2-hydroxyethyl methacrylate) beads	(45)
Poly(imide)-based porous crosslinked beads	(46)
Porous styrene-glycidyl methacrylate copolymers	(47)
Poly(butyl acrylate-*co*-stearyl methacrylate-*co*-styrene-*co*-divinylbenzene)	(48)
Porous poly(2-hydroxyethyl methacrylate)	(49)
Styrene-divinylbenzene copolymers	(50, 51)
Poly(methyl methacrylate)	(52)
N-Vinylcarbazole-divinylbenzene copolymer	(53)
Poly(*i*-butyl methacrylate) microspheres	(54)

2.6 Multistage Polymerization Techniques

Emulsion polymerization is widely used as the first stage of seed polymerization methods, i.e., multistage heterogeneous polymerization techniques. In this two-stage approach, first nonporous *seed* particles are produced, which are the initial relatively small monodisperse particles. These seed particles are subsequently enlarged in a second stage.

A mathematical model has been developed and was also validated for a seeded suspension polymerization process that manufactures PS beads starting from pretreated polyolefin seeds (55). For the manufacture of such a product, the prediction of the dynamics of particle growth, polymerization rate and particle average molecular weight, is of importance.

A hybrid approach was proposed for the parameter subset selection and estimation. An excellent agreement was found with experimental results for different product grades. The monomer feeding rate, which is a key controlling factor of bead properties, was determined based on the proposed model and an optimum semi-batch feeding policy was determined (55).

2.7 Azo Coupling

A series of polyphenol porous polymers were derived from biomass polyphenols via a facile azo coupling method (56).

Batch experiments demonstrated their potentialities for adsorptive separation of metal ions from aqueous solution. Porous polymers prepared with gallic acid as starting material could adsorb Cs^+ at wide pH value range effectively.

2.8 Precipitation Polymerization

The stages of a precipitation polymerization for the production of porous particles are as follows (1):

1. First, only crosslinking and initiator molecules are in the medium.
2. Oligomers and nuclei form because of radical polymerization.
3. As the reaction continues, nuclei grow by adding monomers and oligomers from the medium. However, there is a swollen layer of oligomers around the nuclei.

So, a precipitation polymerization starts as a homogeneous solution of crosslinking agent and initiator in a medium composed of a near Θ-solvent, for the crosslinking agent and the porogens.

When the polymerization starts, oligomers and nuclei are formed (Step 2). Although the oligomers are still soluble in the medium, the nuclei precipitate results in a heterogeneous mixture.

The nuclei are swollen by the porogen and the medium becomes less rich with respect to the porogen. Although if no stabilizer is used, the nuclei are stabilized by a layer of oligomers that are swollen by the medium.

The polymerization continues at the particle-continuous phase interface (57).

The nuclei do not overlap but only grow by adding fresh monomer and oligomers from the medium (Step 3).

The porogen that is initially absorbed by the growing nuclei phase separates from the particle and forms the pores. As a result of the fact that the nuclei do not overlap, highly monodisperse particles, generally in the size range of 1–5 μm, are obtained (58).

As one of the unique properties of this technique, the particles constantly grow and the polymerization can be stopped when the desired particle size is reached.

The near Θ-solvent later becomes the continuous phase as the precipitation of particles starts. Precipitation polymerization requires a high amount of crosslinking agent. In many cases DVB is polymerized alone (59, 60)

Commercial DVB is composed of either 55% or 80% DVB with the rest being mostly an ethyl styrene monomer.

Polymerized particles with precipitation polymerization are rich in remaining double bonds that can be efficiently used for post-functionalization techniques (1).

2.9 Microfluidics

Droplet-based microfluidics, which generates and manipulates discrete droplets through immiscible multiphase flows inside microchannels, has emerged as a versatile tool for the generation of well-ordered porous materials. This method has been reviewed (61). In particular, the pore structure and the properties of microfluidic scaffolds were discussed.

A basic advantage of a microfluidic template is the monodispersity and also that the generation of these homogeneous templates is precisely and effectively controlled and tuned in a passive or active manner (62). This is important for the preparation of well-ordered porous scaffolds.

The synthesis of porous polymer particles via a microfluidics has been reported, including the methods of design and applications (63).

The feasibility of a new direction for microfluidics was studied by demonstrating a set of new methods to fabricate microfluidic systems (64). Similarly to microfluidic paper-based analytical devices, porous materials are being used. However, alternative porous materials and different printing methods are used here to give the material the necessary pattern to act as a microfluidic system.

Microfluidic systems were produced by the following three separate methods (64):

1. By curing a porous monolithic polymer sheet into a necessary pattern with photolithography,
2. By screen printing silica gel particles with gypsum, and
3. By dispensing silica gel particles with poly(vinyl acetate) binder using a modified 3D printer.

Various parameters of the printed chips were determined to evaluate whether the printed chips were suitable for use in microfluidics, i.e., the strength of the printed material, printing accuracy, printed material height, wetting characteristics, and repeatability. All three of these approaches were found to be suitable, and therefore the novel approach to microfluidics was successfully demonstrated (64).

In addition, a practical way to synthesize a hierarchical porous covalent organic polymer monolith was presented (65). Here, a droplet-based microfluidic method was proposed.

The interconnected macropores, with a size of up to ~130 nm, were created within intrinsic microporous covalent organic polymer structures by the further condensation of covalent organic polymer particles in the confined microdroplets.

Concentration, flow rate ratio, catalyst, etc., were screened to figure out the superior condition for producing different hierarchical porous covalent organic polymer monolith in microfluidic (65). Rod-like and spherical monolith with gradient sizes were controllable synthesized to satisfy the requirements of different packing density for practical applications.

A single microreactor was able to produce the spherical hierarchical porous covalent organic polymer monolith with a high space-time yield of 413 $kg\,m^{-3}d^{-1}$ under optimized conditions. This provides a promising scale-up method to synthesize covalent organic polymer monoliths (65).

2.10 Photocatalysis

The history of photocatalysis dates back to the 1960s (66, 67). The pioneering work occurred in 1972, in which water splitting was achieved on a TiO_2 electrode under UV light irradiation (68).

Porous organic polymers are a class of multi-dimensional porous network materials, which are built via strong covalent linkages between various organic building blocks with different geometries and topologies, and have recently become a rising research field in porous materials (69). Porous organic polymers can be generally divided into two categories based on their degree of long-range order, including amorphous and crystalline.

Owning to their advantages of light weight, superior inherent porosity, excellent stability, pre-designable and tunable structures and functions, porous organic polymers have received increasing attention and research interest for their tremendous potential applications in gas storage/separation, heterogeneous catalysis, photoelectric conversion, chemical-sensing and biosensing, energy storage and conversion.

Their porous structure, pore size, specific surface areas and functions can be directly designed and facilely tuned by introducing specific functional building blocks.

Certain representative advances in the field of porous organic polymers, their design and synthetic strategies, with an emphasis on their specific applications in photocatalysis, including photocatalytic chemical transformation, photodetoxification of contaminants from water, and water splitting, have been reviewed (69).

Photocatalytic systems based on poly(benzothiadiazoles) and their derivatives have been reviewed (70). As a strong electron-withdrawing unit, benzothiadiazole is often used for the construction of conjugated microporous polymers (CMPs), and thus BT-based CMPs have become dominant in CMP-photocatalytic chemical transformation.

Some representative CMPs and covalent organic frameworks

have been explained, focusing on their recent progress in photocatalytic transformations (69).

In order to be easily understood, the reactions or chemical transformations are divided into two types, i.e., oxidation reactions and other reactions, according to whether oxygen is involved,

By combining the high internal phase emulsion polymerization technique and micropore engineering by incorporating and removing a cleavable functional group in the polymer network, CMPs with increased surface area and porosity were reported (21). The so obtained CMPs showed an effective photocatalytic activity for the highly selective oxidation of organic sulfides under visible light irradiation.

Visible-light-driven photocatalysis requires only the sunlight as a driving force and an appropriate photocatalyst to conduct the catalytic reactions (71). This method is believed to be one of the most promising technologies. So, the development of efficient photocatalysts with superior photocatalytic properties is of fundamental importance (72, 73).

2.11 Thermal Drawing

The process of thermally drawing a fiber preform into a fiber has been exploited for a wide range of applications, including the production of optical fibers, fiber-based sensors, and fiber-based radiation tools (74). The thermal drawing process is a low-cost and intrinsically scalable fabrication method that enables production of extended lengths of fiber material with well-controlled geometry and electromagnetic transport characteristics over such extended lengths.

Beyond conventional optical transmission fibers, thermal fiber drawing has been shown to be capable of producing combined optical and electrical transmission, as well as microscale microelectronic and optoelectronic device operation, along a fiber length and/or across a fiber's cross section. Many optoelectronic applications have been particularly well-addressed by fibers produced with thermal drawing methodology.

A fiber is provided that includes a fiber material, which is disposed along a longitudinal axis fiber length. There is a porous domain that has a porous domain length along at least a portion of the fiber length, within the fiber material. The porous domain includes solid-phase material regions and fluid-phase interstitial regions that are both along the porous domain length and across the porous domain, for multi-dimensional porosity of the porous domain.

This fiber is produced with a method for forming a fiber including a porous domain. In the method, a fiber preform is assembled, including a solid polymer region having a hollow region extending through the solid polymer along a longitudinal axis of the fiber preform. The hollow region is filled with a liquid solution including a first component and a second component. The first and second liquid solution components are mixed in the liquid solution at temperatures greater than a phase separation temperature and the liquid solution transforms to solid-phase regions and fluid-phase regions at temperatures less than the phase separation temperature.

The fiber preform is drawn into a fiber by applying tension to the fiber preform at a fiber draw temperature that is greater than the phase separation temperature of the liquid solution. The fiber is cooled from the fiber draw temperature to a temperature that is less than phase separation temperature of the liquid solution, to form a thermally drawn fiber having a porous domain, extending the length of the drawn fiber, that includes solid-phase material regions and fluid-phase regions (74).

A schematic view of a fiber draw tower that has four independent heating zones is shown in Figure 2.8.

The process was performed using a cyclic olefin copolymer (COC) preform cladding as a cylindrical reservoir. The COC is an amorphous thermoplastic displaying excellent chemical resistance to many polar organic solvents, and can be thermally drawn between 150°C to 280°C. Before the draw, a cylindrical hollow reservoir is filled with a 1:4 by weight solution of (poly(vinylidene fluoride) (PVDF), Kynar 761, Arkema) in propylene carbonate. This solution has a S-L demixing temperature around 35°C (75), so that it is homogeneous during the draw process but phase separates in the fiber core as it cools down to ambient temperature. Porous PVDF fibers are obtained by selectively dissolving the COC cladding in toluene. The SEM micrograph of the fiber cross section shows a morphology consisting of interconnected PVDF spherulites separated by voids.

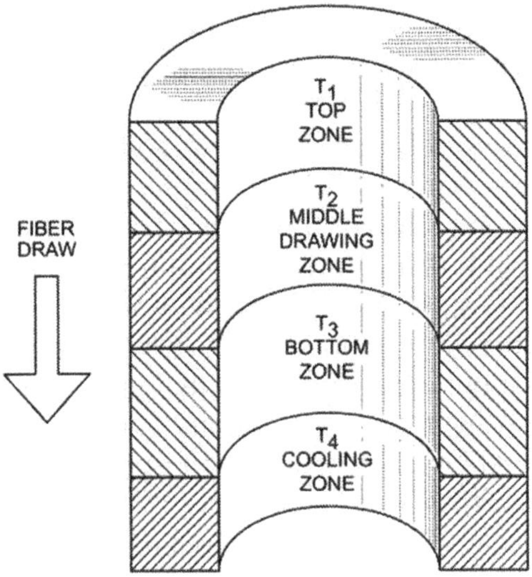

Figure 2.8 Fiber draw tower (74).

Such morphology is common for S-L phase-separated PVDF membranes (76, 77), and arises from nucleation and growth of highly semicrystalline PVDF spherulites (75).

2.12 Biodegradable Foam

A method for forming a dried absorbent foam has been described (78). The foam has an open pore network and pores.

An aqueous dispersion containing an enzymatically biodegradable biopolymer and a foaming agent is used here. Optionally, one or more of gel-forming ions, a plasticizer, a crosslinking agent and a pH modifier, forming, preparing or mixing a foam from the aqueous dispersion and drying the foam to form a dried foam containing open pores.

The gelled composites made from the foams are also described. The method is particularly useful for the synthesis of foams and composites for use in biomedical applications and as an anti-adhesive in tissue regeneration and wound management (78).

2.13 Biocompatible Porous Three-Dimensional Polymer Matrices

Cell therapy is a promising therapeutic strategy for cardiac ischemic syndromes (79). Indeed, the latter are already treated in patients by autologous graft of mesenchymal stem cells (MSCs) directly injected into the damaged organ, but the low cell survival rate (<15%) in the three days following the injection limits the beneficial effects thereof.

In point of fact, the functional recovery of the damaged organ correlates directly with the number of surviving MSCs; recent studies have shown that the regeneration of the infarcted organ is thought to be mainly due to their paracrine effects. In this context, maintaining MSC viability and functionality appears to be a major challenge for improving the benefit/risk ratio of the treatment of ischemias by cell therapy.

The development of porous biomaterials capable of generating a biomimetic environment favorable to maintaining MSCs, while at the same time allowing administration in *patch* form at the surface of the damaged organ, constitutes an advance in this field, and is the subject of intensive research.

A process for preparing a biocompatible and biodegradable polymer matrix comprising a network of open and interconnected pores has been described. This process consists of the following steps (79):

1. Preparation of an aqueous solution comprising at least one biocompatible anionic polysaccharide chosen from alginates and modified alginates; and at least one biocompatible cationic polymer chosen from chitosan, chitosan salts and chitosan derivatives,
2. Mechanically stirring said solution obtained above in the preceding step, in the presence of a foaming agent or of a pressurized gas, so as to form a foam,
3. Freezing the foam obtained above in the preceding step, so as to obtain a frozen foam,
4. Gelling the frozen foam obtained above in the preceding step, by adding, to said foam, at least one gelling agent in solution in a solvent, so as to obtain a gelled foam,
5. Dehydrating the gelled foam obtained above in the preceding step, so as to obtain a dehydrated gelled foam, then
6. Drying the dehydrated gelled foam obtained above in the preceding step, by treatment with supercritical CO_2, so as to obtain said polymer matrix.

2.14 Breath-Figure Method

The fabrication of regularly porous films by the breath-figure method has attracted much attention (80).

This method was first used in 1994 (81) for the preparation of porous polymer membranes for technological applications to control both the size distribution and the relative positions of the pores. This process is based on the rapidly evaporated polymer solutions exerted to humidity (82).

This simple, low-cost technique uses the condensation of water droplets to produce these structures, but the phenomenon itself is complex, requiring control over many interacting parameters that change throughout the process.

Developing a unified understanding of the molecular design of polymers to prepare ordered porous films is challenging.

2.14.1 *Effects of the Chemical Structure of Polymers*

The effects of the chemical structure of polymers in the breath-figure technique were systematically explored using side-chain-type liquid-crystalline star polymers (80).

The formation of porous films was affected by the structure of the polymers. The monomers used for such polymers are shown in Figure 2.9.

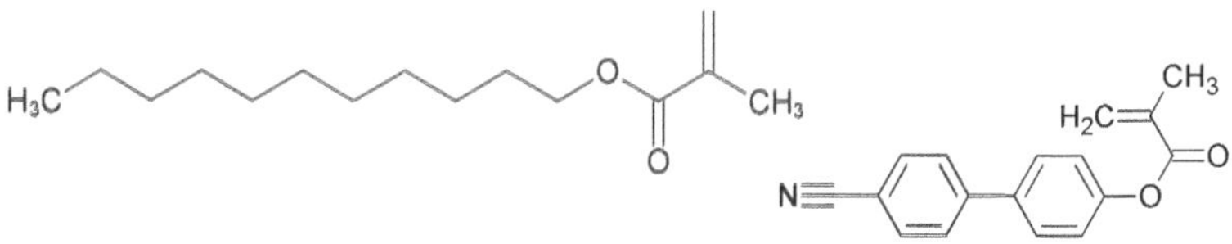

Undecyl methacrylate   4-Cyanobiphenyl methacrylate

Figure 2.9 Monomers used in the breath-figure technique (80).

Although the entire film surface of poly(11-[4-(4-cyanobiphenyl) oxy]undecyl methacrylate) had a hexagonal ordered porous structure over a certain M_n value, regularly arranged holes did not easily form in poly(methyl methacrylate) (PMMA), even though the main chain of PMMA is similar to that of poly(11-[4-(4-cyanobiphenyl) oxy]undecyl methacrylate).

A comparison of poly(11-[4-(4-cyanobiphenyl)oxy]undecyl methacrylate) and poly(11-[(1,1′-biphenyl)-4-yloxy]undecyl methacrylate) without cyano groups) showed that the local polar groups in hydrophobic polymers promoted the formation of ordered porous films. No holes were formed in poly(4-cyanobiphenyl methacrylate) (poly(11-[4-(4-cyanobiphenyl) oxy]undecyl methacrylate) without alkyl spacers) films due to its hydrophilicity.

The introduction of alkyl chains in poly(4-cyanobiphenyl methacrylate) allowed the preparation of honeycomb-structured films by increasing the internal tension. However, alkyl chains in the side chain alone did not result in a porous structure, as in the case of poly(undecyl methacrylate).

Aromatic rings are also required to increase the T_g and to improve the film formability. In the present study, suitable molecular designs of polymers, specifically hydrophobic polymers with local polar groups, were found to form a regularly porous structure. Development of clear guidelines for the molecular design of polymers is the subject of the current research of Naka *et al.*, which will enable the fabrication of porous films using various functional polymers (80).

2.14.2 *Coating Layers with Selective Wettability on Filter Papers*

A comb-like amphiphilic polymer, composed of hydrophobic backbones and hydrophilic side chains, was employed to grow honeycomb coating layers *in-situ* on a filter paper via directly casting a polymer solution and by the subsequent dynamic breath-figure method (83).

Through regulating the hydrophilic polymer side-chain density and the solution concentration, a continuous honeycomb coating layer contoured to the filter paper surface profile, in addition to possessing a water contact angle as high as 146°, was successfully fabricated.

Increasing the hydrophilic side-chain density will turn polymer into a surfactant-like polymer, and thus endow the polymer solution with the capacity of numerous micro/nano-sized water droplets, rather than simply stabilizing the ordered water droplet arrays on the surface of the solution. With vast nano-sized water droplets in it, the once transparent polymer solution changed into a translucent nano-emulsion, which demonstrates a strong Tyndall effect.

While casting such nano-emulsion on a filter paper and then subjecting it to the breath-figure process, the polymeric solute takes both nano-emulsion intrinsic nano-sized water droplets and solvent evaporation-induced water droplets as templates and self-assembles into a bird-nest-like three-dimensional porous microstructure, which possesses micro/nano-sized communicating pores.

By regulating the water content in the nano-emulsion, the bird-nest-like structure can be uniformly formed on the surface of the filter paper, which revealed a water contact angle of 152°. The coated filter papers possess selective wettability, and meanwhile, maintain the inherent permeability of the substrates, which therefore can be directly utilized as oil/water separation materials (83).

2.15 Superabsorbent Polymers

A superabsorbent polymer (SAP) is a type of synthetic polymeric material capable of absorbing 500 to 1000 times its own weight of moisture (84).

In most cases, the SAP is widely used in the field of hygiene products such as diapers and sanitary napkins, and, for this purpose, it is necessary to exhibit a high absorption capacity for moisture. In addition, it is necessary that the absorbed moisture should not leak out even under external pressure. Furthermore, it needs to show excellent permeability by maintaining its shape even in an expanded (swelled) state after absorbing water.

A preparation method of a porous superabsorbent polymer runs as follows (84):

1. Preparing a monomer neutralized solution by adding an internal crosslinking agent, an initiator, an inorganic filler, and a nonionic surfactant to a neutralized solution obtained by neutralizing at least 80 *mol%* of a water-soluble ethylene-based unsaturated monomer containing acidic groups,
2. High-shear mixing of the monomer neutralized solution; preparing a hydrogel polymer from the high-shear mixed monomer neutralized solution by thermal polymerization or photopolymerization, and
3. Preparing a base resin powder by drying, pulverizing, and classifying the hydrogel polymer.

The soluble ethylene-based unsaturated monomer may also include at least one selected from the group consisting of anionic monomer as collected in Table 2.4. Some of these compounds are shown in Figure 2.10.

Examples of internal crosslinking agents are collected in Table 2.5. Some of these compounds are shown in Figure 2.11.

Table 2.4 Additional monomers (84).

Ionic monomer
Acrylic acid
Methacrylic acid
Maleic anhydride
Fumaric acid
Crotonic acid
Itaconic acid
2-Acryloylethane sulfonic acid
2-Methacryloylethane sulfonic acid
2-(Meth)acryloylpropane sulfonic acid
2-(Meth)acrylamide-2-methyl propane sulfonic acid
(Meth)acrylamide
2-Hydroxyethyl(meth)acrylate
2-Hydroxypropyl(meth)acrylate
Methoxypolyethylene glycol(meth)acrylate
Polyethylene glycol(meth)acrylate
(*N,N*)-Dimethylaminoethyl(meth)acrylate
(*N,N*)-Dimethylaminopropyl(meth)acrylamide

Table 2.5 Internal crosslinking agents (84).

Monomers for crosslinking (84)
Poly(ethylene glycol diacrylate)
Glycerine diacrylate
Glycerine triacrylate
Trimethylol propane triacrylate
Trimethylol propane ethoxy triacrylate
Hexanediol diacrylate
Triethylene glycol diacrylate

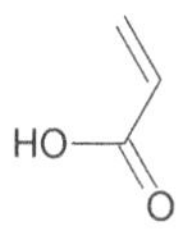

Acrylic acid

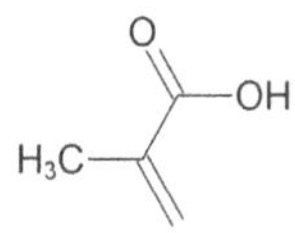

Methacrylic acid

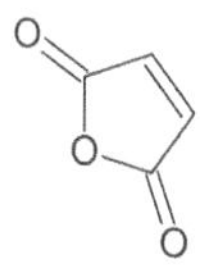

Maleic anhydride

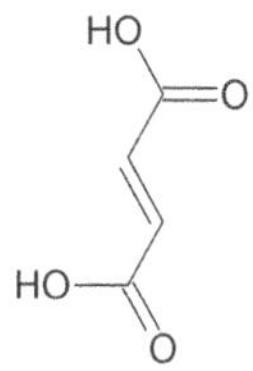

Fumaric acid

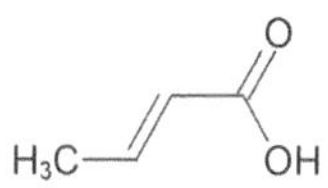

Crotonic acid

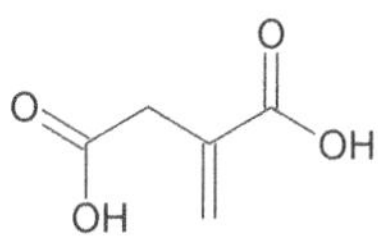

Itaconic acid

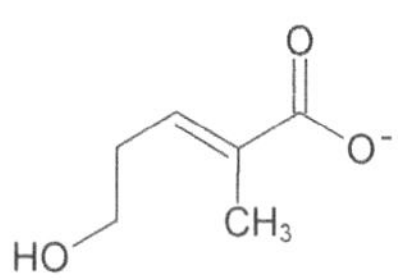

2-Hydroxyethyl methacrylate

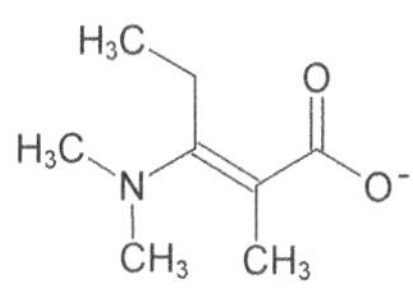

(*N,N*)-Dimethylaminoethyl methacrylate

Figure 2.10 Monomers.

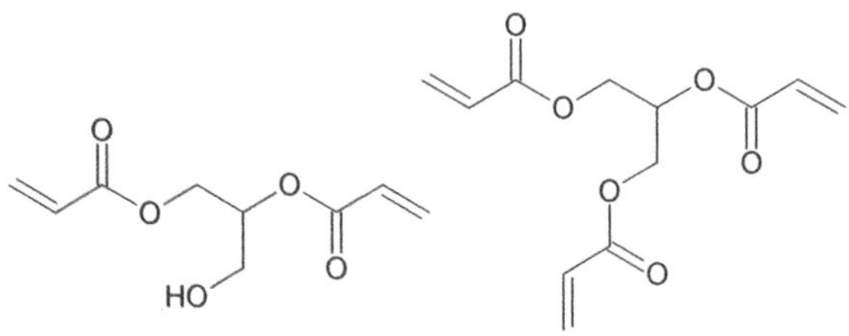

Glycerine diacrylate Glycerine triacrylate

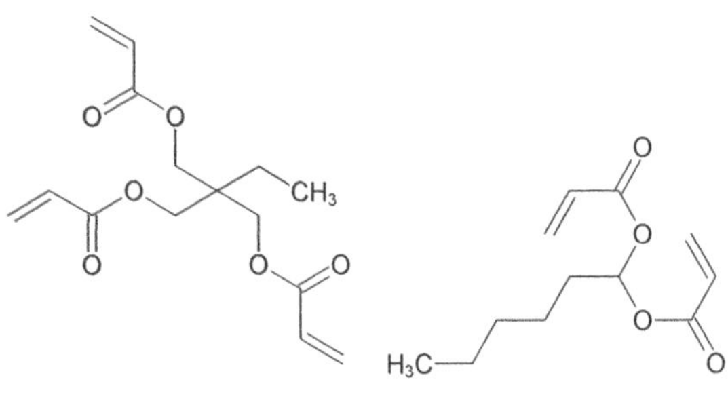

Trimethylol propane triacrylate Hexanediol diacrylate

Figure 2.11 Monomers for crosslinking (84).

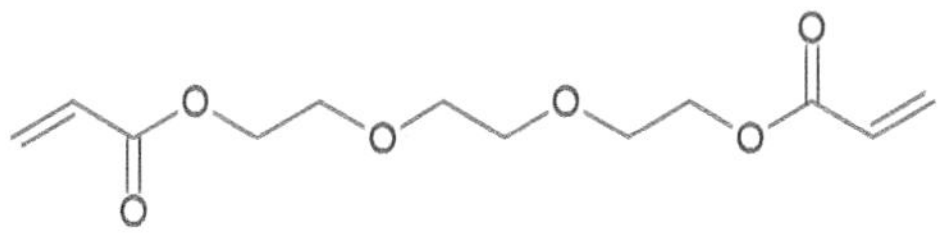

Triethylene glycol diacrylate

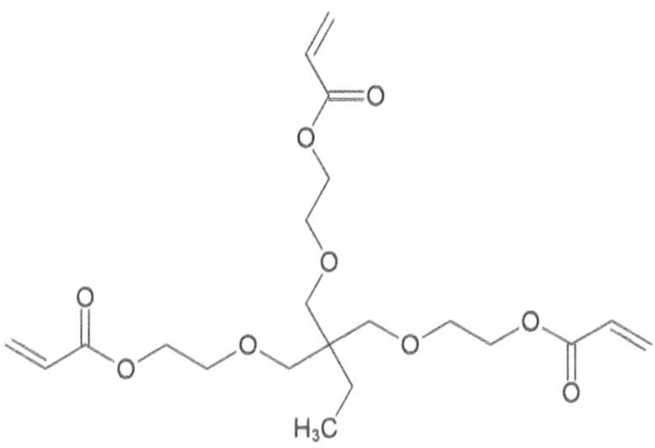

Trimethylol propane ethoxy triacrylate

Figure 2.11 (cont) Monomers for crosslinking (84).

The monomer composition may further include a thermal polymerization initiator or a photopolymerization initiator.

The photopolymerization initiator may be one or more compounds selected from the group consisting of benzoin ether, dialkyl acetophenone, hydroxyl alkylketone, phenylglyoxylate, benzyl dimethyl ketal, acyl phosphine, and α-aminoketone. Furthermore, as the specific example of acyl phosphine, commercial Lucirin TPO, namely, 2,4,6-trimethyl-benzoyl-trimethyl phosphine oxide, may be used. Some of these compounds are shown in Figure 2.12

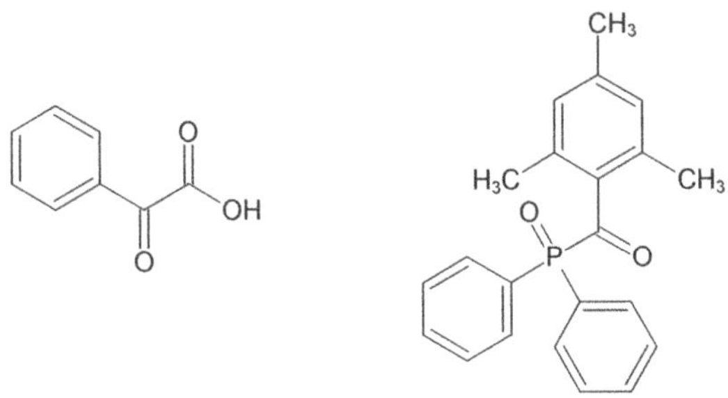

Phenylglyoxylic acid 2,4,6-Trimethyl-benzoyl-trimethyl phosphine oxide

Figure 2.12 Photopolymerization initiators.

Other photopolymerization initiators have been documented (85, 86).

Thermal polymerization initiators are one or more initiators selected from the group consisting of a persulfate-based initiator, an azo-based initiator, hydrogen peroxide, and ascorbic acid may be used. Specifically, sodium persulfate (Na : $2S_2O_8$), potassium persulfate ($K_2S_2O_8$), and ammonium persulfate ($(NH_4)_2S_2O_8$) may be used as examples of the persulfate-based initiators. 2,2-Azobis-(2-amidinopropane)dihydrochloride, 2,2-azobis-(*N*,*N*-dimethylene)isobutyramidine dihydrochloride, 2-(carbamoylazo)isobutylonitril, 2,2-azobis-[2-(2-imidazolin-2-yl) propane]dihydrochloride, 4,4-azobis-(4-cyanovaleric acid), may be used as examples of azo-based initiators. Some of these compounds are shown in Figure 2.13.

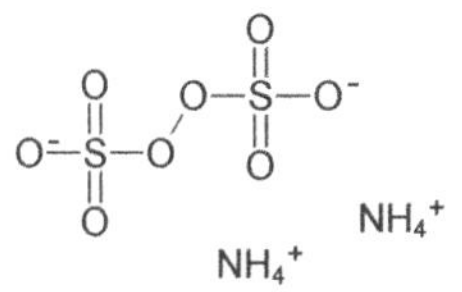

Ammonium persulfate

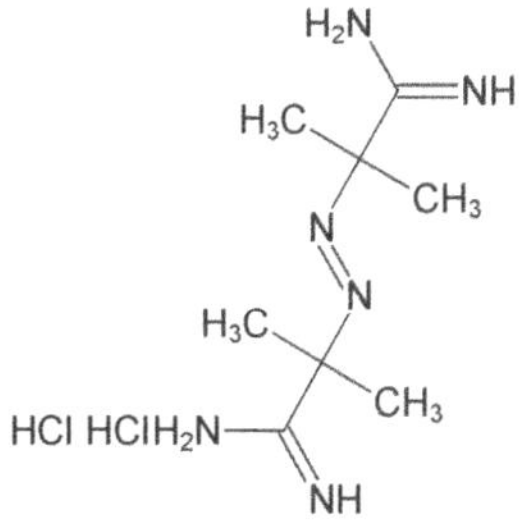

2,2-Azobis-(2-amidinopropane) dihydrochloride

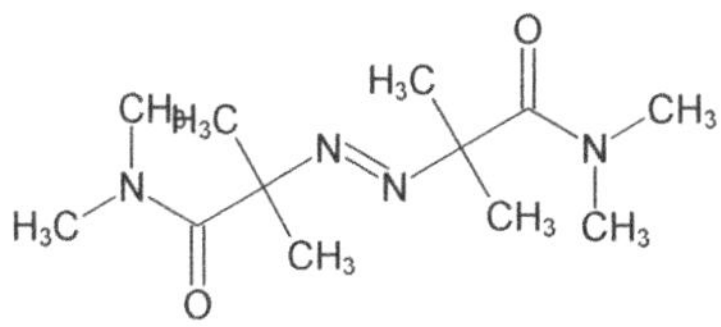

2,2-Azobis-(*N*,*N*-dimethylene) isobutyramidine dihydrochloride

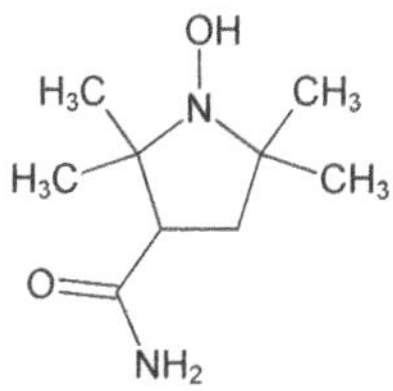

2-(Carbamoylazo) isobutylonitril

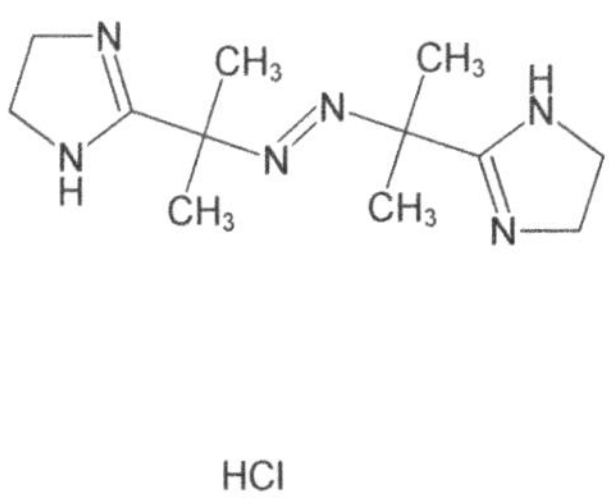

2,2-Azobis-[2-(2-imidazolin-2-yl) propane]dihydrochloride

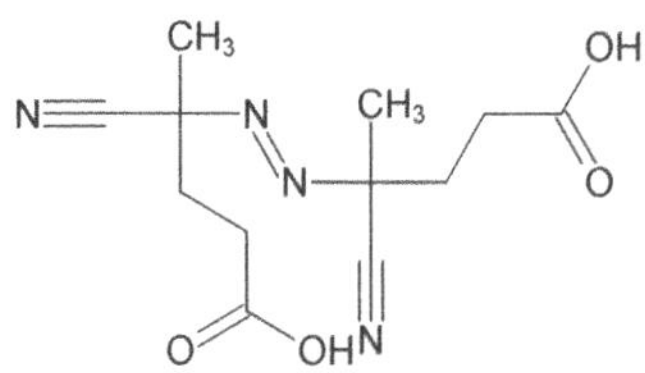

4,4-Azobis-(4-cyanovaleric acid)

Figure 2.13 Thermal polymerization initiators.

More various thermal polymerization initiators are disclosed in a monograph (87).

The preparation may be conducted as follows (84):

Preparation 2–2: First, 35.83 *g* of acrylic acid, 56.54 *g* of sodium hydroxide (NaOH, 30% solution), and 7.76 *g* of water were mixed to prepare a neutralized solution (the neutralization degree of acrylic acid: about 90 *mol*%). Then, 1,000 *ppm* of poly(ethylene glycol diacrylate) (M_w=400) as an internal crosslinking agent, 1.07 *g* of Pluronic® P123 [poly(ethylene oxide)(propylene oxide)(ethylene oxide) 3 block copolymer] and 0.71 *g* of silica particles surface-modified with a hydroxy group (30 wt% in the aqueous solution, colloidal silica having a particle diameter of 12 *nm* with a specific surface area of 200 m^2g^{-1}) were added to the neutralized solution to prepare a monomer aqueous solution having a monomer concentration of 50 wt%.

The monomer aqueous solution was subjected to high-shear mixing at 5,000 *rpm* for about 60 *min* using a mechanical mixer. Thereafter, 810 *g* of the high-shear mixed monomer neutralized solution was first mixed with 30.54 *g* of a 0.18% ascorbic acid solution and 33 *g* of a 1% sodium persulfate solution, and then continuously polymerized with 30.45 *g* of a 0.15% hydrogen peroxide solution while being fed through a feeding part of the polymerization reactor capable of kneading. At this time, the temperature of the polymerization reactor was maintained at 80°C, the maximum temperature of the polymerization was 110°C, and the polymerization time was 1 *min* and 15 *sec*. Thereafter, the polymerization and kneading were carried out for 20 *min* by continuing the kneading.

The size of the prepared polymer was distributed below 0.2 *cm*. Here, the moisture content of the finally prepared hydrogel polymer was 51 wt%. Subsequently, the hydrogel polymer was dried with a hot air drier at 180°C for 30 *min*, and the dried hydrogel polymer was pulverized with a pin mill pulverizer. And then, the polymer having a diameter less than about 150 *μm* and the polymer having a diameter of about 150 *μm* to 850 *μm* were classified by using a sieve. Thereafter, a surface treatment solution containing 5% of 1,3-propanediol and 5% of propylene glycol was sprayed onto the prepared base resin powder, and stirred at room temperature so that the surface treatment solution was evenly distributed on the base resin powder.

Thereafter, a porous superabsorbent polymer having a diameter of about 150 to 850 *μm* was obtained by using a sieve. In the product of the porous superabsorbent polymer, the content of the fine powder having a diameter of about 150 *μm* or less was less than about 2%.

A SEM image of the above prepared porous superabsorbent polymer is shown in Figure 2.14.

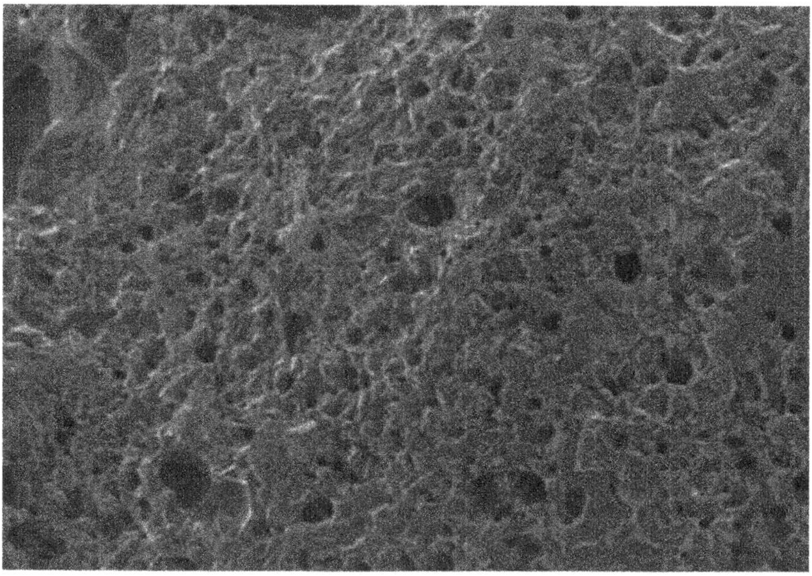

Figure 2.14 SEM image (84).

2.16 Functionalization Methods

2.16.1 Thiol-Ene Click Chemistry

The preparation of porous polymer monoliths with dodecyl and zwitterionic functionalities using the thiol-ene click chemistry of thiol-containing monoliths with both hydrophobic and polar methacrylate monomers has been shown (88).

Here, glycidyl methacrylate, ethylene glycol dimethacrylate, c.f. Figure 4.1, lauryl methacrylate, c.f. Figure 2.15, and [2-(methacryloyloxy)ethyl]-dimethyl-(3-sulfopropyl) ammonium betaine, c.f. Figure 4.20, were used.

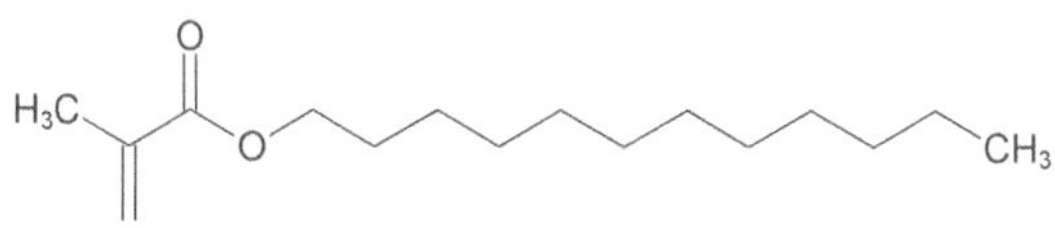

Figure 2.15 Lauryl methacrylate.

Generic monoliths were prepared in vinylized capillaries via *in-*

-*situ* polymerization of a mixture shown in Table 2.6.

Table 2.6 Composition of polymers (88).

Compound	Amount/[%]
Glycidyl methacrylate	24
Ethylene glycol dimethacrylate	16
Cyclohexanol	30
1-Dodecanol	30
2,2´-Azobisisobutyronitrile (initiator)	1

After sonication for 15 *min* and degassing by purging with nitrogen for 5 *min*, the mixture was then introduced into the vinylized capillary. The capillary was sealed at both ends with a rubber septum and immersed in a thermostated water bath at 60°C for 24 *h*. After the polymerization reaction was completed, a few centimeters from both ends of the capillary were cut to liberate the virgin structure, and the monolith was flushed with acetonitrile to remove unreacted components (88).

2.16.2 *Ionic Bond Functionalization*

A grafted polymer was modified by an ionic bond. This material was synthesized by a simple and scalable method based on sulfonated PS and a monoamine-terminated poly(ethylene oxide) derivative (89).

Porous polymer electrolytes containing the grafted polymer and PVDF were prepared by a phase inversion process (90).

The PVDF powder and grafted copolymer were dissolved in a mixture of glycerine, a non-solvent and *N*-methyl-2-pyrrolidone, in a volume ratio of 1:10, to form a homogeneous solution by stirring for 16 *h*. Then the solution was casted onto a glass plate and placed in a vacuum oven at 120°C for 24 *h* to evaporate the solvent. So, a series of freestanding membranes was obtained.

The dense pores distributed in the polymer matrix, which could store a liquid electrolyte and provide enough channels for ionic conduction, and the ionic conductivity of porous polymer electrolytes could be effectively improved.

The effect of the mass fraction of grafted polymer on porosity, electrolyte uptake, and ionic conductivity was investigated. The porous polymer electrolyte membrane with 30% grafted polymer exhibited a better pore distribution and a superior electrolyte uptake in comparison to a pure PVDF membrane (89). This resulted in a remarkable ionic conductivity of 3.05×10^{-3} $S\,cm^{-1}$ at 30°C. In addition, the contact angle decreased significantly owing to the good affinity of grafted polymer with electrolyte.

These results suggest that porous polymer electrolytes containing a grafted polymer via ionic band can be promising electrolyte candidates for high-performance lithium-ion batteries (89).

2.16.3 Pore-Size-Specific Functionalization

The pores of mesoporous silica particles were blocked with fluorenylmethyloxycarbonyl (Fmoc) protected silanes (91). Pore blockage was proved by N_2 sorption measurements. In the next step, a short treatment with a base cleaved only the surface Fmoc groups, which could be then labeled with a fluorophore. Location of the fluorophore was followed by confocal analysis. The usage of polymeric reagents for pore-size-specific functionalization of macroporous particles was developed by Svec and Fréchet (92–94).

When macroporous, epoxy-containing beads were treated with poly(styrene sulfonic acid), only the epoxy groups of the larger pores hydrolyzed, leaving the remaining epoxy groups of the smaller pores to be functionalized with small amines. Epoxy group titration showed that the amount of intact epoxy groups were increasing by the size of hydrolyzing reagent, which was used as a proof of pore-size-specific functionalization.

Recently, the same chemistry was applied to monoliths (95). After the treatment with poly(styrene sulfonic acid), pores smaller than 7 *nm* remained unchanged, half of the ~80 *nm* pores were lost and the number of ~40 *nm* pores increased by 150%. This was followed by inverse size exclusion chromatography as a proof of pore-size-specific functionalization (1).

References

1. M.T. Gokmen and F.E. Du Prez, *Progress in Polymer Science*, Vol. 37, p. 365, 2012. Topical Issue on Nanomaterials.
2. M. Okubo, Y. Konishi, T. Inohara, and H. Minami, *Macromolecular Symposia*, Vol. 175, p. 321, 2001.

3. M. Okubo, Y. Konishi, T. Inohara, and H. Minami, *Journal of Applied Polymer Science*, Vol. 86, p. 1087, 2002.
4. D. Horák, Jiří-Labsky, J. Pilaí, M. Bleha, Z. Pelzbauer, and F. Švec, *Polymer*, Vol. 34, p. 3481, 1993.
5. A. Ferreira, M. Bigan, and D. Blondeau, *Reactive and Functional Polymers*, Vol. 56, p. 123, 2003.
6. K. Kanamori, J. Hasegawa, K. Nakanishi, and T. Hanada, *Macromolecules*, Vol. 41, p. 7186, 2008.
7. F.S. Macintyre and D.C. Sherrington, *Macromolecules*, Vol. 37, p. 7628, 2004.
8. M.J. Benesí, D. Horák, and F. Svec, *Journal of Separation Science*, Vol. 28, p. 1855, 2005.
9. Q. Liu, L. Wang, A. Xiao, H. Yu, Q. Tan, J. Ding, and G. Ren, *The Journal of Physical Chemistry C*, Vol. 112, p. 13171, 2008.
10. S. Dubinsky, J.I. Park, I. Gourevich, C. Chan, M. Deetz, and E. Kumacheva, *Macromolecules*, Vol. 42, p. 1990, 2009.
11. W.-Q. Zhou, T.-Y. Gu, Z.-G. Su, and G.-H. Ma, *Polymer*, Vol. 48, p. 1981, 2007.
12. W.-Q. Zhou, T.-Y. Gu, Z.-G. Su, and G.-H. Ma, *European Polymer Journal*, Vol. 43, p. 4493, 2007.
13. L. Wu and G.-C. Pang, *Chromatographia*, Vol. 66, p. 151, 2007.
14. L. Wu, S. Bai, and Y. Sun, *Biotechnology Progress*, Vol. 19, p. 1300, 2003.
15. N. Ishizuka, K. Konishi, T. Oda, Y. Tsujii, K. Sakakibara, and T. Sato, Porous particle made of organic polymer, method for producing porous particle made of organic polymer, and block copolymer, US Patent 10 858 491, assigned to Kosen National Institute Of Technology, Kyoto University, Emaus Kyoto Inc., December 08, 2020.
16. S.J. Pierre, J.C. Thies, A. Dureault, N.R. Cameron, J.C.M. van Hest, N. Carette, T. Michon, and R. Weberskirch, *Advanced Materials*, Vol. 18, p. 1822, 2006.
17. M. Bokhari, R.J. Carnachan, N.R. Cameron, and S.A. Przyborski, *Biochemical and Biophysical Research Communications*, Vol. 354, p. 1095, 2007.
18. N. Dizge, B. Keskinler, and A. Tanriseven, *Colloids and Surfaces B: Biointerfaces*, Vol. 66, p. 34, 2008.
19. F. Su, C.L. Bray, B. Tan, and A.I. Cooper, *Advanced Materials*, Vol. 20, p. 2663, 2008.
20. K. Zhang, Z. Vobecka, K. Tauer, M. Antonietti, and F. Vilela, *Chem. Commun.*, Vol. 49, p. 11158, 2013.
21. Z.J. Wang, S. Ghasimi, K. Landfester, and K.A. Zhang, *Chemical Communications*, Vol. 50, p. 8177, 2014.
22. Z.J. Wang, K. Landfester, and K.A.I. Zhang, *Polymer Chemistry*, Vol. 5, p. 3559, 2014.

23. R. Foudazi, C. Bezik, D.L. Feke, I. Manas-Zloczower, S.R. Merrigan, and S.J. Rowan, Method for the production of high internal phase emulsion foams, US Patent 9 574 058, assigned to Procter and Gamble Co., February 21, 2017.
24. J.C. Dyer, Flame retardant microporous polymeric foams, US Patent 6 160 028, assigned to Procter and Gamble Co., December 12, 2000.
25. S. Sugiura, M. Nakajima, and M. Seki, *Industrial & Engineering Chemistry Research*, Vol. 41, p. 4043, 2002.
26. S. Sugiura, T. Oda, Y. Izumida, Y. Aoyagi, M. Satake, A. Ochiai, N. Ohkohchi, and M. Nakajima, *Biomaterials*, Vol. 26, p. 3327, 2005.
27. Q.Y. Xu, M. Nakajima, and B.P. Binks, *Colloids and Surfaces A: Physicochemical and Engineering Aspects*, Vol. 262, p. 94, 2005.
28. I. Kobayashi, T. Takano, R. Maeda, Y. Wada, K. Uemura, and M. Nakajima, *Microfluidics and Nanofluidics*, Vol. 4, p. 167, 2008.
29. I. Kobayashi, Y. Wada, K. Uemura, and M. Nakajima, *Microfluidics and Nanofluidics*, Vol. 5, p. 677, 2008.
30. M.M. Dragosavac, M.N. Sovilj, S.R. Kosvintsev, R.G. Holdich, and G.T. Vladisavljević, *Journal of Membrane Science*, Vol. 322, p. 178, 2008.
31. G.T. Vladisavljević, S. Tesch, and H. Schubert, *Chemical Engineering and Processing: Process Intensification*, Vol. 41, p. 231, 2002.
32. A. Trentin, M. Ferrando, F. López, and C. Güell, *Desalination*, Vol. 245, p. 388, 2009. Engineering with Membranes 2008.
33. G.T. Vladisavljević and R.A. Williams, *Advances in Colloid and Interface Science*, Vol. 113, p. 1, 2005.
34. Z. Lian, Y. Ren, J. He, G.Z. Chen, and K.S. Koh, Vol. 22, p. 145, 2018.
35. L. Fang, Z.-H. Guo, C. Wang, Q. Zhang, and H.-C. Zhou, Methanesulfonic acid mediated solvent free synthesis of conjugated porous polymer networks, US Patent 10 934 389, assigned to Texas A&M University System, March 02, 2021.
36. J. March and M.B. Smith, *Advanced Organic Chemistry*, John Wiley & Sons, Inc,, 2007.
37. S.S. Elmorsy, A. Pelter, and K. Smith, *Tetrahedron Letters*, Vol. 32, p. 4175, 1991.
38. F. Cherioux and L. Guyard, *Advanced Functional Materials*, Vol. 11, p. 305, 2001.
39. X.-Y. Cao, W.-B. Zhang, J.-L. Wang, X.-H. Zhou, H. Lu, and J. Pei, *Journal of the American Chemical Society*, Vol. 125, p. 12430, 2003.
40. M. Rose, N. Klein, I. Senkovska, C. Schrage, P. Wollmann, W. Böhlmann, B. Böhringer, S. Fichtner, and S. Kaskel, *J. Mater. Chem.*, Vol. 21, p. 711, 2011.
41. Y.-C. Zhao, D. Zhou, Q. Chen, X.-J. Zhang, N. Bian, A.-D. Qi, and B.-H. Han, *Macromolecules*, Vol. 44, p. 6382, 2011.
42. F.M. Wisser, K. Eckhardt, D. Wisser, W. Böhlmann, J. Grothe, E. Brunner, and S. Kaskel, *Macromolecules*, Vol. 47, p. 4210, 2014.

43. Wikipedia contributors, Suspension polymerization — Wikipedia, the free encyclopedia, https://en.wikipedia.org/w/index.php?title=Suspension_polymerization&oldid=998882934, 2021. [Online; accessed 7-September-2021].
44. J. Wang, Z. Yang, J. Xu, M. Ahmad, H. Zhang, A. Zhang, Q. Zhang, X. Kou, and B. Zhang, *Macromolecular Rapid Communications*, Vol. 40, p. 1800768, 2019.
45. D. Horák, F. Lednický, V. Řehák, and F. Švec, *Journal of Applied Polymer Science*, Vol. 49, p. 2041, 1993.
46. C. Hulubei, C.D. Vlad, I. Stoica, D. Popovici, G. Lisa, S.L. Nica, and A.I. Barzic, *Journal of Polymer Research*, Vol. 21, p. 514, 2014.
47. P.J. Dowding, J.W. Goodwin, and B. Vincent, *Colloids and Surfaces A: Physicochemical and Engineering Aspects*, Vol. 145, p. 263, 1998.
48. X. Cui, H. Shao, Y. Song, S. Yang, F. Wang, and H. Liu, *RSC Advances*, Vol. 9, p. 25730, 2019.
49. J. Jayakrishnan and B.C. Thanoo, *Journal of Biomedical Materials Research*, Vol. 24, p. 913, 1990.
50. F.M.B. Coutinho and R.C.A. Cid, *European Polymer Journal*, Vol. 26, p. 1185, 1990.
51. L. Qin, H. Lin, F.-l. Gong, D.-X. Hao, and G.-H. Ma, *Chinese Journal of Process Engineering*, Vol. 7, p. 530, 2007.
52. D.N. Uy Lan, S. Hadano, A. Abu Bakar, B. Azahari, Z.M. Ariff, and Y. Chujo, *Polymer Testing*, Vol. 30, p. 841, 2011.
53. E.U. Rehman, R. Naheed, M.A. Hamid, and M. Fuzail, *Journal of the Chemical Society of Pakistan*, Vol. 21, p. 1, 2011.
54. H. Bodugoz-Senturk and O. Güven, *Hacettepe Journal of Biology and Chemistry*, Vol. 42, p. 81, 2014.
55. W. Lin, L.T. Biegler, and A.M. Jacobson, *Chemical Engineering Science*, Vol. 65, p. 4350, 2010.
56. S. Chen, J. Hu, Y. Guo, and T. Deng, *Scientific Reports*, Vol. 10, p. 8221, 2020.
57. R.G. Gilbert, *Emulsion Polymerization: A Mechanistic Approach*, Academic Press, 1995.
58. R. Takekoh, W.-H. Li, N.A.D. Burke, and H.D.H. Stöver, *Journal of the American Chemical Society*, Vol. 128, p. 240, 2006.
59. W.-H. Li and H.D.H. Stöver, *Journal of Polymer Science Part A: Polymer Chemistry*, Vol. 36, p. 1543, 1998.
60. W.-H. Li and H.D.H. Stöver, *Macromolecules*, Vol. 33, p. 4354, 2000.
61. P. Zhao, J. Wang, Y. Li, X. Wang, C. Chen, and G. Liu, *Polymers*, Vol. 12, p. 1863, August 2020.
62. A.S. Utada, E. Lorenceau, D.R. Link, P.D. Kaplan, H.A. Stone, and D.A. Weitz, *Science*, Vol. 308, p. 537, 2005.

63. F. Du Prez, M. Gökmen, A.P. Ramaswamy, S.A.F. Bon, T. Kaufmann, and B.J. Ravoo, Porous polymer particles via microfluidics: Design and applications, in *Abstracts of Papers of the American Chemical Society*, Vol. 245, New Orleans, LA, USA, 2013.
64. H. Evard, H. Priks, I. Saar, H. Aavola, T. Tamm, and I. Leito, *Micromachines*, Vol. 12, p. 671, June 2021.
65. Y. Zhao, Z. Liao, and Z. Xiang, *Chemical Engineering Science*, Vol. 195, p. 801, 2019.
66. W. Doerffler and K. Hauffe, *Journal of Catalysis*, Vol. 3, p. 171, 1964.
67. F. Steinbach, *Nature*, Vol. 221, p. 657, 1969.
68. A. Fujishima and K. Honda, *Nature*, Vol. 238, p. 37, 1972.
69. T. Zhang, G. Xing, W. Chen, and L. Chen, *Materials Chemistry Frontiers*, Vol. 4, p. 332, 2020.
70. R. Li, J. Byun, W. Huang, C. Ayed, L. Wang, and K.A.I. Zhang, *ACS Catalysis*, Vol. 8, p. 4735, 2018.
71. G.-B. Wang, S. Li, C.-X. Yan, F.-C. Zhu, Q.-Q. Lin, K.-H. Xie, Y. Geng, and Y.-B. Dong, *Journal of Materials Chemistry A*, Vol. 8, p. 6957, 2020.
72. W. Tu, Y. Zhou, and Z. Zou, *Advanced Materials*, Vol. 26, p. 4607, 2014.
73. C. Wang, Z. Sun, Y. Zheng, and Y.H. Hu, *J. Mater. Chem. A*, Vol. 7, p. 865, 2019.
74. Y. Fink and B.J.-B. Grena, Thermally-drawn fiber including porosity, US Patent Application 20160155534, assigned to Massachusetts Institute of Technology, June 02, 2016.
75. Y. Su, C. Chen, Y. Li, and J. Li, *Journal of Macromolecular Science Part A: Pure and Applied Chemistry*, Vol. 44, p. 99, 2007.
76. D.R. Lloyd, K.E. Kinzer, and H. Tseng, *Journal of Membrane Science*, Vol. 52, p. 239, 1990. Selected papers presented at the Third Ravello Symposium on Advanced Membrane Science and Technology.
77. P. van de Witte, P.J. Dijkstra, J.W.A. van den Berg, and J. Feijen, *Journal of Membrane Science*, Vol. 117, p. 1, 1996.
78. T. Andersen, S. Joslin, and A. Kristiansen, Biodegradable foam, WO Patent 2007103208, assigned to Therese Andersen, Fmc Biopolymer As, Fmc Corp, Sara Joslin, Are Kristiansen, December 24, 2008.
79. S.G. Fullana, B. Sallerin, R. Bushkalova, and C. Ceccaldi, Process for preparing biocompatible and biodegradable porous three-dimensional polymer matrices and uses thereof, US Patent 10781293, assigned to Centre Hospitalier Universitaire de Toulouse, Centre National de la Recherche Scientifique CNRS, September 22, 2020.
80. Y. Naka, H. Takayama, T. Koyama, K.V. Le, and T. Sasaki, *Langmuir*, Vol. 34, p. 6210, 2018.
81. G. Widawski, M. Rawiso, and B. François, *Nature*, Vol. 369, p. 387, 1994.

82. Wikipedia contributors, Breath-figure self-assembly — Wikipedia, the free encyclopedia, https://en.wikipedia.org/w/index.php?title=Breath-figure_self-assembly&oldid=1033675855, 2021. [Online; accessed 7-August-2021].
83. X. Zhang, G. Sun, H. Liu, and X. Zhang, *RSC Advances*, Vol. 11, p. 14276, 2021.
84. K.Y. Yoon, H.S. Joo, G.C. Kim, H. Choi, and J.E. Kim, Preparation method of porous super absorbent polymer, US Patent 10 562 992, assigned to LG Chem Ltd., February 18, 2020.
85. N.S. Allen, *Journal of Photochemistry and Photobiology A: Chemistry*, Vol. 100, p. 101, 1996.
86. R. Schwalm, *UV Coatings: Basics, Recent Developments and New Applications*, Elsevier, 2006.
87. G. Odian, *Principles of Polymerization*, John Wiley & Sons, 2004.
88. Y. Lv, Z. Lin, and F. Svec, *Analyst*, Vol. 137, p. 4114, 2012.
89. M. Guo, B. Zhou, J. Hu, J. Wang, D. He, X. Xie, and Z. Xue, *Journal of Membrane Science*, Vol. 564, p. 663, 2018.
90. F. Deng, X. Wang, D. He, J. Hu, C. Gong, Y.S. Ye, X. Xie, and Z. Xue, *Journal of Membrane Science*, Vol. 491, p. 82, 2015.
91. K. Cheng and C.C. Landry, *Journal of the American Chemical Society*, Vol. 129, p. 9674, 2007.
92. V. Smigol, F. Svec, and J.M.J. Frechét, *Macromolecules*, Vol. 26, p. 5615, 1993.
93. V. Smigol, F. Svec, and J.M.J. Fréchet, *Analytical Chemistry*, Vol. 66, p. 2129, 1994.
94. F. Svec and J.M.J. Frechét, *Science*, pp. 205–211, 1996.
95. R. Bandari, T. Höche, A. Prager, K. Dirnberger, and M.R. Buchmeiser, *Chemistry – A European Journal*, Vol. 16, p. 4650, 2010.

3

Properties

3.1 Special Materials

3.1.1 Porous Polymer Pressure Sensors

Electrically conductive polymer composites based on porous polymeric matrices of PEBA 4033 (polyether block amide) incorporating different types of graphene platelets were prepared and tested (1).

The PEBA 4033 polymer was mixed with different graphene platelets: grade H5, grade MX and grade C–750. Porous morphology was obtained for the electrically conductive polymer composites prepared with graphene platelets grade MX and H5, while for the grade C–750 only a dense electrically conductive polymer composite was obtained.

The porous electrically conductive polymer composite loaded with 15% by volume of graphene M5 exhibited a linear piezoresistive response in a log-log plot. Some limitations were detected for this electrically conductive polymer composite, namely hysteresis and drift due to the poor mechanical properties.

The porous electrically conductive polymer composite loaded with 15% M5 was further crosslinked, which improved its mechanical properties but the piezoresistive effect became negligible. The incorporation of carbon black in this electrically conductive polymer composite formulation was shown to be much more efficient than crosslinking to improve mechanical properties, but the piezoresistive response became poor (1).

3.1.2 *Crack Propagation Behavior*

The crack propagation behavior of porous polymer membranes was investigated in order to identify the criterion for crack propagation (2).

A notch was introduced into the membrane as an initial crack (pre-crack), and a uniaxial loading was applied to produce stable crack propagation. During the test, the crack propagation was observed using a charge-coupled device camera and the digital image correlation method. The strain around the pre-crack tip at the onset of crack propagation was measured experimentally using the digital image correlation method.

It was found that large-scale yielding developed before crack propagation. Stable crack propagation was observed for all tensile tests. In parallel, a homogeneous model that mimicked the porous polymer membrane was created using the finite element method (FEM) to investigate the stress/strain distribution around the crack tip. This study adopted the yield criterion proposed by Deshpande and Fleck (3–5).

The computed strain distribution was compared with the experimental results, and the distributions were found to be similar. The strain distribution obtained by the digital image correlation method and FEM computation yielded a J-integral value, which could be used to investigate the criterion for crack propagation. Regardless of the initial crack length, the J-integral value at the onset of crack propagation was a constant value for all tests.

It was concluded that the criterion for crack propagation in the porous polymer membrane could be successfully determined. This study is useful for identifying the crack propagation behavior of a porous polymer membrane and for determining the criterion for crack propagation (2).

3.2 Standard Test Methods

The ASTM standard test methods that are suitable for the characterization of porous polymeric materials are summarized in Table 3.1.

Table 3.1 ASTM standard test methods for porous polymers.

Standard	Short description	Reference
ASTM F2450-18	Microstructure of Polymeric Scaffolds	(6)
ASTM F2603	Interpreting Images of Polymeric Tissue Scaffolds	(7)
ASTM F3504	Cell Proliferation in 3D Scaffolds	(8)
ASTM E128-99	Maximum Pore Diameter and Permeability	(9)
ASTM E1441-19	Standard Guide for Computed Tomography	(10)
ASTM F316-03	Pore Size Characteristics of Membrane Filters	(11)
ASTM F3191-16	Water Absorption	(12)
ASTM F1929-15	Leaks in Porous Medical Packaging	(13)
ASTM F1608-16	Microbial Ranking of Porous Packaging Materials	(14)
ASTM F755-19	Porous Polyethylene in Surgical Implants	(15)
ASTM F3384	Polydioxanone Polymer Resins for Surgical Implants	(16)
ASTM F1925	Semi-Crystalline Poly(lactide) Polymer and Copolymer Resins for Surgical Implants	(17)
ASTM F2902	Assessment of Absorbable Polymeric Implants	(18)
ASTM E3160-18	Antibacterial Properties	(19)
ASTM E1428	Evaluation of the Performance of Antimicrobials in Polymeric Solids	(20)
ASTM E2015	Preparation of Polymeric Specimens for Microstructural Examination	(21)
ASTM E2611-19	Acoustical Properties	(22)
ASTM F2228-13	Detection of Leaks in Packaging	(23)
ASTM D4404-18	Mercury Intrusion Porosimetry	(24)
ASTM F726-17	Sorbent Performance of Adsorbents	(25)
ASTM F716-18	Sorbent Performance of Absorbents	(26)

3.2.1 *Polymeric Scaffolds*

3.2.1.1 *Microstructure*

The microstructure of the scaffold can be defined by the existence, type, size distribution, interconnectivity, and directionality of pores, all of which are critical for cell migration, growth, and proliferation. Optimizing the design of tissue scaffolds is a complex task, given the range of available materials, different manufacturing routes, and processing conditions.

Tissue scaffolds are typically assessed using an overall value for scaffold porosity and a range of pore sizes, though the distribution of sizes is rarely quantified (6). Published mean pore sizes and distributions are usually obtained from electron microscopy images and quoted in the micrometer range.

Tissue scaffolds are generally complex structures that are not easily interpreted in terms of pore shape and size, especially in three dimensions. Therefore, it is difficult to quantifiably assess the batch-to-batch variance in microstructure or to make a systematic investigation of the role that the mean pore size and pore size distribution has on influencing cell behavior based solely on electron micrographs

A method for the measurement of the specific area for the characterization of polymer-based porous scaffolds has been described (6).

An overview of the test methods that may be used to obtain information relating to the dimensions of pores, the pore size distribution, the degree of porosity, interconnectivity, and measures of permeability for porous materials used as polymeric scaffolds has been given. Such information can be the key to optimizing the structure for a particular application, thus developing robust manufacturing routes, and providing reliable quality control data.

3.2.1.2 *Interpretion of Images*

The factors that need to be considered in obtaining and interpreting images of tissue scaffolds ,including technique selection, instrument resolution and image quality, quantification and sample preparation have been detailed (7). Such information is intended to be applicable to porous polymer-based tissue scaffolds, including naturally derived materials such as collagen.

3.2.1.3 *Cell Proliferation*

How to conduct a nondestructive proliferation test for mammalian cells based on metabolic activity is described in standard ASTM F3504, (8). This test can be used to assess the number of viable cells within three-dimensional (3D) scaffolds for regenerative medicine and in tissue-engineered medical products.

The resazurin cell metabolic activity method was assessed in terms of reagent concentrations, incubation times, cell culture media composition, calibration curve, controls, assay linearity, and limitations of the assay. Also, factors are described that can interfere with accurate cell proliferation assessment (8).

3.2.2 *Leaks in Porous Medical Packaging*

Harmful biological or particulate contaminants may enter a medical package through leaks (13). These leaks frequently can be found at seals between package components of the same or dissimilar materials. Leaks may also result from a pinhole in the packaging material.

The presented test method defines materials and procedures that will detect and locate a leak equal to or greater than a channel formed by a 50 μm wire in package edge seals formed between a transparent material and a porous sheet material.

A dye penetrant solution is applied locally to the seal edge to be tested for leaks. After contact with the dye penetrant for a specified time, the package is visually inspected for dye penetration (13). Three dye application methods are covered in this test method (13):

1. Injection,
2. Edge dip, and
3. Eyedropper.

3.2.3 *Pore Diameter and Permeability*

A test method for the determination of the maximum pore diameter and permeability of rigid porous filters used in the laboratory for filtration or diffusion has been described (9). The method is applicable to filters made of sintered glass, ceramic, metal, or plastic.

Furthermore, this test method establishes a uniform designation for the maximum pore diameter and also provides a means of detecting and measuring changes which occur through continued use (9). The maximum pore diameter is determined by immersing the filter in a suitable test liquid and applying air pressure until the first bubble of air passes through the filter.

Then, the maximum pore diameter is calculated from the surface tension of the test liquid and the applied pressure. The permeability is determined by measuring the flow of air through the filter when subjected to a pressure differential (9).

3.2.4 *Mercury Intrusion Porosimetry*

The ASTM D4404-18 test method was originally intended for use in determining the volume and the volume distribution of pores in soil and rock with respect to the apparent diameter of the entrances of the pores (24). However, the test can also be used for organic materials.

The pores of polymer-based monolithic capillary columns were investigated by inverse size-exclusion chromatography and mercury-intrusion porosimetry (27). The mesopore size distributions of the monolithic materials determined with inverse size-exclusion chromatography were compared with those measured by mercury-intrusion porosimetry. Both techniques were found to be complementary.

3.2.5 *Pore Size of a Membrane Filter*

Membrane filters have discrete pores from one side to the other of the membrane, similar to capillary tubes (11). The bubble-point test (28) is based on the principle that a wetting liquid is held in these capillary pores by capillary attraction and surface tension, and the minimum pressure required to force liquid from these pores is a function of pore diameter.

The pressure at which a steady stream of bubbles appears in this test is the bubble-point pressure. The bubble-point test is significant not only for indicating maximum pore size, but may also indicate a damaged membrane, ineffective seals, or a system leak (11).

3.2.6 *Computed Tomography*

Computed tomography (CT) is a radiographic reconstruction method that provides a sensitive technique whenever the primary goal is to locate and size planar and volumetric detail in three dimensions (10).

Computed tomography is a radiographic examination technique that generates digital images of an object that generates digital images, including the interior structure. Because of the relatively good penetrability of X-rays, CT permits the nondestructive physical and, to a limited extent, chemical characterization of the internal structure of materials.

Also, since the method is X-ray based, it applies equally well to metallic and non-metallic specimens, solid and fibrous materials, and smooth and irregularly surfaced objects (10).

3.2.7 *Water Absorption*

The ability of a substrate surface to absorb water is a key indicator in determining how to correctly install many types of flooring adhesives, primers, self-leveling underlayments, and other products (12).

Installing flooring products over low or non-absorptive substrates, sometimes referred to as *nonporous* substrates, such as densely machine-troweled concrete, mature and well-hydrated concrete, existing resilient flooring, polymer terrazzo and others, may require adjustments to the surface preparation method or product selection to ensure a successful installation.

A substrate surface porosity test has been described in the ASTM standeard F3191-16 (12),

A qualitative assessment of substrate water absorption (porosity) and whether or not that substrate should be regarded as porous/absorptive or nonporous/non-absorptive as these terms relate to the installation of resilient floor coverings, adhesives, self-leveling underlayments, primers, and other products.

In a test, the size, shape, and color of the water drop may indicate the presence of contaminants (12).

3.2.8 Microbial Ranking of Porous Packaging Materials

The exposure-chamber method is a quantitative procedure for determining the microbial-barrier properties of porous materials. The data obtained from this test are useful for assessing the relative potential of a particular porous material in contributing to the loss of sterility to the contents of the package versus another porous material.

The maintenance of the sterility in a particular packaging application will depend on a number of factors, such as (14):

1. The bacterial challenge (number and kinds of microorganisms) that the package will encounter in its distribution and use. This may be influenced by factors such as shipping methods, expected shelf life, geographic location, and storage conditions.
2. The package design, including factors such as adhesion between materials, the presence or absence of secondary and tertiary packaging, and the nature of the device within the package.
3. The rate and volume exchange of air that the porous package encounters during its distribution and shelf life. This can be influenced by factors including the free-air volume within the package and pressure changes occurring as a result of transportation, manipulation, weather, or mechanical influences.
4. The microstructure of a porous material which influences the relative ability to adsorb or entrap microorganisms, or both, under different air-flow conditions.

A round-robin study was conducted with eleven laboratories participating. Each laboratory tested duplicate samples of six commercially available porous materials to determine the log reduction value. Materials tested under the standard conditions described in this test method returned average values that range from a log reduction value of 1.7 to 4 (14).

This standard and its uses were also described in a monograph (29).

3.2.9 *Antibacterial Properties*

Porous articles, often textiles, are often treated with antimicrobial agents to reduce the growth of microorganisms during use, in storage, or while waiting to be laundered, or both (19). In addition, antimicrobial agents are added to reduce or control the overall microbial growth on porous articles that may affect the odor of the material, visual, chemical or physical integrity, or both.

The here described test method is appropriate for porous materials such as textiles, paper, or similar porous materials. It is intended to measure the antibacterial properties of these materials.

To determine the bactericidal or bacteriostatic properties of porous articles treated with an active biocidal agent, samples of porous treated materials, such as textiles or paper, are inoculated with a defined suspension of microorganisms and then incubated.

The changes in numbers of the bacterial populations on the treated article are compared with untreated articles either over a designated time or they are compared to the initial bacterial population at *zero time* for the treated article to measure the antibacterial properties (19).

3.2.10 *Performance of Antimicrobials*

This test method is intended to assess susceptibility of flat two-dimensional vinyl films and other solid polymer products as well as products that may directly contact vinyl to pink-staining by the actinomycete bacteria *Streptomyces* species (20). This test method may not be suitable for highly textured or porous substrates.

3.2.11 *Surgical Implants*

3.2.11.1 *Porous Poly(ethylene)*

The requirements of the raw material and the final product were detailed (15). Also, the associated test methods for the selection of porous high-density and ultrahigh molecular weight poly(ethylene)s intended for use in surgical implants were described.

Porous poly(ethylene) (PE) may be used as a freestanding product or as a coating on a substrate in nonloaded applications.

The physical properties such as pore size and volume, as well as tensile, compressive, flexural, and shear properties should be determined.

Here, evaluation of the tissue response to a porous PE must be completed. A guidance in establishing the biocompatibility can be found in a list of references in this ASTM standard (15).

Clinical experience and animal studies have shown that the tissue will grow into the open pores of porous PE. The tissue ingrowth into the pores may facilitate the establishment of the fixation of the implant (15).

3.2.11.2 *Semicrystalline Poly(lactide)*

The ASTM F1925 standard uses a virgin poly(*l*-lactic acid) (PLLA) resin for use in surgical implants. The specification does not cover stereoisomeric compositions based on various D, L, or DL copolymer ratios (17).

This specification addresses material characteristics of virgin PLLA resin and does not apply to packaged and sterilized finished implants fabricated from this material. The virgin polymer shall be a homopolymer of *L*-lactide. The molecular mass of the virgin polymer shall be indicated by relative solution viscosity (in chloroform). In addition, the weight average molecular mass and molecular mass distributions may be determined by gel permeation chromatography.

The virgin polymer shall be identified as a poly(lactide) by infrared or 1H-NMR spectroscopy and shall have a specific optical rotation (in dichloromethane) and residual monomer content within the prescribe values, and shall conform to the chemical and physical property requirements specified for residual solvent, residual water, residual tin, heavy metals, and sulfated ash. The following test methods shall be used (17):

1. Karl Fischer titration, and
2. Atomic absorption-emission spectroscopy, or
3. Inductively coupled plasma spectroscopy.

3.2.11.3 Poly(dioxanone) Polymer Resins

Virgin poly(dioxanone) homopolymers have been described that are intended for use in surgical implants (16). These polymers can be fully solvated at 30°C by fluorinated solvents such as hexafluoroisopropanol or hexafluoroacetone. Furthermore, these materials are known to be semicrystalline.

3.2.11.4 Absorbable Polymeric Implants

General guidelines have been described for the chemical, physical, mechanical, biocompatibility, and preclinical assessments of implantable synthetic polymeric absorbable devices (18). Here, evaluation methods are also described that are potentially useful and should be considered when assessing absorbable implants or implant components.

These evaluations may assist a manufacturer in establishing the safety and effectiveness of an absorbable implant device. The assessment methods may also be utilized to assist in establishing substantial equivalence to an existing commercially marketed device.

This guide is intended to cover all forms of absorbable polymeric components and devices, including solid and porous forms.

Absorbable polymers used as a matrix to control the *in-vivo* release of bioactive agents (drugs, antimicrobials, and others) may be evaluated according to many of the methods described herein. Composites of absorbable polymers with ceramics and/or metals may be evaluated according to many of the methods described herein (18).

3.2.12 Acoustical Properties

In this approach, the sound impinges on the specimen in a perpendicular direction (*normal incidence*) only, compared to the random incidence of traditional methods. In addition, reverberation room methods specify certain minimum sizes for test specimens which may not be practical for all materials.

Even though this method may not replicate the reverberant room methods for measuring the transmission loss of materials, it can provide comparison data for small specimens, something that cannot be done in the reverberant room method.

Normal incidence transmission loss may also be useful in certain situations where the material is placed within a small acoustical cavity close to a sound source, for example, a closely-fitted machine enclosure or portable electronic device.

The test method covers the use of a tube, four microphones, and a digital frequency analysis system for the measurement of normal incident transmission loss and other important acoustic properties of materials by the determination of the acoustic transfer matrix (22).

The transfer matrix method is a common technique used to calculate the input impedance or reflectance for 1D acoustic systems that can be approximated in terms of concatenated cylindrical and/or conical sections, as well as side branches (30). Each section is represented by a transfer matrix that relates frequency-domain expressions of input to output in terms of pressure and volume velocity.

Methods to derive transfer matrix method characterizations for more arbitrarily shaped geometries and extensions for tonehole interactions have also been briefly discussed (30).

3.2.13 *Detection of Leaks in Packaging*

Harmful biological or particulate contaminants may enter a package through incomplete seals or imperfections such as pinholes or cracks in the trays (23).

A nondestructive test method has been described that may be performed in either laboratory or production environments. Here, CO_2 is used as tracer material. This testing may be undertaken on either a 100% or a statistical sampling basis.

The nondestructive test method can detect leaks in nonporous rigid thermoformed trays, as well as the seal between the porous lid and the tray. The test method detects channel leaks in packages as small as 100 μm diameter in the seal as well as 50 μm diameter pinholes, or equivalently sized cracks in the tray, subject to trace gas concentration in the package, package design and manufacturing tolerances (23).

A system and a method for determining the leakproofness of an object has been described (31). The system includes a closed chamber having a second cavity arranged to envelope the object, an evacuator for lowering the pressure inside one of the cavities, a supply for supplying a tracer gas into one of the cavities to render the higher pressure, a detector sensitive to the tracer gas and an introducer for introducing a transport gas into the of the cavities to render the lower pressure.

Furthermore, the evacuator is arranged to compress arriving gas to the ambient pressure of the chamber and to communicate with the detector, which is suited for operation at the ambient pressure of the chamber (31).

3.2.14 Sorbent Performance of Adsorbents

The test method for the sorbent performance of adsorbents can be used as a basis for comparison of adsorbents in a consistent manner (25). The test method covers laboratory tests that describe the performance of adsorbents in removing crude oils, nonemulsified oils and other floating, immiscible liquids from the surface of water.

The ASTM F716-18 method is used to evaluate the sorbent performance of absorbents for use on chemical and light hydrocarbon spills (26). Such test methods cover the development of laboratory test data that describe the performance of absorbent materials, and their retention ability, for use on chemical and light hydrocarbon spills.

References

1. V. Gonçalves, L.B. ao, and A. Mendes, *Polymer Testing*, Vol. 37, p. 129, 2014.
2. Y. Kodaira, T. Miura, S. Ito, K. Emori, A. Yonezu, and H. Nagatsuka, *Polymer Testing*, Vol. 96, p. 107124, 2021.
3. V.S. Deshpande and N.A. Fleck, *Journal of the Mechanics and Physics of Solids*, Vol. 48, p. 1253, 2000.
4. V.S. Deshpande and N.A. Fleck, *Acta Materialia*, Vol. 49, p. 1859, 2001.
5. A. Yonezu, K. Hirayama, H. Kishida, and X. Chen, *Polymer Testing*, Vol. 50, p. 1, 2016.
6. Standard guide for assessing microstructure of polymeric scaffolds for use in tissue-engineered medical products, ASTM Standard F2450-18, ASTM International, West Conshohocken, PA, 2018. https://www.astm.org/Standards/F2450.htm.
7. Standard guide for interpreting images of polymeric tissue scaffolds, ASTM Standard F 2603, ASTM International, West Conshohocken, PA, 2020. https://www.astm.org/f2603-06r20.html.

8. Standard practice for quantifying cell proliferation in 3d scaffolds by a nondestructive method, ASTM Standard F 3504, ASTM International, West Conshohocken, PA, 2021. https://www.astm.org/f3504-21.html.
9. Standard test method for maximum pore diameter and permeability of rigid porous filters for laboratory use, ASTM Standard E128-99, ASTM International, West Conshohocken, PA, 2019. https://www.astm.org/Standards/E128.htm.
10. Standard guide for computed tomography (CT), ASTM Standard E1441-19, ASTM International, West Conshohocken, PA, 2019. https://www.astm.org/Standards/E1441.htm.
11. Standard test methods for pore size characteristics of membrane filters by bubble point and mean flow pore test, ASTM Standard F316-03, ASTM International, West Conshohocken, PA, 2019. https://www.astm.org/Standards/F316.htm.
12. Standard practice for field determination of substrate water absorption (porosity) for substrates to receive resilient flooring, ASTM Standard F3191-16, ASTM International, West Conshohocken, PA, 2016. https://www.astm.org/Standards/F3191.htm.
13. Standard test method for detecting seal leaks in porous medical packaging by dye penetration, ASTM Standard F1929-15, ASTM International, West Conshohocken, PA, 2015. https://www.astm.org/Standards/F1929.htm.
14. Standard test method for microbial ranking of porous packaging materials (exposure chamber method), ASTM Standard F1608-16, ASTM International, West Conshohocken, PA, 2016. https://www.astm.org/Standards/F1608.htm.
15. Standard specification for selection of porous polyethylene for use in surgical implants, ASTM Standard F755-19, ASTM International, West Conshohocken, PA, 2019. https://www.astm.org/Standards/F755.htm.
16. Standard specification for polydioxanone polymer resins for surgical implants, ASTM Standard F 3384, ASTM International, West Conshohocken, PA, 2021. https://www.astm.org/f3384-21.html.
17. Standard specification for semi-crystalline poly(lactide) polymer and copolymer resins for surgical implants, ASTM Standard F 1925, ASTM International, West Conshohocken, PA, 217. https://www.astm.org/f1925-17.html.
18. Standard guide for assessment of absorbable polymeric implants, ASTM Standard F 2902, ASTM International, West Conshohocken, PA, 2017. https://www.astm.org/f2902-16e01.html.
19. Standard test method for quantitative evaluation of the antibacterial properties of porous antibacterial treated articles, ASTM Standard E3160-18, ASTM International, West Conshohocken, PA, 2018. https://www.astm.org/Standards/E3160.htm.

20. Standard test method for evaluating the performance of antimicrobials in or on polymeric solids against staining by streptomyce species (a pink stain organism), ASTM Standard E 1428, ASTM International, West Conshohocken, PA, 2016. https://www.astm.org/e1428-15a.html.
21. Standard guide for preparation of plastics and polymeric specimens for microstructural examination, ASTM Standard E 2015, ASTM International, West Conshohocken, PA, 2021. https://www.astm.org/e2015-04r21.html.
22. Standard test method for normal incidence determination of porous material acoustical properties based on the transfer matrix method, ASTM Standard E2611-19, ASTM International, West Conshohocken, PA, 2019. https://www.astm.org/Standards/E2611.htm.
23. Standard test method for non-destructive detection of leaks in packaging which incorporates porous barrier material by CO_2 tracer gas method, ASTM Standard F2228-13, ASTM International, West Conshohocken, PA, 2021. https://www.astm.org/Standards/F2228.htm.
24. Standard test method for determination of pore volume and pore volume distribution of soil and rock by mercury intrusion porosimetry, ASTM Standard D4404-18, ASTM International, West Conshohocken, PA, 2018. https://www.astm.org/Standards/D4404.htm.
25. Standard test method for sorbent performance of adsorbents for use on crude oil and related spills, ASTM Standard F726-17, ASTM International, West Conshohocken, PA, 2017. https://www.astm.org/Standards/F726.htm.
26. Standard test methods for sorbent performance of absorbents for use on chemical and light hydrocarbon spills, ASTM Standard F716-18, ASTM International, West Conshohocken, PA, 2018. https://www.astm.org/Standards/F716.htm.
27. J. Urban, S. Eeltink, P. Jandera, and P.J. Schoenmakers, *Journal of Chromatography A*, Vol. 1182, p. 161, 2008.
28. A. Hernández, J. Calvo, P. Prádanos, and F. Tejerina, *Journal of Membrane Science*, Vol. 112, p. 1, 1996.
29. P. Perez, *Reducing Levels of Medical Device Contamination through Package Redesign, Seal Geometry and Opening Technique*, Michigan State University, 2018.
30. G. Scavone, *The Journal of the Acoustical Society of America*, Vol. 149, p. A93, 2021.
31. P. Jenneus and F. Enquist, System and method for determining the leakproofness of an object, US Patent 7 448 256, assigned to Sensistor Technologies AB, November 11, 2008.

4

Medical Uses

Porous polymers have gained increased interest in the field of tissue engineering. This is attributed to their unique physicochemical properties, such as an interconnected pore structure, large surface area, and small pore size (1).

4.1 Medical Diagnostics

4.1.1 Extracellular Vesicles

When diagnosed early, most cancers are more apt to be prevented from metastasizing, which gives rise to a higher survival rate for patients (2). However, the high diversity of the cancer-related factors contributes to the difficulty of an early diagnosis.

Studies have been directed towards biomarkers that are available for the early diagnosis of cancer. One of the most interesting biomarkers is cancer cell ribonucleic acid (RNA). For use in diagnosis, the RNA has been obtained from cancer cells separated from excised cancer tissues or from circulating tumor cells (CTCs) in the bloodstream. However, the excision is accompanied by an operation that is invasive to the subject, and the finding of CTC is poor in effectiveness because it is too rare, e.g., one cell per 109 blood cells, to detect.

Most cells shed extracellular vesicles (EVs) into the body fluid. An EV serves as a transporter carrying various materials, including proteins, lipids, amino acids, etc. These substances have recently been revealed to be important factors that are indicative of properties of the EV sources, that is, the cells shedding the EV. Thus, the importance of EV has been detected. In particular, cancer cells shed a number of vesicles containing metastasis factors. Cancer cell-shed vesicles are present at a higher level in blood and thus are easier to apply to the diagnosis of cancer, compared to CTC. The vesicles shed from cancer cells can, therefore, be used for early diagnosis of cancer in a noninvasive manner.

An apparatus has been described for the isolation of extracellular vesicles from human fluid. This apparatus contains a channel formed on a microchip and a porous polymer monolith filter connected to the channel (2).

The microchip is made from a material selected from the group consisting of poly(methyl methacrylate) (PMMA), a cyclic olefin copolymer, poly(carbonate) (PC), poly(styrene) (PS), and poly(dimethyl siloxane) (PDMS).

On a microchip, two channels are formed in a cross form with a porous polymer monolith (PPM) filter located at the intersection of the crossflow channels. However, the configuration of the channels and the number of PPM filters may be varied, and are not limited to the embodiment. For instance, the microchip structure may be fabricated by connecting only one channel, but not crossflow channels, to a PPM filter.

A method for fabricating an apparatus for the isolation of extracellular vesicles, consists of (2):

1. Forming a channel and a groove connected to the channel,
2. Injecting a porous polymer monolith prepolymer solution into the groove through the channel, and
3. Polymerizing the porous polymer monolith prepolymer solution at the groove with UV light.

The fabrication of a PMMA microchannel associated with a PPM can be done as (2):

Fabrication method 4–1:

Formation of PMMA Microchannels: Microchannels were fabricated using PMMA as a base material. PMMA is suitable as a material for channel formation because it has high durability and is easy to pattern, is permeable to UV light and is compatible with other material. Using a MICO CNC (TinyCNC-S, TinyRobo, Korea), crossflow channel with a depth of 150 μm and a width of 500 μm was carved in the form of an

X-shape on a PMMA plate, while a rectangular groove for a PPM filter was formed with dimensions of 2.5×4.0 *mm* at the intersection of the crossflow channels. The PMMA plate on which the channels and the groove were formed was washed with 100% ethanol before a cover plate was attached to the PMMA plate using 5 *μl* of methyl methacrylate (purchased from Aldrich) and fixed in an oven for 1 *h*, at 65°C. Afterwards, 23-gauge steel hypodermic needle segments were connected to the ends of the channels at the four corners of the PMMA plate and fixed in an oven for 2 *h* at 65°C.

Preparation of PPM Filter Prepolymer Solution: A PPM prepolymer solution was prepared by mixing 1.8 *ml* of glycidyl methacrylate (GMA) monomer, 1.3 *ml* of ethylene glycol dimethacrylate (EGDM), 40 *mg* of 2,2-dimethoxy-2-phenylacetophenone (DMPA), 10 *mg* of hydroquinone (HQ), and 4.3-6.6 *ml* of methanol (58% to 68% of total volume). All of the materials were purchased from Aldrich. To deoxygenate the solution, it was purged for 2 *min* with nitrogen gas. In the solution, GMA serves as a reaction initiator to negatively charge the surface of the channel, thereby preventing the adsorption of negatively charged vesicles into the channel. In addition, this monomer has an acryl group chemically similar to PMMA, functioning to mediate the firm fixation of the monolith into the channel surface after polymerization. EGDM was used as a crosslinking agent, DMPA was used as a photoinitiator, and methanol was used as a porogenic solvent. HQ served as a polymerization initiator, which played a role in preventing the non-specific polymerization attributable to the transmission of heat or the diffusion of polymerization during UV-initiated polymerization.

Fabrication of PPM Filter: The channel plate patterned in 1-1 was covered with a shadow mask designed to limit the width of a central site where a PPM filter is to be formed (left upper and lower panels of Figure 4.2) to 0.5 *mm*. Thereafter, the needle segments at the four corners of the channel plate were connected with a flexible plastic Tygon tube (Saint-Gobain Performance Plastics, USA) (i.d. 0.508 *mm*) through which the PPM prepolymer solution prepared in 1-2 was then infused. Next, the tube was tied to block the introduction of oxygen, followed by exposure to UV light at 365 *nm* from a UV light source (SUV 2010-S, UV Sources & Manufacturing Technology, Korea) (17 *min*). After completion of the polymerization, ethanol and water were sequentially allowed to flow to remove the unreacted solution. The completed channel was used immediately or stored with water introduced thereinto.

Some of the compounds mentioned in method 4–1 are shown in Figure 4.1.

The process of the fabrication of a PPM filter system in which a channel carved in a crossflow pattern on a microchip is filled with

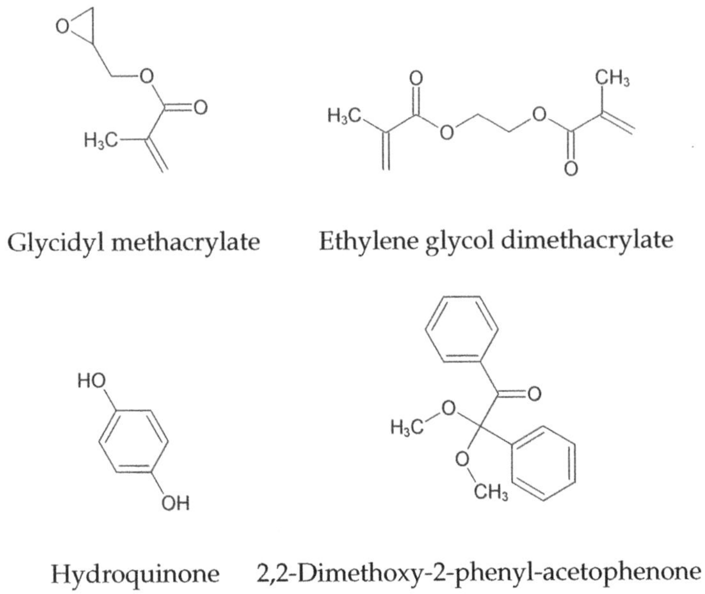

Figure 4.1 Compounds mentioned in method 4–1.

a PPM prepolymer solution and only a central region is exposed to UV to form a PPM filter is shown in Figure 4.2.

The porosity of the PPM filter varies depending on the composition of the PPM prepolymer solution, and the content of the porogenic solvent is particularly known to have direct influences on the porosity of the filter. A PPM filter becomes porous as monolith beads are formed in a net structure.

In order to determine the porosity of the filter fabricated in the fabrication method 4–1, fluorescent PS beads of various sizes were passed through the filter. Because the filter did not pass beads larger than the pore size thereof, the pore size was determined by that of the fluorescent PS beads (Duke Scientific) filtrated into the collection channel. The porosity of the PPM filter is shown in Figure 4.3.

This apparatus can be used for the diagnosis of disease, including cancer from vesicular nucleic acid in a noninvasive manner.

A method for diagnosing a disease using the here detailed apparatus can be done by:

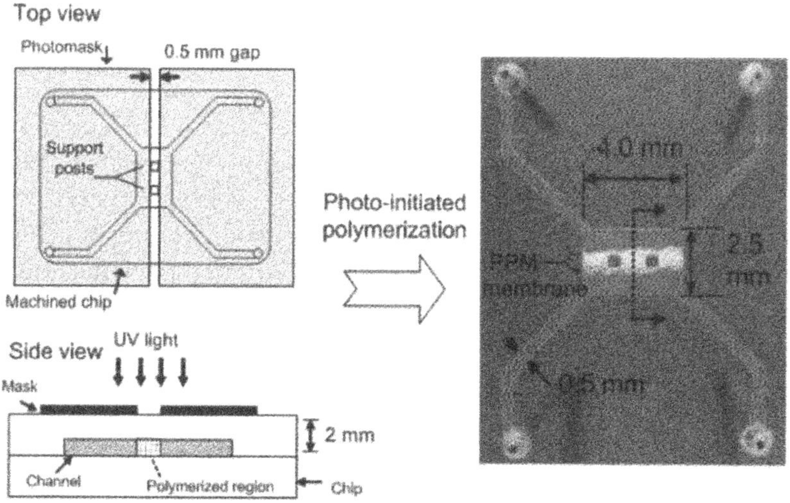

Figure 4.2 Fabrication of a PPM filter (2).

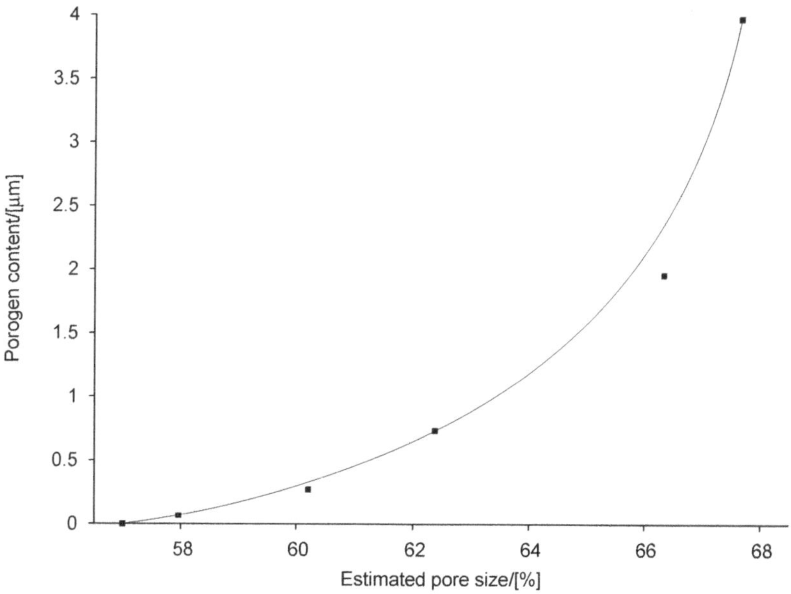

Figure 4.3 Porosity of the filter (2).

1. Isolating extracellular vesicles from body fluid using the apparatus, and
2. Analyzing a nucleic acid prepared from the extracellular vesicles.

The isolation of the vesicles can be done using a PPM filter system-pressure-driven mode and a PPM filter system-electro-driven mode (2). The apparatus, which is capable of isolating and purifying a large quantity of vesicles from a small amount of a body fluid sample within a short time, is expected to be advantageously and widely applied in the medical research and clinical diagnosis of diseases, including cancer.

For example, vesicles were isolated from 25 *μl* of blood. The filtrate was incubated with 0.8 *ml* of TRIzol (Life Technologies) for 5 *min* to dissolve the vesicle membrane. The brand name of TRIzol is guanidinium thiocyanate, c.f. Figure 4.4. The addition of 0.2 *ml* of chloroform separated RNA from other materials (proteins and deoxyribonucleic acid) (5 *min*, 4°C). Distinct layers were formed by centrifugation at 13,500×g for 10 *min*, and the supernatant (RNA layer) was mixed with one volume of isopropyl alcohol to precipitate the RNA (20 *min*, -20°C).

Following the recentrifugation at 13,500×g for 10 *min*, the pellet was washed with 75% ethanol. This was centrifuged again at 13,500×g for 10 *min* to form a RNA precipitate. Ethanol was aspirated, and the pellet was dried at room temperature for 5 *min* or longer and diluted in distilled water. The concentration of the RNA solution was determined using a spectrophotometer (NanoDrop SD2000, BioPrince). Visualization was achieved with an ethidium bromide dye, c.f. Figure 4.4. To compare the quantity of RNA with the positive control (embryonic stem cells), the housekeeping gene actin RNA was employed.

4.2 Medical Devices

Medical articles with porous polymeric structures and methods of forming them were described (3). The porous polymeric structure of a medical device can be made partially or completely from a biodegradable or bioabsorbable material.

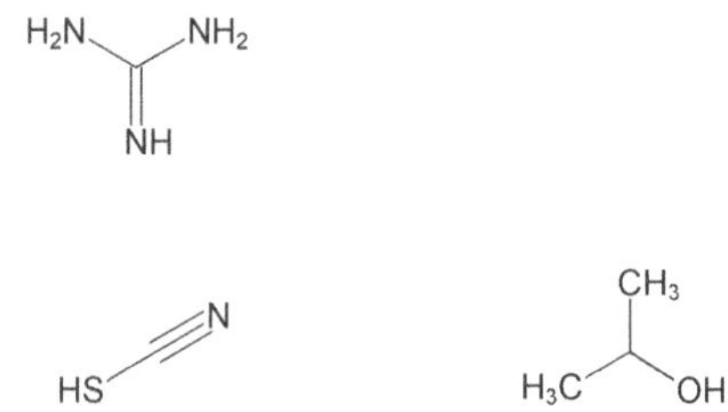

Guanidinium thiocyanate isopropyl alcohol

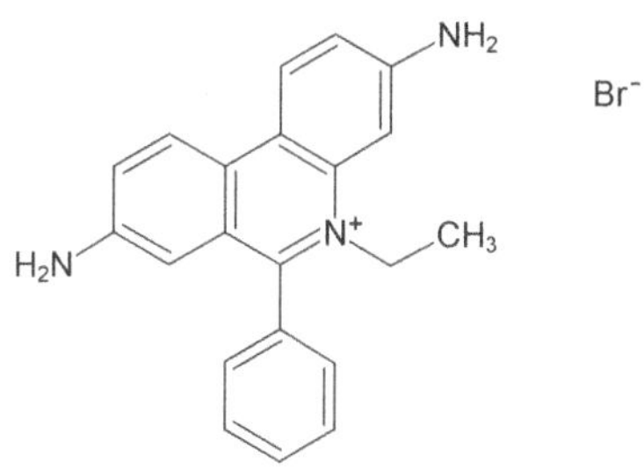

Ethidium bromide

Figure 4.4 Compounds for vesicle characterization.

4.2.1 *Stent Grafts*

Such medical devices can be radially expandable prostheses, such as stents and stent grafts, catheters, and leads for pacemakers and implantable cardioverter defibrillators.

Bioabsorbable polymers that are candidates for their use in medical devices degrade through hydrolytic mechanisms. Common examples of such biodegradable polymers are poly(hydroxy acids) such as poly(*L*-lactide) (PLLA), poly(*D,L*-lactic acid) (PDLA), poly(glycolide), poly(caprolactone), polyanhydrides, polyarylates, and copolymers thereof such as poly(glycolide-*co*-lactide) (PLGA).

Bioabsorbable elastomeric polymers are poly(caprolactone), poly(dioxanone), poly(hydroxy butyrate), and poly(trimethylene carbonate). The monomers are shown in Figure 4.5.

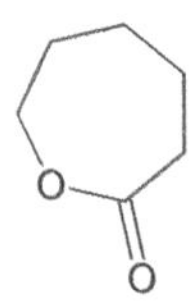

Caprolactone

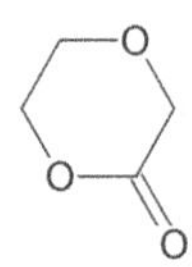

Dioxanone

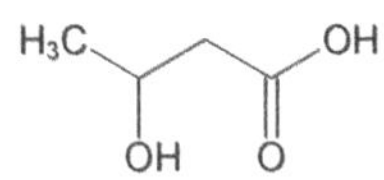

Hydroxy butyric acid

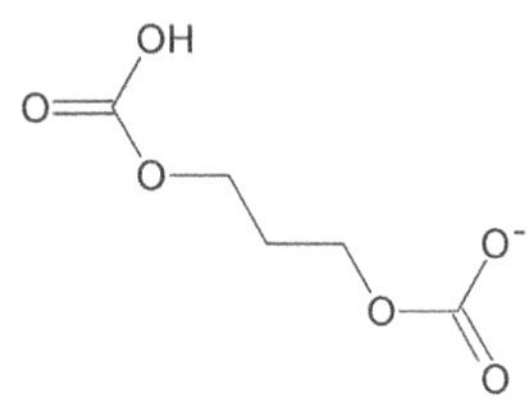

Trimethylene carbonate

Figure 4.5 Monomers for bioabsorbable elastomeric polymers.

Porous bioabsorbable polymeric structures can be useful for controlling the drug release and controlling the degradation rate of a structure. With respect to drug release, the drug release rate and content can be a function of the porosity. When the porosity increases, the amount of drug that can be loaded into a structure increases. In addition, the drug release rate can be influenced by the porosity.

Furthermore, the degradation rate of a polymer structure can be controlled by porosity, since the degradation rate depends on the moisture content or uptake of the structure. Thus, the porosity of a porous structure can be modified to obtain selected drug release rate from drug loading in a bioabsorbable polymeric structure and its degradation rate.

The pore size can be controlled by adjusting or optimizing processing conditions during pore formation. Also, the porous structures can be nanoporous. Nanoporous structures refer to pore sizes in the nanometer range, or less than 1000 *nm* or 1 *μm*. In particular, nanoporous structures have pores with sizes of less than 2 *nm* (microporous), 2 *nm* to 50 *nm* (mesoporous), and 50 *nm* to 1000 *nm* (macroporous).

A majority, all, or substantially all of the pores can be in the nanoporous range. The definitions of microporous, mesoporous, and macroporous are set out by the International Union of Pure and Applied Chemistry (IUPAC). The porous structures can additionally have pores greater than 1 *μm*, in particular, 1–10 *μm*, 10–100 *μm*, or greater than 100 *μm*.

A method of fabricating a stent can be performed as follows (3):

1. Disposing a bioabsorbable polymer tube in a chamber,
2. Contacting the tube with a fluid at supercritical conditions, wherein the fluid is impregnated in the polymeric tube,
3. Adjusting conditions in the chamber so that the fluid is subcritical to form a porous structure in the tube, and
4. Reducing or preventing radial expansion of the tube during formation of the porous structure.

In some cases, a therapeutic agent can be disposed or impregnated within the porous coating. Upon implantation of the stent, the agent can be released into a bodily lumen either through diffusion or degradation of the porous coating. Therapeutic agents can include anti-proliferative agents, anti-inflammatory agents, antibiotics, and antithrombotic agents.

Anti-proliferative agents include everolimus, rapamycin, and/or their derivatives, c.f. Figure 4.6. Everolimus is available under the trade name Certican™, Novartis Pharma AG, Germany. An anti-inflammatory agent can include, clobetasol, which is available under the trade name Temovate™, GlaxoSmithKline, UK. Antithrombotic agents include heparin.

Other materials are a catheter with a porous polymer layer. In general, a catheter is a tube that can be inserted into a body cavity, duct, or vessel and is used in a number of medical applications. The presence of a catheter in the body can lead to various complications or disorders including endothelialization, infection, inflammation, and thrombosis. A catheter can include a porous polymer layer including a therapeutic agent impregnated in the pores of the porous layer. The porous layer can be formed using a bioabsorbable or biostable polymer (3).

In general, a porous polymeric structure or foam can be formed from a nonporous polymer composition using physical and/or chemical blowing agents. A particularly advantageous method of forming a porous polymeric structure is the use of a supercritical fluid as a blowing agent. In such a technique, a polymeric foam structure is formed by impregnating a polymer with a supercritical fluid and inducing a thermodynamic instability through a temperature or pressure change to a subcritical state.

The instability results in the formation of clusters or bubbles of higher density causing formation of pores in the polymer. Other methods known to those with skill in the art may also be used to form a porous polymer structure in the disclosed medical devices (3).

The formation of porous polymeric structures in medical devices can be performed in batch or continuous modes of operation. A batch operation can be done by disposing a medical article having a polymeric portion in a chamber. For example, the chamber can be part of an autoclave that enables precise control of temperature and pressure within the chamber (3).

In the batch operation, the method is done by contacting the medical article in the chamber with a fluid at supercritical conditions. In particular, the supercritical fluid can be injected into the chamber. Prior to contacting the medical article with the supercritical fluid, the temperature and pressure in the chamber may be adjusted to or above supercritical temperature (T_c) and pressure (P_c) of the fluid. The temperature of the chamber can be adjusted to a target temperature or temperature range above T_c that allows or facilitates impregnation of the supercritical fluid in the polymer of the medical article.

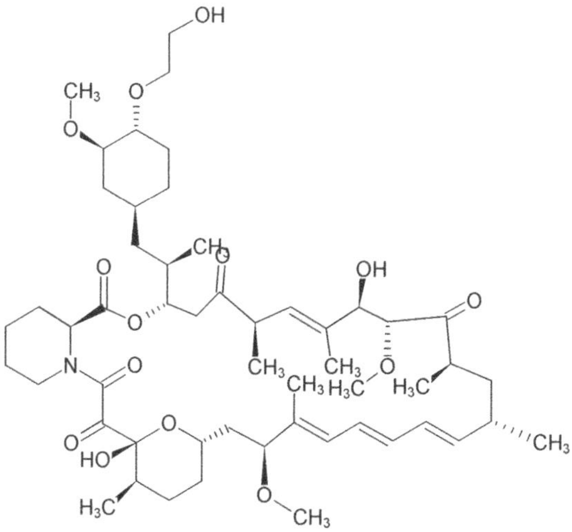

Everolimus

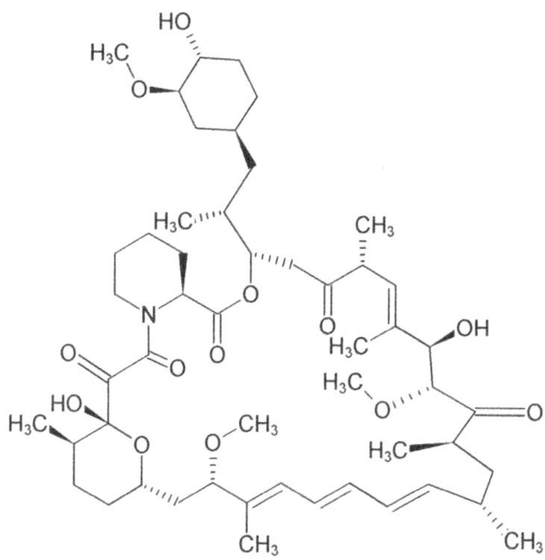

Rapamycin

Figure 4.6 Anti-proliferative agents.

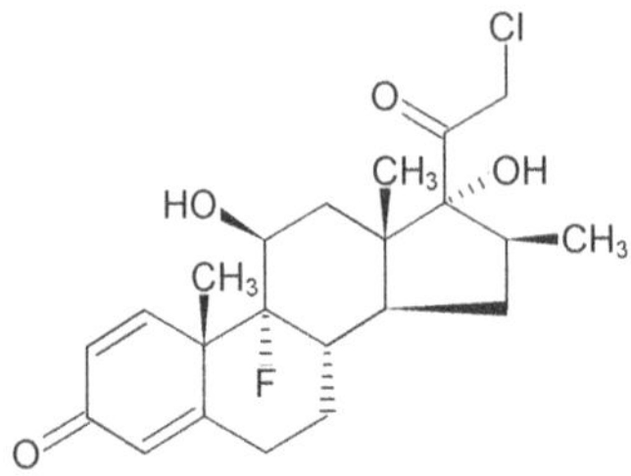

Clobetasol

Figure 4.6 (cont) Anti-inflammatory agent.

The pressure can be adjusted to a target pressure or range up to a saturation pressure above P_c of the supercritical fluid in the polymer at the target temperature. The target pressure and temperature can be maintained for a selected time period which depends on the desired properties of the porous structure to be formed.

To facilitate the impregnation of the supercritical fluid in the polymer, the target temperature can be above the glass transition temperature (T_g) or the softening point (T_s) of the polymer. The target temperature can be above or below the T_g or T_s of the polymer. Here, the supercritical fluid may be dissolved in the polymer so that the polymer and supercritical fluid are a single-phase mixture.

The glass transition temperatures of some polymers that can be used are collected in Table 4.1.

Table 4.1 Melting points and glass transition temperatures of some polymers (3).

Polymer	Melting point/[°C]	Glass transition temperature/[°C]
Poly(glycolic acid)	225 – 230	35 – 40
Poly(*L*-lactide)	173 – 178	60 – 65
Poly(*D,L*-lactic acid)	–	55 – 60
Poly(caprolactone)	58 – 63	-65 – -60
Poly(dioxanone)	–	-10 – 0

A porous polymer structure of a medical article can also be formed through contact of a supercritical fluid with a polymer melt. This means fabrication of a medical article in a continuous operation using extrusion (3).

Extrusion refers to the process of conveying a polymer melt through an extruder and forcing the polymer melt through a die that imparts a selected shape to the polymer exiting the extruder. In the case of tubing extrusion, the polymer melt (extrudate) forced through the die forms a cylindrical film in the shape of a tube. The film is cooled and drawn axially to form the tubing. An extruder generally includes a barrel through which a polymer melt is conveyed from an entrance to an exit port.

The polymer is fed to an extruder barrel near its proximal end as a melt or in a solid form, for example, as a pellet from a hopper. The solid polymer is melted as it is conveyed through the barrel and mixed, for example, through interaction with rotating screws. The polymer in the extruder barrel is heated to temperatures above the melting temperature (T_m) of the polymer and exposed to pressures that are generally far above ambient. Since the viscosity decreases with temperature, the temperature is at a level that allows a desired flow rate of polymer through the extruder.

An extruder barrel also generally includes a breaker plate positioned at a distal end of the extruder barrel. A breaker plate is a thick metal disk with holes drilled through it. The breaker plate serves to create a back pressure in the barrel which facilitates uniform melting and mixing (3).

The polymer melt exits the distal end of the extruder barrel into a die. A die generally refers to a device having an orifice with a specific shape or design geometry that it imparts to a polymer melt pumped from an extruder. In the case of tubing extrusion, the die has a circular shaped orifice that imparts a cylindrical shape to the polymer melt exiting the die. The function of the die is to control the shape of the polymer melt by delivering polymer melt to the orifice.

The polymer melt can be delivered at a selected temperature and pressure. The formed product exiting the die is cooled by passing it through a bath of fluid such as water. Representative examples of extruders for use with the present invention may include single screw extruders, intermeshing co-rotating and counter-rotating twin-screw extruders, and other multiple screw masticating extruders.

Here, a supercritical fluid is contacted with a polymer being conveyed through an extruder. In addition, a supercritical fluid can be injected into the extruder and as the polymer is conveyed along the extruder barrel, the supercritical fluid is mixed or dispersed within the polymer. The supercritical fluid can be injected into the extruder an any point along the length of the extruder barrel that allows uniform or substantially uniform mixing of the fluid with the polymer. In general, a pressure and temperature gradient can exist in the extruder barrel between the proximal end and the distal end and through the extruder die.

The mixture of the polymer and supercritical fluid can be a single or two-phase composition. In some embodiments, the pressure and temperature can be maintained above T_c and P_c along at least a portion of the extruder barrel and die to keep the fluid in a supercritical state. In other embodiments, a gas or liquid can be injected into the extruder barrel and conditions in the extruder barrel can transform the fluid to a supercritical state. When a supercritical fluid is injected into the extruder, a pressure gradient is created within the extruder. In some embodiments, the proximal or upstream pressure can be 1,000-3,500 *psi* while the distal or downstream pressure can be 500-3,000 *psi*. The back plate or extruder die can control and maintain the pressure within the extruder from about 500-3,500 *psi*.

Under some conditions, a single-phase solution of supercritical fluid and polymer may be formed. The supercritical fluid tends to reduce the viscosity of the polymer. Thus, the polymer and supercritical fluid mixture can be processed at a lower temperature. The lower temperature processing is advantageous since molecular weight degradation of the polymer during processing is reduced.

The T_m of some polymers that can be used are given in Table 4.1. The method further includes inducing formation of pores in the polymer and fluid mixture. As above, a thermodynamic instability is created in the polymer and fluid mixture through a temperature, pressure, or both a temperature and pressure change in the polymer and fluid mixture to a subcritical state of the fluid. The instability results in the formation of clusters or bubbles of fluid within the polymer causing formation of pores in the polymer. In some embodiments, the temperature and pressure of the polymer and fluid mixture can be maintained at supercritical conditions throughout the extruder barrel and die.

In such embodiments, the polymer and fluid mixture is exposed to subcritical conditions as it exits the die, resulting in the formation of pores or cells in the polymer. In other embodiments, the polymer and fluid mixture in the extruder barrel is at supercritical conditions throughout and the polymer and fluid mixture is exposed to subcritical conditions in the extruder die. This results in the formation of pores in the polymer while the polymer is in the die. In additional embodiments, the polymer and fluid mixture can be exposed to supercritical conditions in the extruder barrel at a location along the axis of the barrel distal to the point of injection of the supercritical fluid. As a result, pores can form in the polymer while within the extruder barrel (3).

4.2.2 *Vascular Grafts*

The research for an ideal blood vessel substitute has been focused on biological tissues and synthetics (4). Despite intensive efforts to improve the nature of blood vessel substitutes, many problems remain such as increasing failure rate with decreasing caliber of the blood vessel substitute, a high failure rate when infection occurs, and aneurysm formation.

The major need for vascular grafts is for adequate supply of blood to organs and tissues whose blood vessels are inadequate either through defects, trauma or diseases. Vascular grafts are also needed to provide access to the bloodstream for individuals undergoing hemodialysis. The three major types of vascular grafts are peripheral, arterial-to-venous access, and endovascular (4).

Peripheral grafts are those used in the neck and extremities, with the most common being used in the leg. This results in supply problems being that some intermediate and most small diameter arteries are replaced or bypassed using an autologous saphenous vein, the long vein extending down the inside of the leg, with a secondary source being the radial veins of the arms. In a given patient, suitable veins may be absent, diseased or too small to be used, and removal of the vein is an additional surgical procedure that carries attendant risk.

Arterial-to-venous access grafts are used to access the circulatory system during hemodialysis. Vascular grafts used in connection with hemodialysis are attached to an artery at one end and sewn to a vein at the other. Two large needles are inserted into the graft. One needle removes the blood where it flows through an artificial kidney machine and is then returned to the body via the second needle. Normal kidney function is destroyed by several acute and chronic diseases, including diabetes and hypertension.

Patients suffering from kidney failure are maintained by dialysis three times a week for approximately four hours per session. Due to the constant punishment these grafts undergo, there is a high occurrence of thrombosis, bleeding, infections, and pseudoaneurysm.

Endovascular grafts are used to reline diseased or damaged arteries, particularly those in which aneurysms have formed, in a less invasive manner than standard vascular surgical procedures. Various surgical techniques and materials have been developed to replace and repair blood vessels. Ideally, the thickness of the prosthesis is minimized, so that it can be delivered to the implantation site using a percutaneous procedure, typically catheterization, and kept in place utilizing stents. Problems associated with this type of implantation include thrombosis, infection and new aneurysm formation at the location of the stent.

Vascular prostheses have been coated with bioabsorbable substances such as collagen, albumin, or gelatin during the manufacture process instead of preclotting at surgery.

Compliance problems with woven poly(ethylene terephthalate) and drawn out poly(tetrafluoroethylene) (PTFE) prompted interest in thermoplastic elastomers for their use as blood conduits. Medical grade poly(urethane) (PU) copolymers are an important member of the thermoplastic elastomer family. PUs are generally composed of short, alternating polydisperse blocks of soft and hard segment units. The soft segment is typically a polyester, polyether or a poly(tetrahydrofuran), e.g., poly(tetramethylene oxide). The hard segment is formed by the polymerization of either an aliphatic or aromatic diisocyanate with a chain extender, such as a diamine or a glycol.

The resulting product containing the urethane or urea linkage is copolymerized with the soft segment to produce a variety of PU formulations. PUs have been tested as blood conduits for over 30 years. Medical grade PUs, in general, have material properties that make it an excellent biomaterial for the manufacture of vascular grafts as compared to other commercial plastics. These properties include excellent tensile strength, flexibility, toughness, resistance to degradation and fatigue, as well as biocompatibility.

Despite these positive qualities, it became clear in the early 1980s that conventional ether-based PU elastomers presented long-term biostabilty issues as well as some concern over potential carcinogenic degradation products. Furthermore, in contrast to excellent performance in animal trials, clinically disappointing results with PU-based grafts diminished the attractiveness of the material for this application.

Recent developments in new-generation PUs, however, have made this biomaterial, once again, a promising choice for a successful long-term vascular prosthesis.

The production of a porous polymeric material using the gel-enhanced phase separation technique has been documented (4).

4.2.2.1 Porous (NanoCliP) Gel

A self-assembled amphiphilic nanogel-crosslinked porous (NanoCliP) gel was developed that could trap proteins, liposomes, and cells (5). The NanoCliP gel was prepared by the Michael addition of a self-assembled nanogel of acryloyl group-modified cholesterol-bearing pullulan to pentaerythritol tetra(mercaptoethyl) poly(oxyethylene), followed by freezing-induced phase separation. Pullulan is shown in Figure 4.7. Dynamic rheological analysis revealed that the storage modulus of the NanoCliP gel was approximately 10 times greater than that of a nonporous nanogel-crosslinked gel.

Two-photon excitation deep imaging revealed that the NanoCliP gel comprises interconnected pores of several hundred micrometers in diameter. The NanoCliP gel trapped proteins and liposomes via hydrophobic interactions because its amphiphilic nanogels exhibit a chaperone-like activity. Mouse embryonic fibroblasts penetrated the interconnected pores and adhered to the porous surface of fibronectin-complexed NanoCliP gel. *In-vivo*, the NanoCliP gel enhanced the cell infiltration, tissue ingrowth, and neovascularization without requiring exogenous growth factors, suggesting that the NanoCliP gel is a promising scaffold for tissue engineering (5).

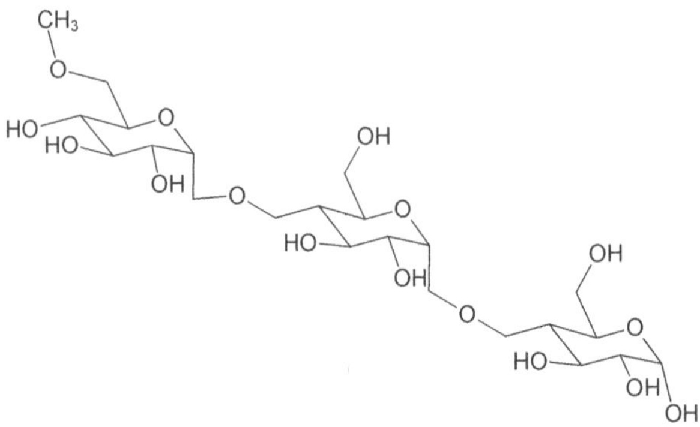

Figure 4.7 Pullulan.

4.3 Medical Applications

4.3.1 *Porous Polymer Microneedles*

Porous polymer microneedles have great potential in transdermal medical applications due to their three-dimensional (3D) porous structures and high porosity. A facile and effective phase inversion route has been described to prepare polymer microneedles with highly porous and interconnected pore structures (6).

The fabrication process is simple and mild without involving high temperatures or irradiation, and can be applied to a broad spectrum of commonly used polymers, e.g., cellulose acetate (CA), poly(sulfone) polysulfone (PSF), poly(ethersulfone), poly(lactic acid).

Thanks to the capillary effect and large cavity given by highly porous and interconnected structures, the resulting porous polymer microneedles show the capability of rapidly extracting dermal interstitial fluid (ISF) and efficiently loading/releasing drug compounds.

The use of these porous CA microneedles could be demonstrated in the highly efficient extraction of ISF for glucose-level detection and administration of insulin for hyperglycemia. So, this study provides a new opportunity for the fabrication of microneedle-based devices for transdermal ISF extraction and drug delivery (6).

4.3.2 *Flexible Pressure Sensors*

The use of microporous PDMS foams as a highly deformable film to improve the sensitivity of a flexible capacitive pressure sensor dedicated to wearable use has been detailed (7).

PDMS is a nontoxic material, with dielectric properties as well as relevant mechanical properties. This allows a deformation of up to 100% without damage (8).

A fabrication process allowing the mechanical properties of foams to be adjusted was proposed (7):

Preparation 4–2: PDMS is formed by mixing of a silicon elastomer (Sylgard 184A) and a curing agent (Sylgard 184B) at a ratio of 10 to 1. Sugar particles are added within the PDMS in volume ratios ranging from 4 to 6 (sugar) per 1 PDMS. The preparation is poured into a 5 *cm* diameter Petri dish and compacted to form a block of the mixture. The mixture is cured in a stove for 2 *h* at 90°C.

After curing, the mixture is removed from the Petri dish and soaked in a water bath to dissolve the sugar and thereby obtain the microporous PDMS foam. Once the sugar is completely dissolved, the PDMS foam is again placed in the stove for 1 *h* 30 *min* at 75°C to remove the excess water.

The capacitance changes versus applied load characteristics of the fabricated sensors were obtained by means of a dedicated PC-controlled electromechanical test bench. This bench is composed of:

1. An indenter applying loads to the sensor under test through an ISEL iMC-S8 robotic arm,
2. A Futek FSH00105 110N range force sensor connected to an HP34401A multimeter measuring the applied load, and
3. A HP4192A impedance analyzer measuring the sensor capacitance changes at a frequency of 1 *MHz*.

In order to quantify the sensor performances, a first order exponential behavioral model was proposed, which reads:

$$\frac{\Delta C}{C_0} = \frac{\Delta C_{max}}{C_0}\left(1 - e^{-P/P_c}\right) \tag{4.1}$$

Here: P (*kPa*) is the applied pressure, C_{max} is the maximum capacitance change and P_c is the characteristic pressure at which 63% of C_{max} is reached. This model is adjusted to the experimental data and allows the sensitivity *S* of the sensor to be defined as the slope of the tangent to the curve at the origin, and the pressure measurement range to be defined as the pressure at which 95% of the C_{max} is reached.

The sensors fabricated and characterized showed that the sensitivity and the measurement range can be adjusted from 0.14% *kPa* up to 13.07% *kPa*, and from 594 *kPa* to 183 *kPa*, respectively, while the porosity of a PDMS film ranges from 0% up to 85% (7).

4.3.3 Bone Regeneration

The design, workflow, and manufacture of highly porous, resorbable additively manufactured, three-dimensional (3D) custom scaffolds for the regeneration of large volume alveolar bone defects was investigated (9).

Porous biodegradable scaffolds greatly benefit from the combination with products that can deliver bioactive cues to the surrounding progenitor cells.

The additive manufacturing technology of 3D printing can be used to form custom, porous, and volumetrically stable scaffolds for guided bone regeneration. A layer-by-layer printing approach was applied throughout the entire volume, resulting in the formation of a porous scaffold.

Patient-specific scaffolds were 3D printed using an EnvisionTEC 3D-Bioplotter® Developer Series (EnvisionTEC GmbH) printer, equipped with a low temperature printing head, a high temperature printing head, and a pressure booster.

A medical grade PC08 poly(ε-caprolactone) (PCL) (Corbion N.V.) was loaded into the high temperature head, and EnvisionTEC low temperature support hydrogel RG (EnvisionTEC GmbH) was loaded into the low temperature head after dissolution.

Computed tomography (CT) scans of 5 posterior mandibular vertical bone defects were done. Surface masks (3D surface contours) of the recipient site were first isolated using a contrast threshold, transformed into 3D objects, and used to guide the formation of custom implant template models (9).

In order to determine the accuracy of the model and the accuracy of the fitting, the gap and overlap between the patient geometry models and the idealized template 3D models were quantified.

Models were 3D printed from medical grade PCL into porous scaffolds. For scaffold dimensional quantification, scaffolds were scanned using a micro-CT scanner. The design and printing processes each achieved dimensional errors of <200 μm on average. The average gap between the template implant model and the scanned scaffold model was found to be 74±14 μm. The printed scaffold was confirmed as having a porosity of 83.91%, a mean polymer or filament thickness of 200±46 μm, and a mean pore size of 590±243 μm.

The here described approach is straightforward, adaptable to a range of patient geometries, and results in the formation of reproducible, dimensionally accurate custom implants. These highly porous 3D structures manufactured from resorbable medical grade material represent a potentially transformative technology toward the clinical implementation of scaffold-guided bone regeneration procedures (9).

4.3.3.1 *Bone Tissue Engineering*

The employment of tissue engineering scaffolds in the reconstruction of the damaged bone tissues has shown remarkable promise since they significantly facilitate the healing process (10). The fabrication of highly porous biocompatible scaffolds with sufficient mechanical strength is still challenging. Polymers have been widely utilized to construct three-dimensional porous scaffolds due to their excellent processability and biocompatibility.

The fabrication of polymer/metal composite 3D porous scaffolds by additive manufacturing techniques is the most convenient approach to construct complex polymer-based parts; their structural, mechanical, biodegradation, biocompatibility, osteogenic and angiogenic features have been reviewed (10).

Synthetic polymer scaffolds are candidates for the regeneration of host tissues damaged by injury or disease (11). However, the relatively small pore size (30 μm to 60 μm) within the polymerized high internal phase emulsions limits the *in-vivo* application in bone tissue engineering due to the lack of vasculature formed in the scaffolds.

High internal phase emulsion templating represents a route to produce highly porous polymer scaffolds capable of supporting cell in-growth (11).

Strategies for creating larger channels in the highly porous structures with the help of 3D printing technologies have been presented. These strategies are expected to be beneficial for bone tissue engineering and related applications (11).

Here, 3D printing technology can be utilized as a solution to generate larger pores or channels in the synthetic scaffolds with the aim of promoting vascularization and bone regeneration *in-vivo*.

The fused deposition modeling (FDM) technology is a well-known 3D printing technique for manufacturing complex geometrical parts in the market. This method was first devised by Crump in 1989 (12). Also, in recent times the fabrication of new instruments using the FDM technology have been presented. (13–16).

A combination of high internal phase emulsion templating and 3D printing has been investigated to create large channels in highly porous polymer scaffolds. PCL was 3D-printed into a *woodpile* porous structure using a FDM technology to assess the biological response of human osteosarcoma cells (MG63) to pore sizes of 250 and 500 μm. MG63 cells were observed to infiltrate into the center of the scaffolds, attach to the surface of the printed PCL struts and proliferate in the scaffolds. An *in-vivo* evaluation was conducted on the PCL scaffolds using the rat mandible critical size defect model.

The results demonstrated that vascularization and bone regeneration happened in both of the PCL scaffolds with two different pore sizes (11). High internal phase emulsion templating was subsequently incorporated using indirect 3D printing to generate interconnected channels in the highly porous polymer scaffolds. A poly(vinyl alcohol) (PVA) negative mold was printed into a woodpile porous structure using a FDM technology in advance, followed by infiltration of high internal phase emulsion into the interstice between adjacent PVA struts.

After polymerization, the PVA negative mold was dissolved in water and highly porous polymer scaffolds with interconnected channels (3D-highly porous polymer scaffolds) were obtained. The mechanical properties of 3D-highly porous polymer scaffolds were found to decrease after the incorporation of the channels (11).

It was found that MG63 cells could infiltrate into the center of the 3D-highly porous polymer scaffolds with the help of the interconnected channels, and migrate from the channels to the internal voids and proliferate therein. Also, the possibility of printing highly porous polymer scaffolds using digital light processing was explored. Here, internal phase emulsion formulations were optimized, and the influence of internal phase emulsion preparation condition and printing parameters was investigated.

The results demonstrate that internal phase emulsions are indeed printable with digital light processing. Nevertheless, challenges still remain in printing the porous structures with a sufficiently high printing resolution. Overall, this study presents strategies for creating larger channels in polymerized internal phase emulsion structures with the help of 3D printing technologies. These strategies are expected to be beneficial for bone tissue engineering and related applications (11).

4.3.4 Release of Therapeutic Agents

Implantable or insertable medical devices were described that have at least one nanoporous polymeric region. These methods include (17):

1. Providing a phase-separated polymeric region comprising a stable polymeric phase and a disintegrable polymeric phase of nanoscale dimensions, and
2. Selectively removing the disintegrable polymeric phase, thereby producing the nanoporous polymeric region.

Porous polymeric regions can be formed *ex-vivo* by methods in which at least one disintegrable phase is selectively removed from at least one stable polymeric phase. Examples of conditions which can be used to selectively remove at least one disintegrable polymeric phase include (17):

1. Exposure to elevated temperatures,
2. Exposure to aqueous and/or organic solvents at temperatures and pressures that are effective to dissolve the disintegrable polymeric phases,
3. Exposure to reactive chemical species (including biological fluids) at concentrations, temperatures and pressures that result in the degradation of at least one disintegrable polymeric phase (e.g., chemical lysis, including hydrolysis, catalytic breakdown, etc.),

4. Exposure to radiation levels sufficient to cause chain scission within at least one disintegrable polymeric phase.

The key to the success of each technique is the ability to provide a phase-separated region which contains at least one polymer phase that is stable under such conditions, and which contains at least one other polymer phase that is not stable under such conditions and thus can be removed.

Therapeutic agents may be used in these medical devices. The therapeutic agents may be provided within or beneath phase-separated polymeric regions that contain at least one biostable polymeric phase and at least one biodisintegrable polymeric phase, in which case the therapeutic agents may be released into the biological milieu. It is believed that the release of the therapeutic agent in these materials will occur by one or more of the following mechanisms:

1. By diffusion of the therapeutic agent within an intact polymeric phase,
2. By swelling/hydration of a polymeric phase followed by diffusion of the therapeutic agent within the swelled/hydrated polymeric phase, and
3. By polymer degradation.

Exemplary non-genetic therapeutic agents for use in connection with these medical devices are (17):

1. Antithrombotic agents such as heparin, heparin derivatives, urokinase, and PPack (dextrophenylalanine proline arginine chloromethyl ketone), c.f. Figure 4.8,
2. Anti-inflammatory agents such as dexamethasone, prednisolone, corticosterone, budesonide, estrogen, sulfasalazine and mesalamine (5-aminosalicylic acid), c.f. Figure 4.10,
3. Antineoplastic/anti-proliferative/antimiotic agents such as paclitaxel, 5-fluorouracil, cisplatin, vinblastine, vincristine, epothilones, endostatin, angiostatin, angiopeptin, monoclonal antibodies capable of blocking smooth muscle cell proliferation, and thymidine kinase inhibitors,
4. Anesthetic agents such as lidocaine, bupivacaine and ropivacaine, c.f. Figure 4.11,
5. Anticoagulants such as D-Phe-Pro-Arg chloromethyl ketone, an RGD peptide-containing compound, heparin, hirudin,

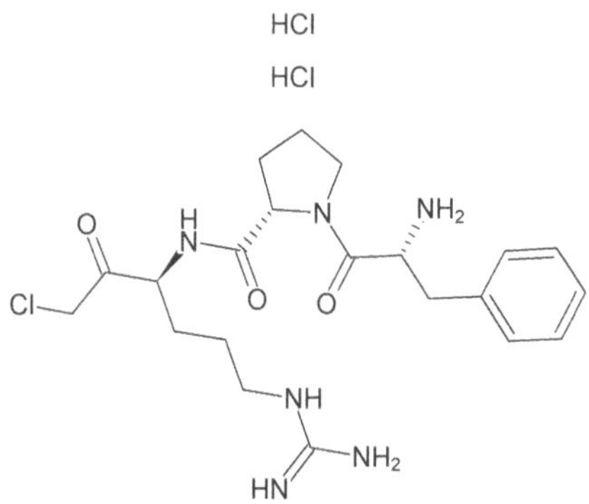

Figure 4.8 *D*-Phenylalanyl-*N*-[(3*S*)-6-carbamimidamido-1-chloro-2-oxo-3-hexanyl]-*L*-prolinamide dihydrochloride.

antithrombin compounds, platelet receptor antagonists, antithrombin antibodies, antiplatelet receptor antibodies, aspirin, prostaglandin inhibitors, platelet inhibitors and tick antiplatelet peptides,

6. Vascular cell growth promoters such as growth factors, transcriptional activators, and translational promoters,
7. Vascular cell growth inhibitors such as growth factor inhibitors, growth factor receptor antagonists, transcriptional repressors, translational repressors, replication inhibitors, inhibitory antibodies, antibodies directed against growth factors, bifunctional molecules consisting of a growth factor and a cytotoxin, bifunctional molecules consisting of an antibody and a cytotoxin,
8. Protein kinase and tyrosine kinase inhibitors (e.g., tyrphostins, genistein, quinoxalines, c.f. Figure 4.12),
9. Prostacyclin analogs,
10. Cholesterol-lowering agents,
11. Angiopoietins,
12. Antimicrobial agents such as triclosan, cephalosporins, aminoglycosides and nitrofurantoin, c.f. Figure 4.13.
13. Cytotoxic agents, cytostatic agents and cell proliferation effectors,
14. Vasodilating agents,
15. Agents that interfere with endogenous vasoactive mecha-

nisms,

16. Inhibitors of leukocyte recruitment, such as monoclonal antibodies,
17. Cytokines,
18. Hormones,
19. Inhibitors of HSP 90 protein (i.e., heat shock protein, which is a molecular chaperone or housekeeping protein and is needed for the stability and function of other client proteins/signal transduction proteins responsible for growth and survival of cells), including geldanamycin, c.f. Figure 4.9,
20. Beta blockers,
21. βARKct inhibitors,
22. Phospholamban inhibitors, and
23. SERCA2 gene/protein.

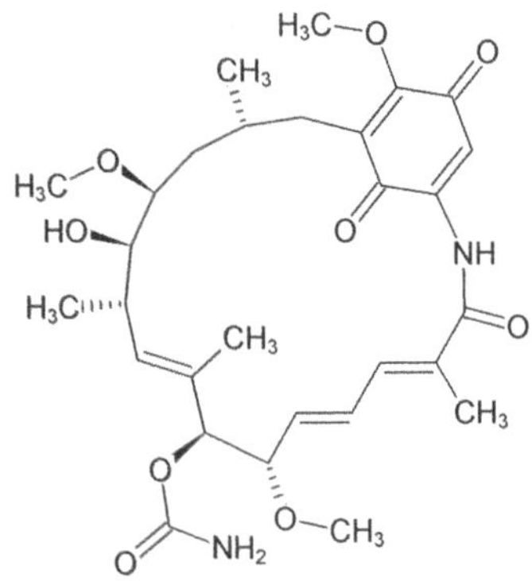

Figure 4.9 Geldanamycin.

Included among prostacyclin analogs are ciprostene, epoprostenol, carbacyclin, iloprost and beraprost (17). Some compounds are shown in Figure 4.14.

4.3.5 *Implant Dentistry*

The ability to place an implant in an edentulous space in the maxillary and/or mandibular arch is often hindered by the surrounding tissues in the oral cavity, which alters the planning of treatment

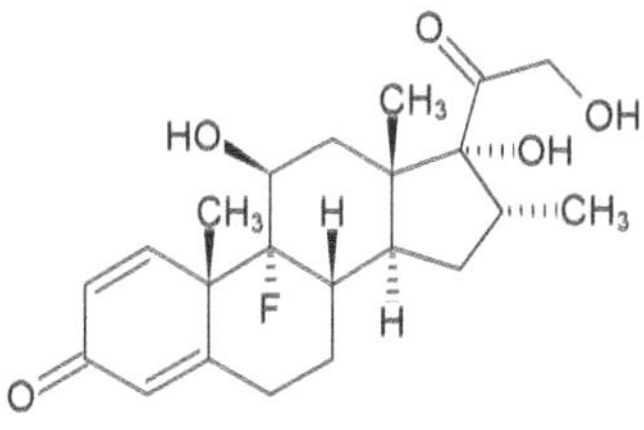

Dexamethasone

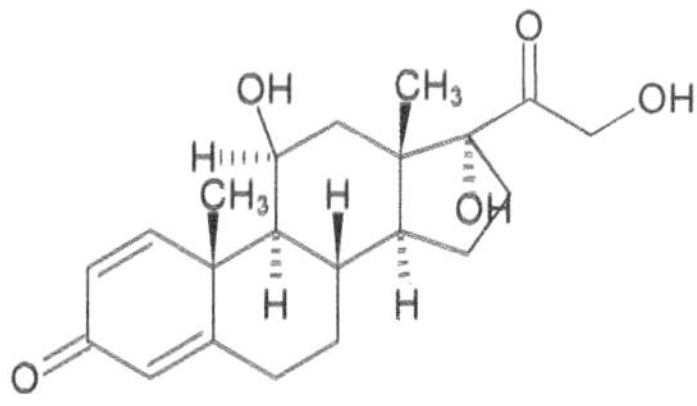

Prednisolone

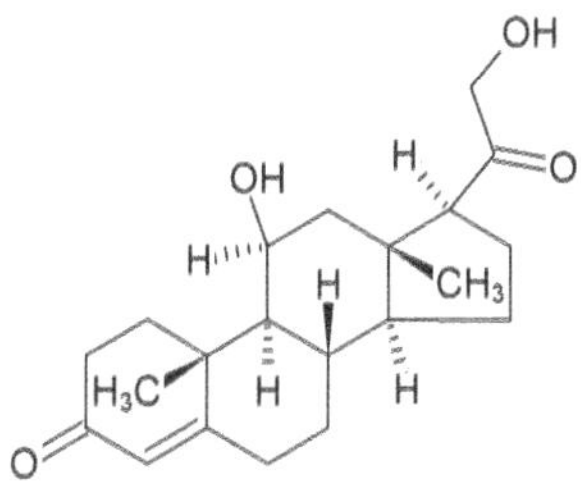

Corticosterone

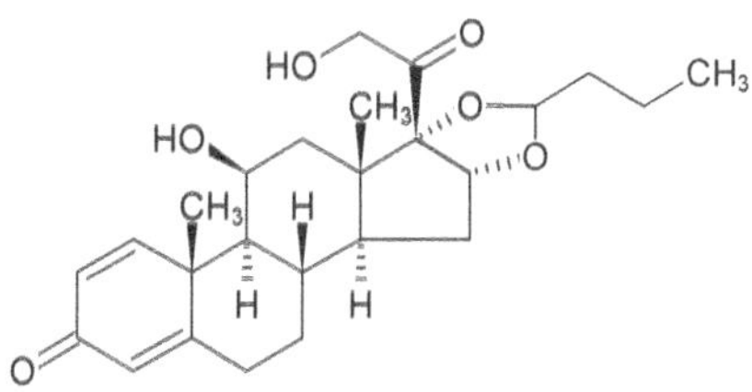

Budesonide

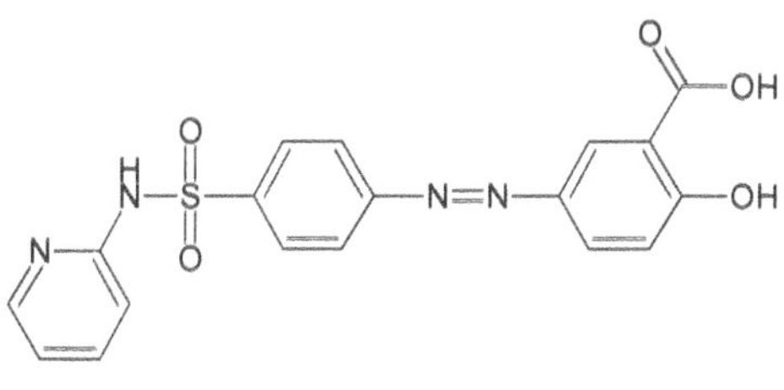

Sulfasalazine

Figure 4.10 Anti-inflammatory agents.

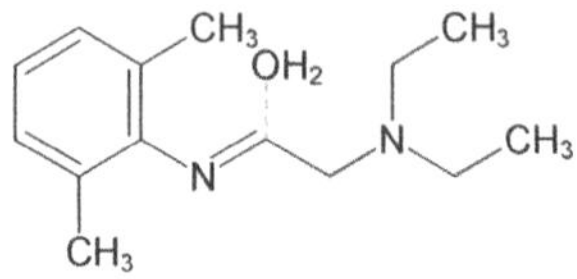

Lidocaine

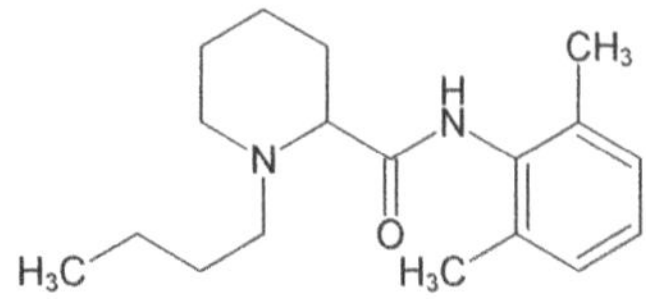

Bupivacaine

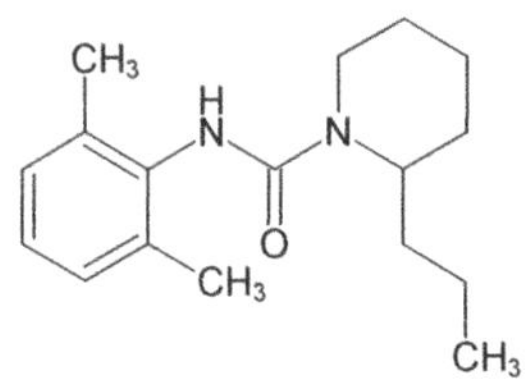

Ropivacaine

Figure 4.11 Anesthetic agents.

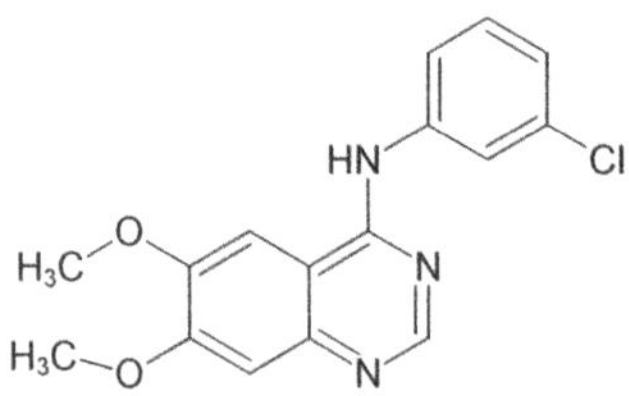

Tyrphostin

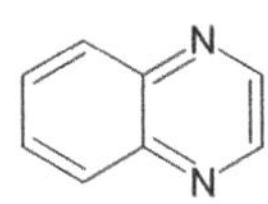

Quinoxaline

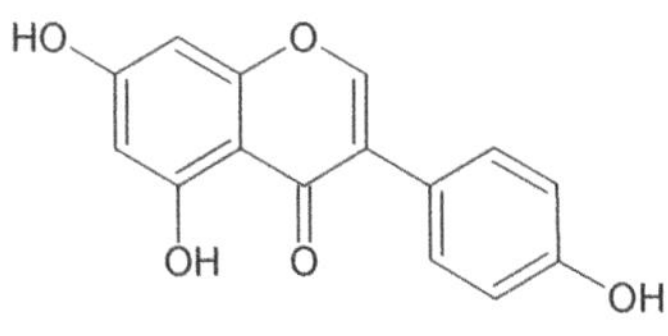

Genistein

Figure 4.12 Protein kinase and tyrosine kinase inhibitors.

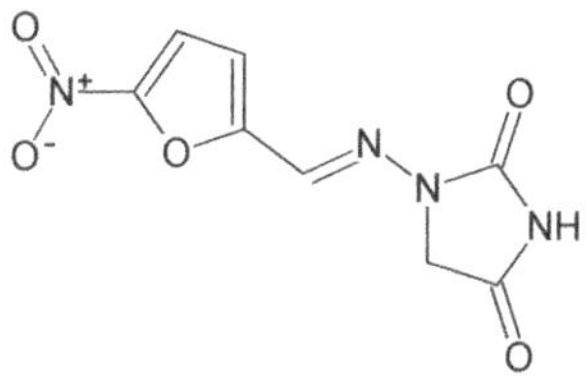

Nitrofurantoin

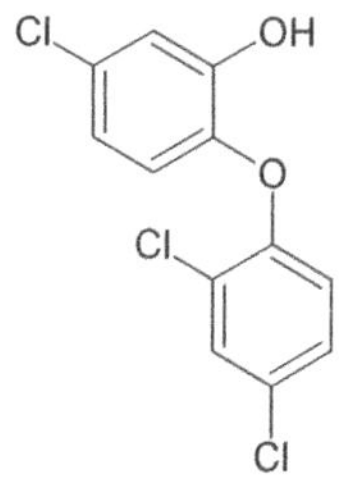

Triclosan

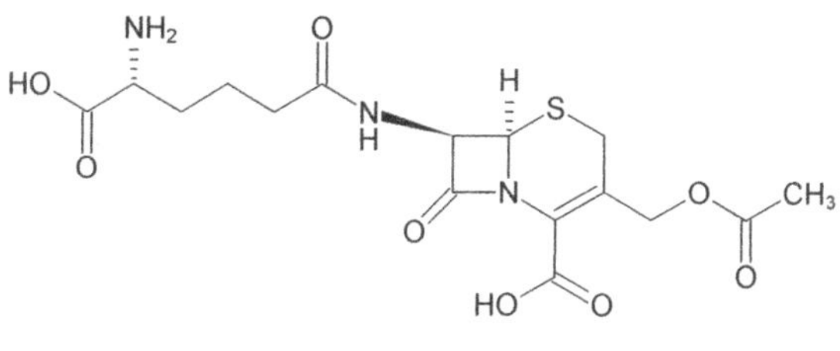

Cephalosporin C

Figure 4.13 Antimicrobial agents.

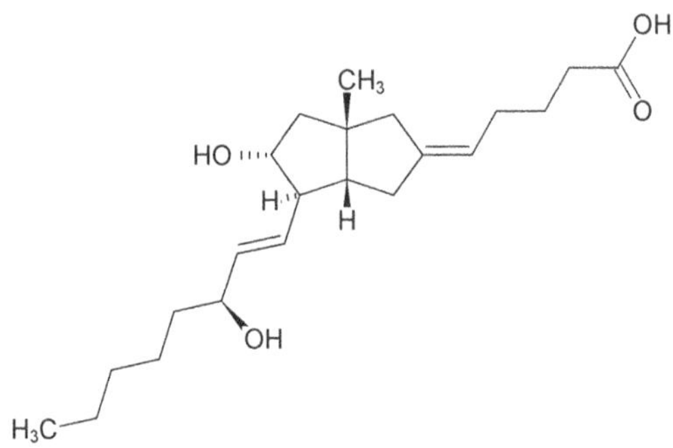

Ciprostene

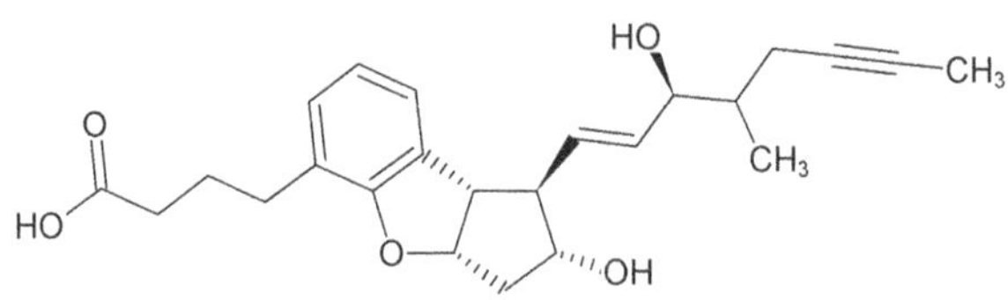

Beraprost

Figure 4.14 Prostacyclin analogs.

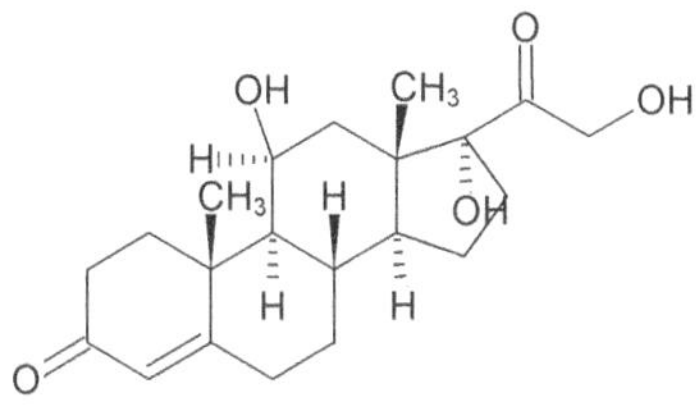

Cortisol

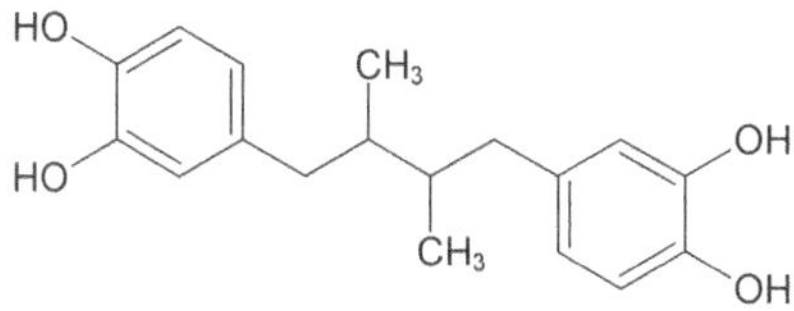

Nordihydroguaiaretic acid

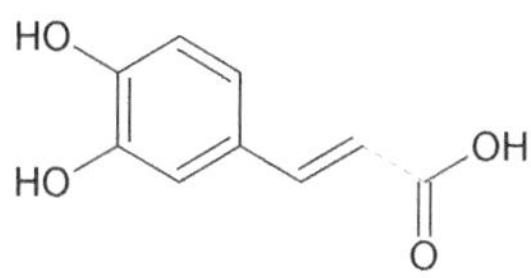

Caffeic acid

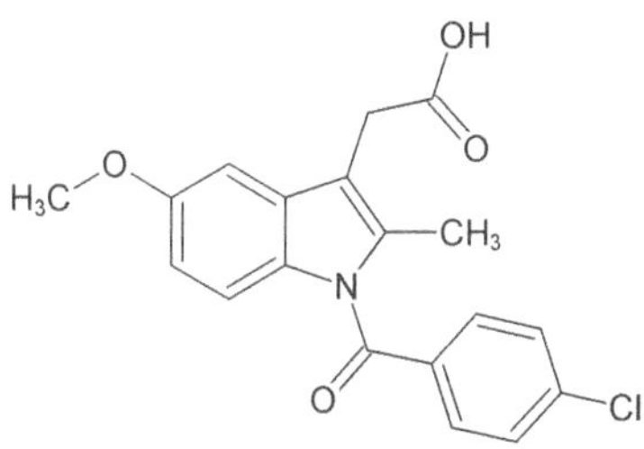

Indomethacin

Figure 4.14 (cont) Drugs.

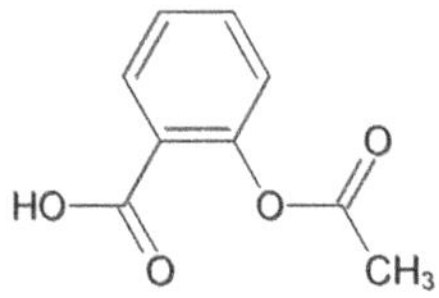

Aspirin

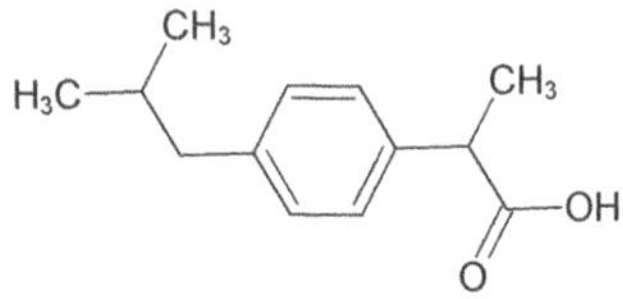

Ibuprofen

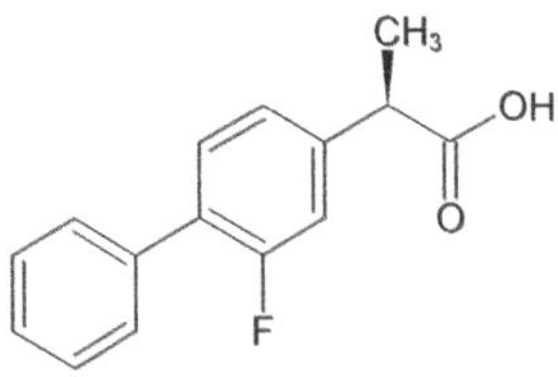

Flurbiprofen

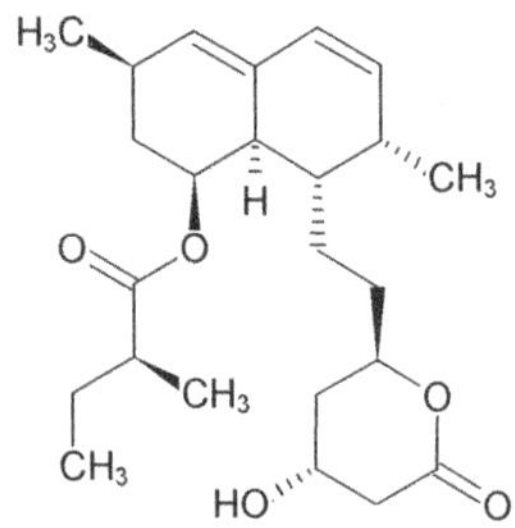

Lovastatin

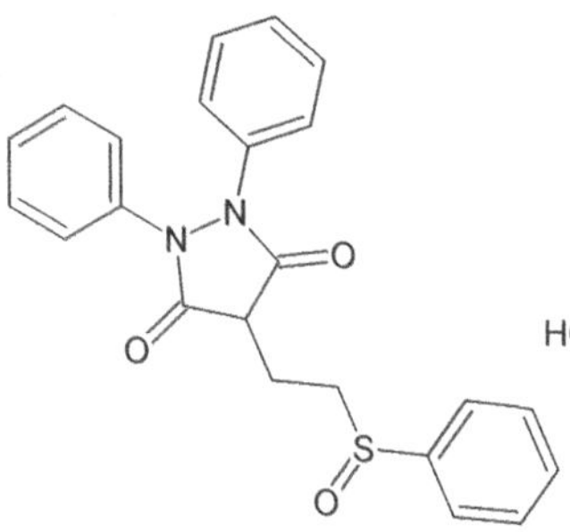

Sulfinpyrazone

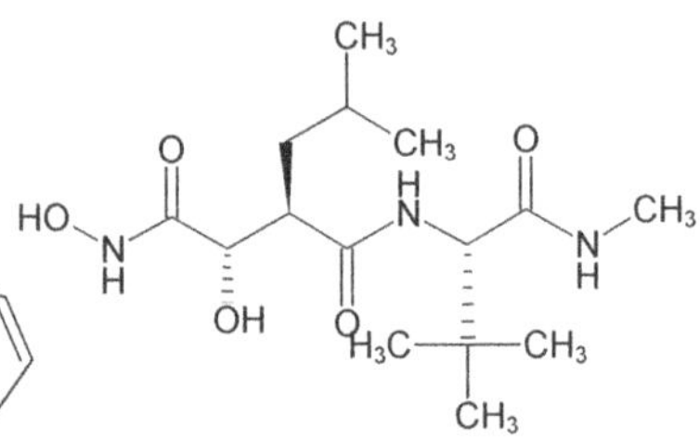

Marimastat

Figure 4.14 (cont) Drugs.

for oral rehabilitation (18). Ideally, oral rehabilitation should regain normal physiological functions such as mastication, phonetics, and aesthetics, so these functions should be major considerations of treatment planning.

Various methods for three-dimensional reconstruction of jawbones include alveolar distraction, guided bone regeneration, and bone grafting (18). Such methods are used most commonly for the reconstruction of the posterior region of the mandible. On the other hand, the posterior region of the maxilla is reconstructed using sinus lifting or ridge augmentation.

Guided bone regeneration is the most widely used procedure for jaw reconstruction, and provides high predictability. It uses a selectively permeable membrane, which is either resorbable or non-resorbable, that acts as a barrier, so holding in place particulate bone grafts. This also maintains the space that allows for an uniform growth of the bone.

Non-resorbable membranes, bone block grafts and distraction osteogenesis can also be viable treatment options for the posterior region of the mandible, but these demand a large amount of autologous bone from intra or extra-oral origins, which can be challenging due to inadequate availability and a morbidity of the patient (18).

On the other hand, the posterior region of the maxilla has the maxillary sinus, which can be encroached into with a variety of bone grafts to attain an adequate thickness of bone. The procedure of sinus lifting is a convenient approach with a good prognosis as the site is naturally sheltered by the palatine bone and the sinus walls.

Bone tissue engineering has emerged as an alternative treatment modality for severely resorbed ridge areas. It uses biomaterials that have an osteoconductive property, in combination with bioactive molecules, and with progenitor cells that initiate osteogenesis and form healthy bone at the site (18).

Much work has explored the interface between these biomaterials and fluids in the oral cavity such as blood, plasma, and saliva. Many researchers, clinicians, and healthcare professionals have a keen interest in this area. Current work focuses on methods to attain bone regeneration working from the molecular level to yield a meaningful quality and quantity of bone. More clinical trials are needed to verify progress, as newer materials are challenged by older techniques (18).

4.3.5.1 Periodontal Tissue Engineering

The management of periodontal and peri-implant diseases constitutes a significant healthcare burden with the potential to greatly improve the quality of life of affected patients (19).

Such microbially-mediated inflammatory diseases are destructive in nature and culminate in progressive loss of tooth-supporting alveolar bone. Besides periodontitis and peri-implantitis, orofacial soft and hard tissue deficiencies can also be caused by post-extraction atrophy, trauma, tumor resection, and congenital or developmental conditions such as cleft lip and palate (20).

Periodontal tissue engineering involves a multi-disciplinary approach towards the regeneration of periodontal ligament, cementum and alveolar bone surrounding teeth, whereas bone regeneration specifically applies to ridge reconstruction in preparation for future implant placement, sinus floor augmentation and regeneration of peri-implant osseous defects (21). A successful periodontal regeneration is based on verifiable cementogenesis on the root surface, oblique insertion of periodontal ligament fibers and formation of new and vital supporting bone.

Ultimately, regenerated periodontal and peri-implant support must be able to interface with surrounding host tissues in an integrated manner, withstand biomechanical forces resulting from mastication, and restore normal function and structure. Current regenerative approaches utilized in everyday clinical practice are mainly guided tissue/bone regeneration based.

Although these approaches have shown positive outcomes for small and medium-sized defects, predictability of clinical outcomes is heavily dependent on the defect morphology and clinical case selection. In many cases, it is still challenging to achieve predictable regenerative outcomes utilizing current approaches. Periodontal tissue engineering and bone regeneration aims to improve the state of patient care by promoting reconstitution of damaged and lost tissues through the use of growth factors and signaling molecules, scaffolds, cells and gene therapy.

The growth factors and the signaling molecules used in periodontal tissue engineering and bone regeneration are proteins with the capacity to promote chemotaxis, proliferation, differentiation, extracellular matrix synthesis and angiogenesis (22). The biologic functions of these molecular mediators vary widely, but their selection as candidates for a regenerative therapy is based on their important roles in periodontal tissue development and wound healing.

Promising results from preclinical and clinical trials led to the subsequent introduction of various growth factors into the commercial market for the purposes of periodontal, and peri-implant soft and hard tissue regeneration (23).

The major issues are now discussed (21).

Enamel matrix derivatives. An enamel matrix derivative (EMD) is composed of a combination of peptides harvested from immature enamel of developing porcine tooth buds, and was the first Food and Drug Administration (FDA)-approved biologic for periodontal regeneration (22).

The EMD contains a mixture of enamel matrix proteins within an aqueous solution of propylene glycol alginate, permitting application via a syringe due to its shear-thinning properties (24). During tooth development, enamel matrix proteins secreted by Hertwig's epithelial root sheath (HERS) play an important role in odontogenesis. Hertwig's epithelial root sheath was discovered by Oskar Hertwig in 1874 in an amphibian (25).

It was shown that amelogenin is the major enamel matrix protein expressed in human teeth during root formation, and in response to EMD exposure, hard tissue formation could be stimulated, resulting in tissues histologically similar to extrinsic acellular cementum (26).

These results were among the first to show the potential importance of EMD proteins during cementogenesis. Although early *in-vivo* studies demonstrated promising results, in later preclinical and human studies, EMD has inconsistently induced acellular cementum formation, leading to some questions over the precise mechanism of action (27).

A review of *in-vitro* investigations of EMD indicated that in addition to enhanced PDL and osteoblastic gene expression, protein synthesis, mitogenesis and differentiation, EMD could also be involved in angiogenesis and inhibition of epithelial proliferation (28). In addition, it was found that the regenerative capacity of EMD was not completely matched by full-length amelogenin, shorter amelogenin or ameloblastin peptides. These findings point towards a synergistic effect amongst amelogenin and non-amelogenin peptides found within EMD in a ratio approximating 9:1.

Further studies are needed to delineate the biologic effects of EMD on osteoblastic and cementoblastic mesenchymal cells in order to develop a more precise understanding of mechanism of action (21).

The clinical and radiographic outcomes after regenerative surgery and open flap debridement (OFD) for the treatment of deep, greater than 3 *mm*, intrabony periodontal defects were compared (29). Both guided tissue regeneration (GTR) and EMD were found to be superior to OFD alone.

Although the use of EMD was associated with higher radiographic bone filler compared to OFD alone, no statistically significant differences in clinical and radiographic outcomes were found between EMD vs. GTR or EMD vs. bone filler alone. The authors concluded that EMD and GTR with absorbable membranes are the gold standard for treatment of deep intrabony periodontal defects.

Also, a review with regard to biology-based regenerative approaches for periodontal soft tissue engineering has been presented (30). Here it was documented that consistent high-quality evidence supports the adjunctive use of EMD for root coverage procedures.

Furthermore, human clinical studies demonstrated a lot of promising results for the adjunctive use of EMD during the surgical treatment of peri-implantitis (31, 32). Increased bone levels after 12 months and decreased prevalence of gram-negative microbes were found (33).

In summary, EMD has shown similar regenerative outcomes relative to the use of membranes and bone filler biomaterials, and is simpler to use, resulting in fewer post-operative complications (34). These findings could be achieved by evaluating the Cochrane Oral Health Group's Trials Register, CENTRAL, MEDLINE and EMBASE.

Recombinant human platelet-derived growth factors. The recombinant human platelet-derived growth factor (PDGF)-BB (rh-PDGF-BB) is one of the most studied growth factors that is implemented in periodontal tissue engineering. Since its introduction in the late 1980s (35), rh-PDGF-BB has become commercially available for use in the form of Growth-Factor Enhanced Matrix (GEM) 21S (Lynch Biologics), which utilizes β-tricalcium phosphate as a scaffold (36).

PDGF is also one of the most abundant growth factors found in platelet-rich plasma (PRP) (37) and plasma rich in growth factors (PRGF) (38), both of which are concentrates prepared from centrifugation of patient blood.

Of note, the concentration of PDGF in GEM 21S is about 750-fold and 6,400-fold higher relative to the concentration of PDGF in PRP and PRGF, respectively.

The PDGF family is composed of four gene products capable of forming five different dimeric isoforms (39). These isoforms include PDGF-AB, -AA, -BB, -CC, and -DD variants, each of which exhibit different affinities for PDGF z and β tyrosine kinase receptors (40). PDGF-BB is known as the *universal PDGF* due to its ability to bind to all receptor isotypes.

An *in-vitro* investigation into the response of human PDL cells to the various PDGF isoforms was done (41). Here it was discovered that the PDGF-BB isoform exhibited the strongest mitogenic and chemotactic effects. Stored in the z-granules of platelets and released upon activation, PDGF has also been shown to induce DNA synthesis, chemotaxis, as well as collagen and glycosaminoglycan production in fibroblasts (42).

In addition to platelets, activated macrophages and fibroblasts are also known to secrete PDGF (43). PDGF has been demonstrated to stimulate chemotactic and mitogenic responses in mesenchymal cells (in particular PDL cells), in preclinical animal models (44, 45)

In-vitro analysis has also demonstrated that PDGF-BB is a powerful stimulator of stem cell marker expression and proliferation in mesenchymal stem cells (MSCs) isolated from human PDL tissues (46).

In addition, PDGF can modulate bone formation and regeneration through its actions on pericytes (cells of mesenchymal origin located on the perivascular surface believed to harbor populations of MSCs with osteogenic potential).

It has been shown that PDGF plays an important role in mitotic expansion of pericytes (47). It is a major molecular player in pericyte-MSC-osteoblast dynamics, acting as an osteoprogenitor cell mitogen and regulating activity towards osteogenic growth factors such as bone morphogenetic proteins (BMPs).

Fibroblast growth factor-2. The fibroblast growth factor (FGF)-2 is also known as basic fibroblast growth factor. It possesses strong mitogenic properties and has important roles in wound healing, granulation tissue formation, angiogenesis and tissue remodeling (48). Heparin/heparin sulfates are essential for mediating FGF bioactivity (49).

When bound to heparin, the degradation rate of FGF-2 is significantly reduced and the mitogenic activity is greatly enhanced (50).

Heparin sulfate proteoglycans also act as cofactors for FGF2-FGF receptor interactions, allowing for activation of downstream signaling pathways (51).

Growth and differentiation factor. The growth and differentiation factor (GDF)-5 belongs to the TGF-β superfamily of proteins and bears a close structural resemblance to BMPs (52).

As a key regulator of skeletal, tendon and ligament morphogenesis, GDF-5 has also been shown to play important roles in odontogenesis (53) and PDL development (54).

Preclinical studies have demonstrated an enhanced bone formation in response to the local delivery of GDF-5 in rat calvarial defects (55), long bone healing (56) and spinal fusion (57).

GDF-5 also plays important roles in wound healing and periodontal regeneration by regulating extracellular matrix metabolism (58).

In a preclinical investigation it could be demonstrated that exposing human PDL cells to GDF-5 stimulated the synthesis of sulfated glycosaminoglycan and PDL cell proliferation, but not osteoblastic differentiation (59).

Also, the preclinical evidence for the use of GDF-5 in periodontal regeneration was shown, with increased new bone, cementum and PDL formation in small and large animal models (60).

In a review of both preclinical and clinical evidence it was documented that GDF-5 is a safe and effective candidate for periodontal regeneration and alveolar ridge augmentation (61).

4.3.5.2 Tissue Implants

Spinal fusion is a common technique used to treat chronic back pain caused by degenerated or herniated disk (62). The technique involves the removal of a disc between two vertebrae and replacing it with an intervertebral spacer. The intervertebral spacer maintains spacing between the two vertebrae and preferably results in fusion through the spacer.

The intervertebral spacers may be constructed of autogenic bone tissue taken from a patient's own bone. Allogenic spacers are constructed of bone harvested from donors. Artificial spacers are currently the most common spacer type and may be constructed of metallic material such as titanium or stainless steel or polymers such as poly(ether ether ketone) (PEEK).

PEEK has recently become popular due to its biocompatibility and naturally radiotranslucent characteristics, resulting in limited interference with X-ray and CT imaging. However, while PEEK is biocompatible, bone treats it as a foreign body during the remodeling process and isolates it with a fibrous tissue capsule. This fibrous tissue prevents direct bony apposition and adhesion to the implant. Other materials, such as titanium, allow for direct bony apposition and ongrowth, but they are typically not radiotranslucent and it becomes difficult to assess the fusion formation.

Other areas where PEEK is used as an orthopedic biomaterial experience similar fibrous encapsulation. Such indications include custom machined bodies that are used to fill defects in the skull and cranium. With PEEK, magnetic resonance imaging and CT imaging is generally easier as compared to titanium, but the implant is never fully incorporated into the bone and soft tissue does not adhere to the implant (62).

A porous structure for promoting tissue ingrowth contains (62): A first solid portion bonded to a second porous portion, wherein the second porous portion is a network of polymer granules-bonded together at contact points. The network of polymer granules defines a plurality of interstitial spaces between the polymer granules, wherein each polymer granule of the network defines a polymer granule surface (62).

4.3.5.3 Manufacturing Methods

Designing 3D constructs with appropriate materials and structural frameworks for complex dental restorative/regenerative procedures has always remained a multi-criteria optimization challenge (63).

It has long been known that 3D printing is a potent tool for various tissue regenerative applications, however, the preparation of biocompatible, biodegradable, and stable inks is yet to be explored and revolutionized for overall performance improvisation.

The currently used manufacturing processes have been reviewed (63). These processes include the development of engineered self-supporting, easily processable, and cost-effective 3D constructs with target-specific tuneable mechanics, bioactivity, and degradability aspects in the oral cavity for their potential use in numerous dental applications ranging from soft pulp tissues to hard alveolar bone tissues.

A hybrid synergistic approach, consisting of development of multi-layered, structurally stable, composite building blocks with desired physicomechanical performance and bioactivity presents an optimal solution to circumvent the major limitations and develop new-age advanced dental restorations and implants (63).

Electrospinning methods. For example, a Ti-metal-organic framework nanofiber (MOFNF) was prepared using an electrospinning method (64). The basic structure is shown in Figure 4.15.

Its thermodynamic, morphological and mechanical properties were considered using related analysis methods, based on the ideal features of the metal-organic framework and nanofibers using fractional factorial design.

While preparing such nanocoatings with some capabilities, such as environmental friendliness, transparency, and desirable mechanical and physicochemical properties, are significant. The MOFNF was used as nanocoatings with surface area close to 3204 m^2g^{-1}, thermal stability up to 368°C and 559 *Pa* compressive strength in

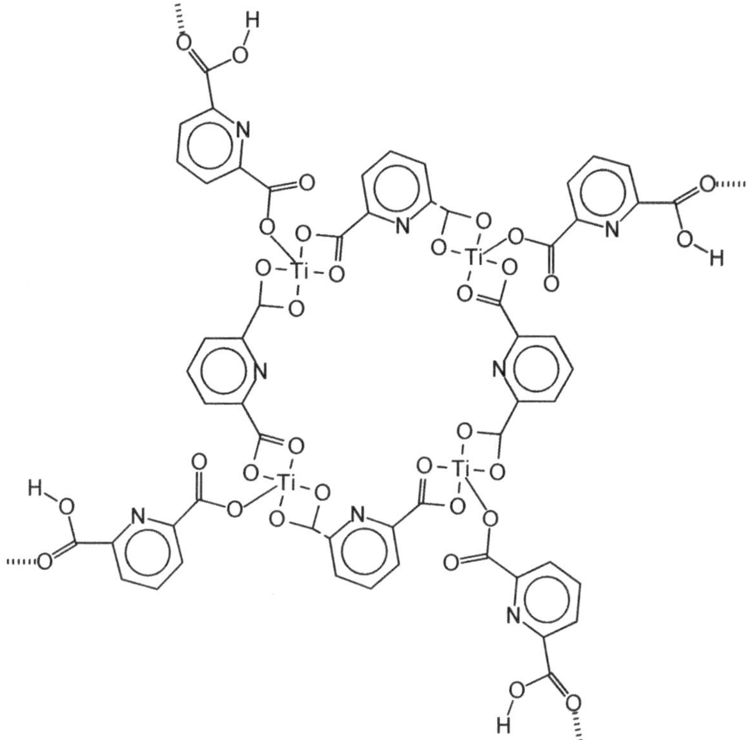

Figure 4.15 Structure of MOFNF (64).

dentistry. This study could be a new strategy for making dental nanocoatings with desired properties (64).

Blending strategies. Hierarchical porous polymer particles from the assembly of polymer chains have attracted increasing interest because of their prospective applications in drug delivery and catalysis (65).

A blending strategy of polymer mixture in mixed solvents was used to fabricate hierarchical porous polymer particles with adjustable pore structures (65). First, using methacrylic acid and methyl methacrylate with precise ratio, a copolymer solution of P(MMA-MAA) was prepared in a mixed solvent of ethanol/water. Then, styrene was added stepwise along with polymerization for achieving a translucent polymer mixture.

Furthermore, a quantitative amount of polymer mixture and six-fold volume of water were rapidly mixed with vortex oscillation. The hierarchical porous polymer particles were eventually nurtured by the assembly of the hydrophilic P(MMA-MAA) copolymers and hydrophobic poly(styrene) homopolymers, as well as the micelles comprised of P(MMA-MAA) copolymer and low polarity solvent.

The size and the number of nanopores were effectively regulated by the solvent polarity, polymerization temperature and the monomer ratios of different attributes. Moreover, the resultant hierarchical porous polymer particles exhibited a hierarchical porous structure, which has a broad application in biomedical and environmental sciences (65).

4.4 Biomedical Applications

Conventional and modern techniques for the synthesis of porous organic polymers have been detailed (66). A huge variety of polymer architectures and functions can be obtained by foaming, phase separation, imprinting or templating approaches.

Also, several applications of porous polymers are discussed, with the focus on biotechnological and biomedical applications such as chromatography, protein synthesis, drug carrier systems, tissue engineering, etc. (66).

4.4.1 *Macroporous Hydrogels*

The development of bulk, three-dimensional (3D), macroporous polymers with high permeability, large surface area and large volume is highly desirable for a range of applications in the biomedical, biotechnological and environmental areas. The experimental techniques currently used are limited to the production of small size and volume cryogel material (67).

A novel, versatile, simple and reproducible method for the synthesis of large volume porous polymer hydrogels by cryogelation has been proposed (67).

By controlling the freezing process of the reagent/polymer solution, large-scale 3D macroporous gels with wide interconnected pores (up to 200 μm in diameter) and large accessible surface area have been synthesized. Here, macroporous gels (of up to 400 *ml* bulk volume) with controlled porous structure could be manufactured, with potential for scale-up to much larger gel dimensions.

This method can be used for production of novel 3D multi-component macroporous composite materials with a uniform distribution of embedded particles. The proposed method provides a better control of freezing conditions and thus overcomes existing drawbacks limiting production of large gel-based devices and matrices (67).

Here, the preparation of gels runs as follows:

Preparation 4–3: Gelatine hydrogel (G gel) sheets were prepared in a glass mold. Glutaraldehyde (3% v/w) was added to the solution of gelatine (6% w/w). The solution was placed in the glass mold and frozen in a Julabo cooling chamber at -12°C for 20 *h*. It was then defrosted and washed with an excess of water. 2-Hydroxyethyl methacrylate (HEMA) gels were prepared by dissolving HEMA (Acros organics 98%) and poly(ethylene glycol) diacrylate (PEGDA, Aldrich, M_n 258) in water (6 w/% by volume solution, HEMA:PEGDA molar ratio 8:1). The reaction mixture was degassed at low pressure for 25 *min* to eliminate dissolved oxygen before gelation. In the conventional method the mixture was cooled to 0°C for 15 *min* and then *N,N,N′,N′*-tetramethyl-ethylene diamine (TEMED) (Fisher Scientific, 99%) and ammonium persulfate (APS, 98%) were added and the mixture allowed to freeze completely. In the pre-freezing method the mixture was cooled with constant mixing in an ethanol cooling bath at -18°C. After ice crystals formed, the mixture was pre-cooled to -2°C with constant mixing.

Next, TEMED and ammonium persulfate (APS, 98%), were added and the mixture allowed to freeze completely. The frozen mixture was kept at -18°C for 20 *h* and then defrosted at room temperature.

The preparation of a composite can be also done as follows:

Preparation 4–4: Iron oxide nanoparticle composites were prepared by dissolving 4.8 *g* acrylamide (AAm) (Sigma-Aldrich, 99%) and 1.6 *g* of *N,N'*-methylenebisacrylamide (MBAA) (Sigma-Aldrich, 96%) in 100 *ml* of water. The solution was degassed for 20 *min*. Then, 10 *g* iron oxide nanoparticles (α-Fe2O3, Arry International Group Limited, particle size 30 *nm*) and 0.01 *ml* of TEMED were added. Next, 10 *ml* of the suspension was placed in a glass tube and 0.007 *g* APS was added. After mixing, the glass tube was placed into a cooling bath at -18°C.

The rest of the suspension of nanoparticles in the monomer solution was pre-cooled in the cooling bath at −18°C with constant mixing until ice crystals formed. Then, 10 *ml* of pre-frozen suspension was added to the glass tube and 0.007 *g* APS was added. The suspension was mixed and allowed to freeze completely at −18°C. The frozen samples were kept at −18°C for 20 *h* and then defrosted at room temperature. The iron oxide cryogel of 140 *ml* was prepared in a similar manner. The reaction solution (150 *ml* of monomer solution plus 15 *g* $\alpha\text{–}Fe_2O_3$) was pre-cooled in the cooling bath at −18°C with constant mixing until ice crystals formed. The reaction suspension was placed into a 150 *ml* syringe and initiator was added. The samples were allowed to freeze at −18°C overnight.

The next day, the sample was defrosted and washed with water. Activated carbon bead acrylamide composites were similarly made by adding to 100 *ml* of AAm and MBAA solution carbon beads (MAST Carbon International Ltd, S-250/500-TE9/16–20°C) pre-wetted in water overnight (3.3 *g* in 15 *ml* of water). For making carbon nanotubes composites, 6 *ml* of carbon nanotube suspension (0.09 *g* of dry weight) was added to 100 *ml* of AAm (4.8 *g*) and MBAA (1.6 *g*) solution.

The solution was degassed for 20 *min* and 0.01 *ml* of TEMED was added. The suspension of nanotubes in the monomer solution was pre-cooled in the cooling bath at −18°C with constant mixing until ice crystals formed. Then, 0.07 *g* of APS was added. The suspension was mixed and allowed to freeze completely at −18°C. The frozen mixture was kept at −18°C for 20 *h* and then defrosted at room temperature.

4.4.2 *Alginate Foams*

Alginates in the format of scaffolds provide important functions as materials for cell encapsulation, drug delivery, tissue engineering and wound healing among others (68). A method for preparation of alginate-based foams is based on homogeneous, ionotropic gelation of aerated alginate solutions, followed by air drying. The method allows higher flexibility and better control of the pore structure, hydration properties and mechanical integrity compared to foams prepared by other techniques.

The main variables for tailoring the hydrogel properties include operational parameters, such as degree of aeration and mixing times and concentration of alginate, as well as macromolecular properties such as the type of alginate (chemical composition and molecular weight distribution). An exposure of the foams to γ-irradiation resulted in a dose-dependent (0–30 kGy) reduction in the molecular weight of the alginate and a corresponding reduction in the tensile strength of the foams (68).

4.4.3 Biodegradable Sponges

A dispensing system has been described that can be used to introduce a gel-forming fluid into or onto the body that will produce a homogeneous gel *in-situ*. The system contains:

1. A first reservoir comprising a first flowable composition in the form of an aqueous solution of ionically crosslinkable alginate,
2. A second reservoir comprising a second flowable composition, said second flowable composition containing one or more polyols,
3. A first outlet from the first reservoir; a second outlet from the second reservoir, said first outlet and said second outlet being arranged in such a way that the first flowable composition is mixed with the second flowable composition when both compositions are simultaneously expelled from the respective reservoirs, thereby forming the gel-forming fluid.

Here, the first and/or second flowable composition contains a crosslinking agent and upon admixture of the first flowable composition with the second flowable composition said crosslinking agent initiates the formation of a solid or semi-solid ionically crosslinked alginate gel. The presence of a substantial amount of polyol in the second flowable composition favorably affects the crosslinking reaction between the crosslinking agent and the alginate (69).

The use of an inactive crosslinking agent in the present system offers the advantage that the rate of alginate gelation can be controlled effectively, depending on the rate of activation of the inactive crosslinking agent. According to a very preferred embodiment, the inactive crosslinking agent is a water-insoluble salt of a multivalent cation. Examples of multivalent cations include multivalent metal cations as well as multivalent cationic substances, such as gentamicin or cationic peptides. Gentamicin is shown in Figure 4.16,

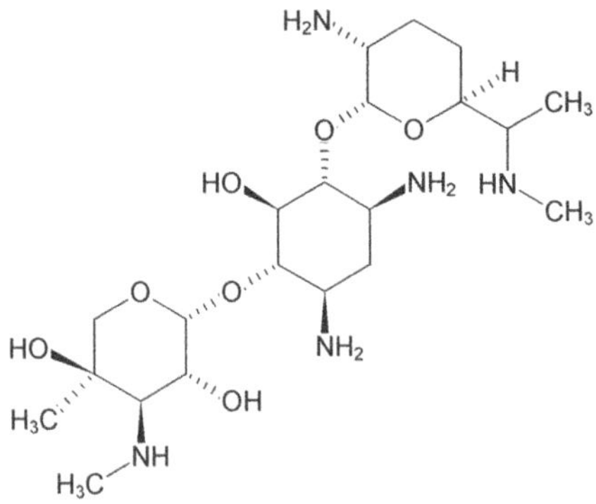

Figure 4.16 Gentamicin C1.

4.4.4 *Biomedical Scaffolds*

Degradable polymers are widely used in tissue engineering and regenerative medicine (70). The maturing capabilities in additive manufacturing coupled with advances in orthogonal chemical functionalization methodologies have enabled a rapid evolution of defect-specific form factors and some strategies for designing and creating bioactive scaffolds.

However, such defect-specific scaffolds, especially when utilizing degradable polymers as the base material, present processing challenges that are distinct and unique from other classes of materials.

In a review, a guide for the fabrication of biodegradable polymer-based scaffolds has been presented that includes a complete pathway starting from selecting materials, choosing the correct fabrication method, and considering the requirements for tissue-specific applications of the scaffold (70).

Scaffolds made from synthetic polymers normally lack a bioactive cell-material interface and undergo a burst release of growth factors, which may hinder their further application in tissue engineering. A metal-phenolic network was interfacially constructed on the pore surface of a porous poly(*d,l*-lactide) scaffold (71).

Based on the molecular gating property of the metal-phenolic network supramolecular structure, the poly(*d,l*-lactide) metal-phenolic network scaffold achieved the sustained release of the loaded molecules. In addition, the metal-phenolic network coating provided a bioactive interface, thus encouraging the migration and osteogenic differentiation of bone marrow mesenchymal stem cells.

The poly(*d,l*-lactide)metal-phenolic network scaffolds exhibited an enhanced bone regeneration in a rat femoral defect model *in-vivo* compared to poly(*d,l*-lactide), which was ascribed to the combined effect of sustained bone morphogenetic protein-2 release and the osteogenic ability of metal-phenolic network. This nanodressing technique provides a viable and straightforward strategy for enhancing the performance of porous polymer scaffolds in bone tissue engineering (71).

4.4.5 Biodegradable Electronic Materials

Biodegradable or transient electronics is an emerging technology whose key characteristic is an ability to dissolve, resorb, physically disappear or disintegrate in physiological environments after fulfilling their functions over prescribed time frames (72).

Biodegradable electronics as temporary implants can be safely absorbed by the body, and as green electronics can effectively alleviate landfill and environmental issues caused by electronic wastes, which are expected to ultimately eliminate associated costs and risks resulting from recycling operations.

The state-of-the-art of the biodegradable polymers, their fabrication techniques, and potential applications in environment friendly electronic skins (e-skins) were comprehensively summarized and systematically discussed (72).

Also, the future outlook for the development of biodegradable e-skins was also discussed. So, vital tools and beneficial strategies are opened for the design of biodegradable e-skins with the functions of diagnosis, therapy and green sensing that are beneficial for human healthcare and environmental protection (72).

4.4.6 *Optical Fibers*

A cost-effective fabrication scheme for porous polymer optical fibers has been proposed (73). Different porous polymer fibers made from PC and PMMA using this method have been thermally drawn and characterized.

The porosity in the fiber cladding is introduced by the absorbed water in one layer of the polymer fiber preforms under heat treatment and/or thermal drawing, and can be controlled by adjusting the water concentration. In addition, it has been shown that the fabricated porous polymer fibers have the potential application in localized drug delivery for cancer treatment (73).

Thermal drawing is the most common method for polymer fiber fabrication (74, 75). Here, a polymer fiber preform resembling the final fiber geometry is fed into a furnace where it is heated. The polymer gradually softens, allowing it to be pulled into a continuous fiber whose diameter is controlled by the preform down feed rate, the fiber capstan rate and the furnace temperature.

One inch diameter PC and PMMA rods were first drawn after a three-day 60°C water bath (73). The resulting fibers had pores elongated along the axial direction. A fiber draw down ratio of greater than 50 can stretch the original submillimeter pores to several meters long. To limit the pore generation in the fiber cladding, vacuum-dried polymer rods were inserted into the water-bathed tubes which would result in a solid-core, porous-cladding fiber. An outer layer made from a vacuum-dried tube or tightly wrapped polymer film layers (vacuum-dried 75 μm thick PC films (McMaster-Carr)) were added to provide additional support for the porous-cladding in the drawn fiber.

Thirty minutes of consolidation of the preforms in a vacuum oven at 140°C for PMMA and 180°C for PC eliminates possible air gaps at the interfaces. Water-induced pores could be readily observed in the PC preforms after consolidation when the pore formation starts at 180°C. Besides the water absorbance method, an additional fabrication scheme was also proposed to mitigate the issue of irregular pore generation in PMMA. Here, vacuum-dried tubes were first drawn into capillary tubes with diameters between 500 μm and 1000 μm.

These capillary tubes were placed between a vacuum-dried core rod and an outer tube to form the cladding region. Consolidation was conducted under atmospheric pressure instead of vacuum to minimize capillary tube collapsing before the preform can be readily drawn into a fiber (73).

Porous polymer fiber materials capable of transmitting light and locally deliver drug are particularly useful for a photodynamic therapy, where the released drug requires light activation. Sustained and controlled drug release could be achieved by loading drug-doped biodegradable polymers, such as Poloxamer 407, into the porosity (73).

4.4.7 *Bead Sorbent*

Packed red blood cell (pRBC) units contain reactive donor antibodies, free hemoglobin, high extracellular potassium levels, and biologically active inflammatory mediators that have the potential to cause adverse effects during blood transfusions (76). Such adverse effects can include non-hemolytic febrile and allergic transfusion reactions, atypical infections, allo-immunization, and potentially fatal reactions, like transfusion-related acute lung injury. Furthermore, transfusion risk increases in patients receiving multiple pRBCs, such as those involved in trauma or undergoing surgery, and in primed susceptible patients, such as those in critical care or undergoing high-risk surgery.

The likelihood of adverse effects increases over time for stored blood or blood products, as concentrations of many biological response modifiers, such as potassium, free hemoglobin, and cytokines, increase with storage duration. Cytokines are produced by residual leukocytes during storage of platelets and pRBCs, and can cause inflammation, fever, and direct vascular and organ injury.

Erythrocytes contain phosphatidyl choline, and cytosolic and membrane phospholipase A2, contributing increasing levels of lysophosphatidylcholine (lysoPC) during storage. Structural and biochemical changes that RBCs undergo are described as *storage lesion* and lead to a progressive loss of hemoglobin and potassium.

Plasma free hemoglobin can rapidly overwhelm the scavenging capability of haptoglobin, resulting in oxidative damage to lipids, proteins, endothelial cells, tissues, and renal proximal tubules, and in depletion of nitric oxide (NO) upon transfusion. Increases in extracellular potassium during storage lead to an increased risk of hyperkalemia and arrhythmia, particularly for large volume or massive transfusions and transfusions in newborns and infants.

Hyperkalemia describes a condition in which the potassium level in the blood exceeds a concentration of 5 mEq/L, where concentrations exceeding 7 mEq/L are considered severe cases. The electrical rhythm of the heart can be altered by moderate hyperkalemia, while severe conditions may cause the heart to stop beating.

In addition to blood transfusions, another major cause of hyperkalemia is tissue destruction that causes dying cells to release potassium into blood circulation. Tissue destruction typically results from trauma, burns, hemolysis, massive lysis of tumor cells, rhabdomyolysis, or major surgery, such as cardiac surgery or cardiopulmonary bypass, where severe tissue destruction leads to more severe cases of hyperkalemia.

In addition to the release of potassium into blood circulation, massive tissue injury is characterized by release of a large amount of myoglobin from damaged muscle tissue, plasma free hemoglobin from hemolyzed red blood cells, damage associated molecular pattern (DAMP) factors from damaged cells, and an up-regulation of pro- and anti-inflammatory mediators such as cytokines. Excessive free myoglobin, free hemoglobin, and other inflammatory mediators, can lead to complications such as renal failure or even death. Abnormal regulation of cytokines, or release of DAMPS, may lead to systemic inflammatory response syndrome and multi-organ dysfunction.

Currently, there are existing technologies for potassium removal or antibody removal from stored blood or blood products. Kawasumi Laboratories has developed a single-pass in-line potassium adsorption filter to reduce the risk of hyperkalemia and improve the safety of blood transfusions. The filter functions by exchanging potassium ions (K^+) for sodium ions (Na^+) to decrease the concentration of K^+ in stored RBC units. In an *in-vitro* study, 10 filters were tested using each of three AS-3 RBC units via gravity filtration.

The mean decrease in potassium was 97.5%, 91.2%, and 64.4% for the first, second, and third units, respectively. Accompanying the decrease in potassium were mean increases of sodium by 33%, magnesium by 151.4%, and total calcium by 116.1%. Plasma hemoglobin was unchanged after filtration.

A journal article (77) describes a study assessing the development of a sodium/calcium/magnesium exchange resin mixture that removes potassium without associated electrolyte abnormalities. At the time the article was written, direct hemoperfusion over an exchange resin was capable of lowering elevated serum potassium levels, but had not been used clinically due to subsequent electrolyte abnormalities. Prior to evaluating the exchange resin in an *in-vivo* model, batch experiments were conducted in vitro to identify an effective ratio of sodium to calcium to magnesium for the resin mixture.

The results from the study demonstrated a reduction of elevated plasma potassium levels from about 6.7 to about 3.5 mEq/L in anephric dogs, without any significant change in levels of sodium, calcium, magnesium, albumin, total protein, or cholesterol, after 2 hours of direct hemoperfusion through an exchange resin column. Pre- and post-hemoperfusion platelet counts and plasma free hemoglobin levels were also measured, where post-hemoperfusion platelet counts were only about 45% of pre-hemoperfusion levels, and there was no significant change in plasma free hemoglobin levels. Cholesterol is shown in Figure 4.17.

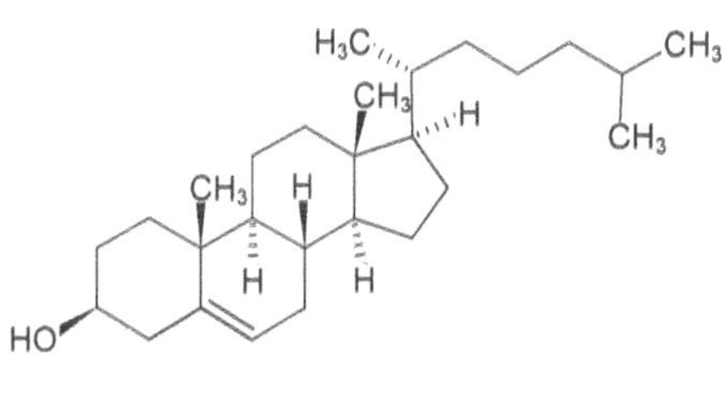

Figure 4.17 Cholesterol.

Differentiating a multifunctional polymer from other filters that remove only reactive proteins or only potassium is its ability to remove both simultaneously without sacrificing binding capacity for either. Additionally, the sorbent is able to remove cytokines and inflammatory protein moieties simultaneously while removing potassium and antibodies. For hemoperfusion applications, it is a requirement that the polymer is hemocompatible. Using the unactivated partial thromboplastin time assay as a measure of thrombogenicity, the polymer described herein exhibits minimal activation, indicating a plasma-like interaction.

This polymer is suited for a wide variety of applications, as many cases of trauma, burn, and major surgery result in hyperkalemia, cytokine storm, and require blood transfusions. The ability to use one multi-application filter has many advantages over using many single-application filters. Given the value of blood and blood products, the use of a single, smaller filter that minimizes cell loss within the retained volume and reduces complexity of material quality assurance is very desirable (76).

A biocompatible polymer system has been developed that contains at least one polymer with a plurality of pores and a sulfonic acid salt functionality. The polymer system is capable of adsorbing a broad range of protein-based toxins having a molecular weight of from less than about 0.5 *kDa* to about 1,000 *kDa* and also positively charged ions.

Here, the preferred polymers are hemocompatible. The polymer system has the form of a solid support. Certain preferred polymer systems have a geometry of a spherical bead. Other polymer systems have the form of a fiber, monolithic column, film, membrane, or semi-permeable membrane (76, 78).

Preferred monomers for such polymers are collected in Table 4.2. Some of these monomers are shown in Figure 4.18.

An organic solvent can be used as the porogen or pore-former, and the resulting phase separation induced during polymerization yields porous polymers. Preferred porogens are shown in Table 4.3. Some porogens are shown in Figure 4.19.

The synthesis of a base sorbent can be done as follows (78):

Preparation 4–5: Reactor Setup: A 4-neck glass lid was affixed to a 3*L* jacketed cylindrical glass reaction vessel using a stainless steel flange clamp and poly(tetrafluoroethylene) (PTFE) gasket. The lid was fitted with a PTFE stirrer bearing, resistance temperature detector (RTD) adapter, and water-cooled reflux condenser. A stainless steel stirring shaft having five 60° agitators was fit through the stirrer bearing and inserted into a digital overhead stirrer. An RTD was fit through the corresponding adapter, and connected to a PolyStat circulating heating and chilling unit. Compatible

Table 4.2 Monomers for hemocompatible polymers (76).

Monomer	Monomer	Monomer
Acrylonitrile	Allyl ether	Allyl glycidyl ether
Butyl acrylate	Butyl methacrylate	Cetyl acrylate
Cetyl methacrylate	Vinyl acetate	Vinylformamide
Vinylnaphthalene	2-Vinyloxirane	Vinyl toluene
Styrene	Ethylstyrene	Vinylbenzyl alcohol
Ethylvinylbezene	Methyl acrylate	Methyl methacrylate
Octyl acrylate	Octyl methacrylate	Trivinylbenzene
1,2-Epoxy-5-hexene	Ethyl acrylate	Ethyl methacrylate
Divinylbenzene	Divinylformamide	Divinylnaphthalene
Divinylsulfone	3,4-Epoxy-1-butene	1,2-Epoxy-9-decene
Glycidyl methacrylate	Trivinylcyclohexane	

Monomer	Monomer
3,4-Dihydroxy-1-butene	Dipentaerythritol diacrylate
Dipentaerythritol dimethacrylate	Dipentaerythritol tetraacrylate
Dipentaerythritol tetramethacrylate	Dipentaerythritol triacrylate
Dipentaerythritol trimethacrylate	Pentaerythritol diacrylate
Pentaerythritol dimethacrylate	Pentaerythritol tetraacrylate
Pentaerythritol tetramethacrylate	Pentaerythritol triacrylate
Pentaerythritol trimethacrylate	Trimethylol propane diacrylate
Trimethylol propane dimethacrylate	Trimethylol propane triacrylate
Trimethylol propane trimethacrylate	Vinylcyclohexene epoxide

Table 4.3 Porogens (76).

Compound	Compound	Compound
Benzyl alcohol	Cyclohexane	Cyclohexanol
Cyclohexanone	Decane	Dibutyl phthalate
Ethyl acetate	*n*-Heptane	*n*-Hexane
n-Octane	Pentanol	Tetraline
Toluene	Xylene	Poly(styrene)
Isoamyl acetate	Isoamyl alcohol	2,2,4-Trimethylpentane

Compound	Compound
Di-2-ethylhexyl phthalate	Di-2-ethylhexylphosphoric acid
2-Ethyl-1-hexanoic acid	2-ethyl-1-hexanol
Poly(propylene glycol)	Poly(styrene-*co*-methyl methacrylate)
Tri-*n*-butyl phosphate	1,2,3-Trichloropropane

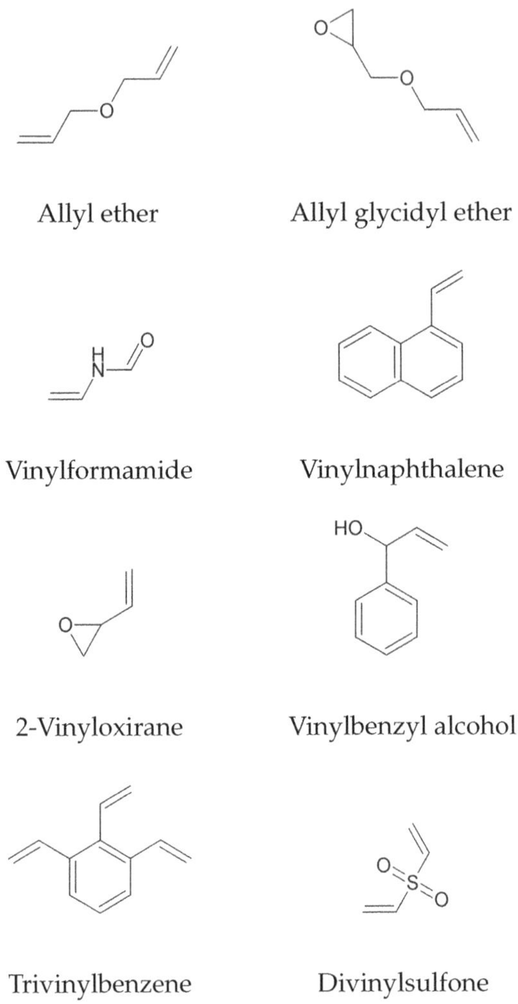

Figure 4.18 Monomers for hemocompatible polymers (76).

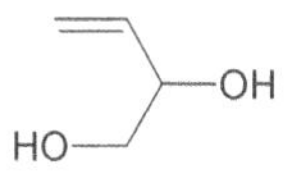

3,4-Dihydroxy-1-butene

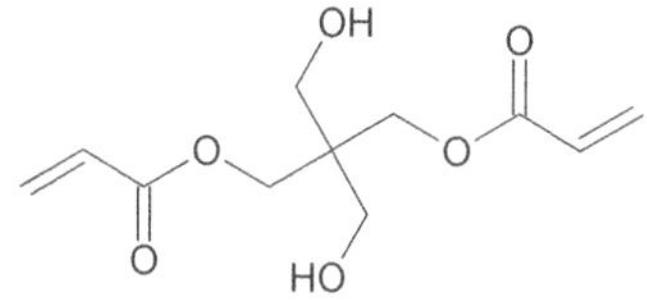

Pentaerythritol diacrylate

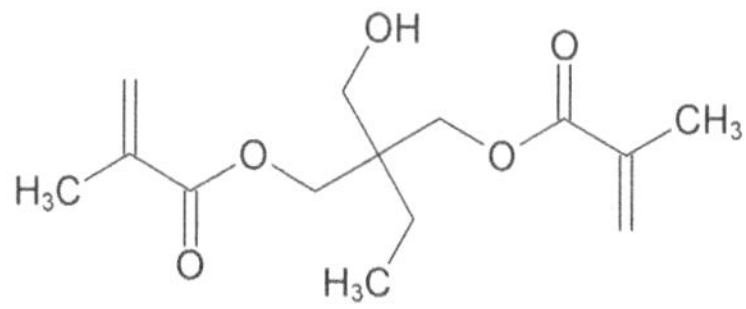

Trimethylol propane dimethacrylate

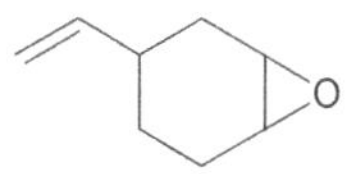

4-Vinyl-1-cyclohexene 1,2-epoxide

Figure 4.18 (cont) Monomers for hemocompatible polymers (76).

tubing was used to connect the inlet and outlet of the reaction vessel jacket to the appropriate ports on the PolyStat. The unused port in the lid was used for charging the reactor and was plugged at all other times.

Polymerization: Aqueous phase and organic phase compositions are shown in Table 4.4 and Table 4.5, respectively. Ultrapure water was split into approximately equal parts in two separate Erlenmeyer flasks, each containing a PTFE-coated magnetic stir bar. PVA, having a degree of hydrolysis of 85.0 to 89.0 *mol%* and a viscosity of 23.0 to 27.0 *cP* in a 4% aqueous solution at 20°C, was dispersed into the water in the first flask and heated to 80°C on a hot plate with agitation. The salts presented in Table 4.4, i.e., monosodium phosphate, disodium phosphate, trisodium phosphate and sodium nitrite, were dispersed into the water in the second flask and heated to 80°C on a hot plate with agitation. Circulation of heat transfer fluid from the PolyStat through the reaction vessel jacket was started, and fluid temperature heated to 60°C. Once PVA and salts dissolved, both solutions were charged to the reactor, one at a time, using a glass funnel. The digital overhead stirrer was powered on and the rpm set to a value to form appropriate droplet sizes upon organic phase addition.

The temperature of the aqueous phase in the kettle was set to 70°C. The organic phase was prepared by adding benzoyl peroxide to the divinylbenzene in a 2*L* Erlenmeyer flask and swirling until completely dissolved. Then, 2,2,4-trimethylpentane and toluene were added to the flask, which was swirled to mix well. Once the temperature of the aqueous phase in

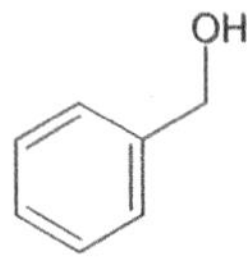

Benzyl alcohol

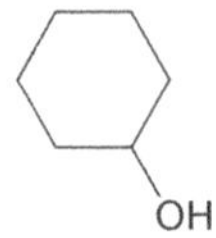

Cyclohexanol

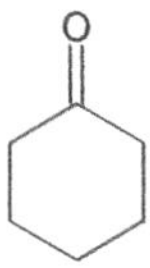

Cyclohexanone

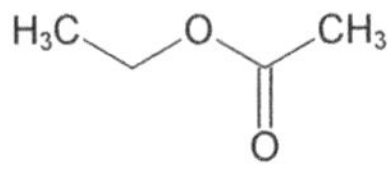

Ethyl acetate

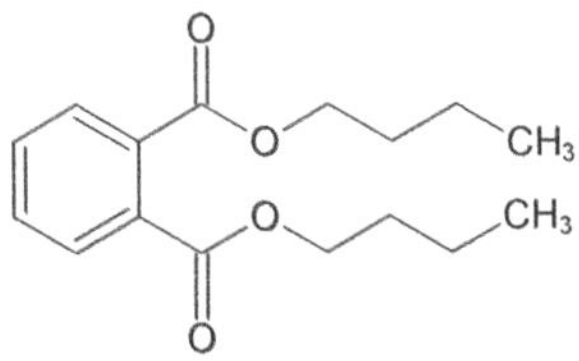

Dibutyl phthalate

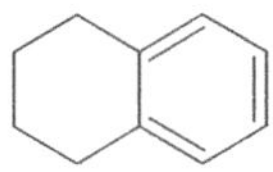

Tetraline

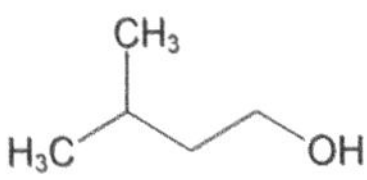

Isoamyl alcohol

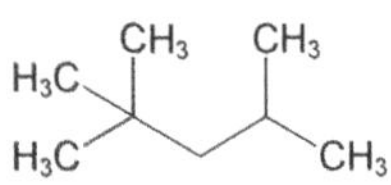

2,2,4-Trimethylpentane

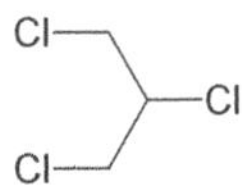

1,2,3-Trichloropropane

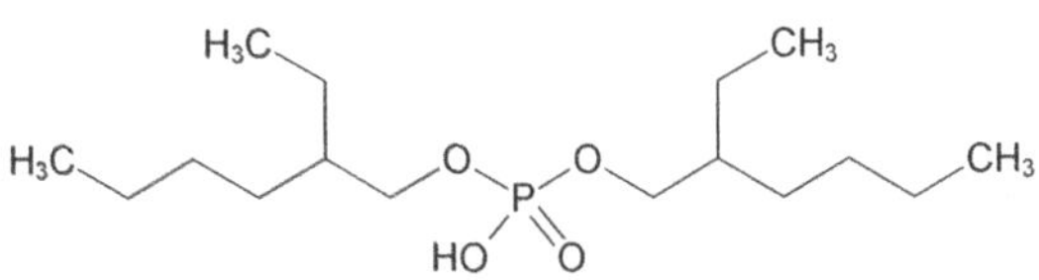

Di-2-ethylhexylphosphoric acid

Figure 4.19 Porogens (76).

the reactor reached 70°C, the organic phase was charged into the reactor using a narrow-necked glass funnel. Temperature of the reaction volume dropped upon the organic addition. A temperature program for the PolyStat was started, heating the reaction volume from 60 to 77°C over 30 *min*, 77 to 80°C over 30 *min*, holding the temperature at 80°C for 960 *min*, and cooling to 20°C over 60 *min*.

Table 4.4 Aqueous phase composition reagents (78).

Component	Amount/[g]
Ultrapure water	1500.000
Poly(vinyl alcohol) (PVA)	4.448
Monosodium phosphate (MSP)	4.602
Disodium phosphate (DSP)	15.339
Trisodium phosphate (TSP)	9.510
Sodium nitrite	0.046
Total	1533.899

Table 4.5 Organic phase compositions (78).

Component	Reagent CY14175/[g]	Reagent CY15077/[g]
Divinylbenzene, 63% (DVB)	508.751	498.383
2,2,4-Trimethylpentane (Isooctane)	384.815	482.745
Toluene	335.004	222.404
Benzoyl peroxide, 98% (BPO)	3.816	3.738
Total (excluding BPO)	1228.571	1203.532

The modification to get polymer CY15154 was done as follows (78):

Preparation 4–6: First, 20.05 *g* dried base polymer CY15077 was added to a 500 *ml* jacketed glass reactor, which was equipped with a Teflon-coated agitator and RTD probe. Dried polymer was rewet into distilled water making 100 *ml* slurry in reactor. Then, 9.00 *g* zwitterionic neutral methacrylate monomer, [(2-methacryloyloxy)ethyl]-dimethyl-3-(sulfopropyl) ammonium hydroxide, c.f. Figure 4.20, and 1.1 *g* ammonium persulfate were dissolved in 100 *ml* distilled water and the solution added to the reactor containing base polymer slurry. The mixture was heated to 75°C and maintained at 75°C for 24 *h* with agitation at 100 *rpm*.

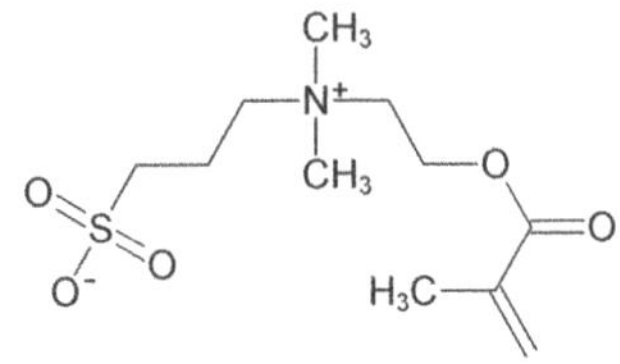

Figure 4.20 [(2-methacryloyloxy)ethyl]-dimethyl-3-(sulfopropyl) ammonium hydroxide.

The endotoxin removal from plasma in a recirculation model can be done as follows (78):

Method 4–7: First, 20 *ml* of heat-inactivated human citrated plasma was spiked with 3 EU/*ml* endotoxin purchased from Associates of Cape Cod, East Falmouth, MA, and then spiked into plasma reservoir. Plasma with endotoxin was mixed on a stir plate for 15 *min* and recirculated through the 1.5 *ml* polymer column at a flow rate of 2.5 $ml\,min^{-1}$. Samples were collected from the reservoir using sterile pipette tips and diluted 1:20 in endotoxin-free water. All diluted samples were tested in the Pierce LAL Endotoxin Assay from Life Technologies Corp., Grand Island, NY. The data for endotoxin removal from plasma were obtained in a dynamic recirculation model. Table 4.6 shows the endotoxin removal data for the polymer CY15154 and its non-modified precursor polymer CY15077.

Table 4.6 *In-vitro* endotoxin removal (78).

Time/[*min*]	Endotoxine remaining in	
	CY15077/[%]	CY15154/[%]
0	100	100
15	113.70	53.20
30	105.60	44.27
60	101.61	39.78
120	101.15	48.56

References

1. X. Liang, D. King, and A. Weimer, Ceramic ultra-thin coatings using atomic layer deposition in R. Banerjee and I. Manna, eds., *Ceramic*

Nanocomposites, Woodhead Publishing Series in Composites Science and Engineering, pp. 257–283. Woodhead Publishing, 2013.

2. J.S. Park, J.H. Kim, and R.T. Davies, Method for diagnosing a disease using extracellular vesicles comprising porous polymer monolith microfilter, US Patent 9 081 012, assigned to Postech Foundation and Postech Academy Industry Foundation, July 14, 2015.
3. K. Yang and N. Fernandez, Method for fabricating medical devices with porous polymeric structures, US Patent 9 061 093, assigned to Abbott Cardiovascular Systems Inc., June 23, 2015.
4. T.A. Ringeisen and S.M. Goldman, Method for making a porous polymeric material, US Patent 8 697 108, assigned to Kensey Nash Corp., April 15, 2014.
5. Y. Hashimoto, S. Mukai, S. Sawada, Y. Sasaki, and K. Akiyoshi, *Biomaterials*, Vol. 37, p. 107, 2015.
6. P. Liu, H. Du, Y. Chen, H. Wang, J. Mao, L. Zhang, J. Tao, and J. Zhu, *J. Mater. Chem. B*, Vol. 8, p. 2032, 2020.
7. S. Bilent, T.H.N. Dinh, E. Martincic, and P.-Y. Joubert, *Proceedings*, Vol. 2, 2018.
8. T.H.N. Dinh, E. Martincic, E. Dufour-Gergam, and P.Y. Joubert, *Journal of Sensors*, Vol. 2017, p. 8235729, 2017.
9. M. Bartnikowski, C. Vaquette, and S. Ivanovski, *Clinical Oral Implants Research*, Vol. 31, p. 431, 2020.
10. M. Mohammadi Zerankeshi, R. Bakhshi, and R. Alizadeh, *Bioprinting*, Vol. 25, p. e00191, 2022.
11. S. Du, *Polymer Composite Scaffolds for Bone Tissue Engineering*. PhD thesis, Monash University, Melbourne, Australia, 2921.
12. S.S. Crump, Apparatus and method for creating three-dimensional objects, US Patent 5 121 329, assigned to Stratasys Inc., June 09, 1992.
13. I.A. Armesto, L.D. Villamil, R.F. Yazicioglu, K. Schlegel, and K. Leyde, Method for implanting a medical device in the body tissue of a human or animal patient, US Patent 20 210 402 194, assigned to Galvani Bioelectronics Ltd., Greatbatch Ltd., December 30, 2021.
14. J.R. Roeder and W.J. Park, Sheathes for surgical instruments, and related devices and methods, US Patent 20 210 401 530, assigned to Intuitive Surgical Operations Inc., December 30, 2021.
15. E.J. Giphart, A.K. Levien, T. Wilmering, and B. Schafer, Method and apparatus for controlling an eye lid during ultrasound imaging, US Patent 20 210 401 401, assigned to Arcscan Inc., December 30, 2021.
16. F. Tourlomousis, R. Chang, and D. Kalyon, Integrated methods for precision manufacturing of tissue engineering scaffolds, US Patent 20 210 395 675, assigned to Stevens Institute of Technology, December 23, 2021.

17. R.E. Richard, F.H. Strickler, M.C. Schwarz, R. Faust, S.V. Ranade, and M.N. Helmus, Medical devices having porous polymeric regions for controlled drug delivery and regulated biocompatibility, US Patent 8 535 702, assigned to Boston Scientific Scimed Inc., September 17, 2013.
18. T. Fernandez-Medina and A. Nanda, Regeneration for implant dentistry in S. Hosseinpour, L.J. Walsh, and K. Moharamzadeh, eds., *Regenerative Approaches in Dentistry: An Evidence-Based Perspective*, pp. 133–150. Springer International Publishing, Cham, 2021.
19. D. Kaigler, J.A. Cirelli, and W.V. Giannobile, *Expert Opinion on Drug Delivery*, Vol. 3, p. 647, 2006.
20. W.V. Giannobile, *Periodontal*, pp. 1151–1154, 2014. AP Centennial Commentary: Theme 9.
21. M. Galli, Y. Yao, W.V. Giannobile, and H.-L. Wang, *Plastic and Aesthetic Research*, Vol. 8, 2021.
22. Z. Lin, H.F. Rios, and D.L. Cochran, *Journal of Periodontology*, Vol. 86, p. S134, 2015.
23. I.B. Darby and K.H. Morris, *Journal of Periodontology*, Vol. 84, p. 465, 2013.
24. S. Gestrelius, S.P. Lyngstadaas, and L. Hammarström, *Clinical Oral Investigations*, Vol. 4, p. 120, 2000.
25. Wikipedia contributors, Epithelial root sheath — Wikipedia, the free encyclopedia, https://en.wikipedia.org/w/index.php?title=Epithelial_root_sheath&oldid=994434140, 2020. [Online; accessed 12-August-2021].
26. L. Hammarström, L. Heijl, and S. Gestrelius, *Journal of Clinical Periodontology*, Vol. 24, p. 669, 1997.
27. D.D. Bosshardt, A. Sculean, P. Windisch, B.E. Pjetursson, and N.P. Lang, *Journal of Periodontal Research*, Vol. 40, p. 158, 2005.
28. H.M. Grandin, A.C. Gemperli, and M. Dard, *Tissue Engineering Part B: Reviews*, Vol. 18, p. 181, 2012.
29. L. Nibali, V.P. Koidou, M. Nieri, L. Barbato, U. Pagliaro, and F. Cairo, *Journal of Clinical Periodontology*, Vol. 47, p. 320, 2020.
30. L. Tavelli, M.K. McGuire, G. Zucchelli, G. Rasperini, S.E. Feinberg, H.-L. Wang, and W.V. Giannobile, *Journal of Periodontology*, Vol. 91, p. 147, 2020.
31. F. Mercado, S. Hamlet, and S. Ivanovski, *Clinical Oral Implants Research*, Vol. 29, p. 583, 2018.
32. C. Isehed, B. Svenson, P. Lundberg, and A. Holmlund, *Journal of Clinical Periodontology*, Vol. 45, p. 744, 2018.
33. C. Isehed, A. Holmlund, S. Renvert, B. Svenson, I. Johansson, and P. Lundberg, *Journal of Clinical Periodontology*, Vol. 43, p. 863, 2016.
34. M. Esposito, M.G. Grusovin, Y.S. Chew, P. Coulthard, and H.V. Worthington, *Eur J. Oral Implantol.*, Vol. 2, p. 91, 2009.

35. S.E. Lynch, R.C. Williams, A.M. Poison, T.H. Howell, M.S. Reddy, U.E. Zappa, and H.N. Antoniades, *Journal of Clinical Periodontology*, Vol. 16, p. 545, 1989.
36. F. Suárez-López del Amo, A. Monje, M. Padial-Molina, Z. Tang, and H.-L. Wang, *BioMed Research International*, Vol. 2015, p. 2, 2015.
37. G. Weibrich, W.K.G. Kleis, and G. Hafner, *International Journal of Oral & Maxillofacial Implants*, Vol. 17, p. 184, 2002.
38. G. Weibrich, W.K.G. Kleis, W.E. Hitzler, and G. Hafner, *International Journal of Oral & Maxillofacial Implants*, Vol. 20, p. 118, 2005.
39. L. Fredriksson, H. Li, and U. Eriksson, *Cytokine & Growth Factor Reviews*, Vol. 15, p. 197, 2004. Platelet-derived Growth Factor.
40. J. Andrae, R. Gallini, and C. Betsholtz, *Genes & Development*, Vol. 22, p. 1276, 2008.
41. L. Boyan, G. Bhargava, F. Nishimura, R. Orman, R. Price, and V. Terranova, *Journal of Dental Research*, Vol. 73, p. 1593, 1994.
42. S.E. Lynch, J.C. Nixon, R.B. Colvin, and H.N. Antoniades, *Proceedings of the National Academy of Sciences*, Vol. 84, p. 7696, 1987.
43. X.E. Dereka, C.E. Markopoulou, and I.A. Vrotsos, *Growth Factors*, Vol. 24, p. 260, 2006.
44. J.-B. Park, M. Matsuura, K.-Y. Han, O. Norderyd, W.-L. Lin, R.J. Genco, and M.-I. Cho, *Journal of Periodontology*, Vol. 66, p. 462, 1995.
45. M.-I. Cho, W.-L. Lin, and R.J. Genco, *Journal of Periodontology*, Vol. 66, p. 522, 1995.
46. Z. Mihaylova, R. Tsikandelova, P. Sanimirov, N. Gateva, V. Mitev, and N. Ishkitiev, *Archives of Oral Biology*, Vol. 85, p. 1, 2018.
47. A.I. Caplan and D. Correa, *Journal of Orthopaedic Research*, Vol. 29, p. 1795, 2011.
48. M. Murakami and M. Simons, *Current Opinion in Hematology*, Vol. 15, p. 215, May 2008.
49. T. Spivak-Kroizman, M.A. Lemmon, I. Dikic, J.E. Ladbury, D. Pinchasi, J. Huang, M. Jaye, G. Crumley, J. Schlessinger, and I. Lax, *Cell*, Vol. 79, p. 1015, 1994.
50. M.A. Caldwell, E. Garcion, M.G. terBorg, X. He, and C.N. Svendsen, *Experimental Neurology*, Vol. 188, p. 408, 2004.
51. N. Quarto and F. Amalric, *Journal of Cell Science*, Vol. 107, p. 3201, 11 1994.
52. U. Klammert, T.D. Mueller, T.V. Hellmann, K.K. Wuerzler, A. Kotzsch, A. Schliermann, W. Schmitz, A.C. Kuebler, W. Sebald, and J. Nickel, *BMC Biology*, Vol. 13, p. 77, 2015.
53. Y. Morotome, M. Goseki-Sone, I. Ishikawa, and S. Oida, *Biochemical and Biophysical Research Communications*, Vol. 244, p. 85, 1998.
54. K. Sena, Y. Morotome, O. Baba, T. Terashima, Y. Takano, and I. Ishikawa, *Journal of Dental Research*, Vol. 82, p. 166, 2003.

55. H. Kuniyasu, Y. Hirose, M. Ochi, A. Yajima, K. Sakaguchi, M. Murata, and J. Pohl, *Clinical Oral Implants Research*, Vol. 14, p. 490, 2003.
56. R.C. Spiro, L.-S. Liu, M.A. Heidaran, A.Y. Thompson, C.K. Ng, J. Pohl, and J.W. Poser, *Biochemical Society Transactions*, Vol. 28, p. 362, August 2000.
57. T.-A. Jahng, T.-S. Fu, B.W. Cunningham, A.E. Dmitriev, and D.H. Kim, *Neurosurgery*, Vol. 54, p. 171, January 2004.
58. Y. Hatakeyama, J. Hatakeyama, Y. Maruya, K. Oka, E. Tsuruga, T. Inai, and Y. Sawa, *Bone and Tissue Regeneration Insights*, Vol. 3, p. BTRI.S8120, 2010.
59. T. Nakamura, M. Yamamoto, M. Tamura, and Y. Izumi, *Journal of Periodontal Research*, Vol. 38, p. 597, 2003.
60. Y.R. Moore, D.P. Dickinson, and U.M.E. Wikesjö, *Journal of Clinical Periodontology*, Vol. 37, p. 288, 2010.
61. J. Lee and U.M.E. Wikesjö, *Journal of Clinical Periodontology*, Vol. 41, p. 797, 2014.
62. S.H. Kerr, A. Recber, T. Pepe, D. Messerli, L. Laurence, R. Walsh, T. Kueenzi, and B. Randall, Porous biocompatible polymer material and methods, US Patent 10 549 014, assigned to DePuy Synthes Products Inc., February 04, 2020.
63. D. Sharma, V.P. Mathur, and B.K. Satapathy, *Annals of Biomedical Engineering*, 2021.
64. O. Azizabadi, F. Akbarzadeh, G. Sargazi, and N.P.S. Chauhan, *Polymer-Plastics Technology and Materials*, Vol. 60, p. 734, 2021.
65. B. Wang, T. Li, W. Guo, P. Cao, R. Wang, P. Song, Y. He, and Y. Li, *Polymer International*, Vol. n/a, p. in press, 2022.
66. H.-P. Hentze and M. Antonietti, *Reviews in Molecular Biotechnology*, Vol. 90, p. 27, 2002. New Polymers.
67. I.N. Savina, G.C. Ingavle, A.B. Cundy, and S.V. Mikhalovsky, *Scientific Reports*, Vol. 6, p. 21154, 2016.
68. T. Andersen, J.E. Melvik, O. Gøaserød, E. Alsberg, and B.E. Christensen, *Biomacromolecules*, Vol. 13, p. 37033710, 2012.
69. J.C.M.E. Bender, M.G.A. Mann, and M. Kamps, In situ gelling alginate systems, WO Patent 2 010 117 266, assigned to Broockeville Corporation NV, October 14, 2010.
70. A. Kirillova, T.R. Yeazel, D. Asheghali, S.R. Petersen, S. Dort, K. Gall, and M.L. Becker, *Chemical Reviews*, 2021.
71. Y. Zhang, K. Li, L. Shen, L. Yu, T. Ding, B. Ma, S. Ge, and J. Li, *ACS Applied Materials & Interfaces*, Vol. 14, p. 268, 2022.
72. X. Peng, K. Dong, Z. Wu, J. Wang, and Z.L. Wang, *Journal of Materials Science*, 2021.
73. L. Yu, H. Xuan, Y. Guo, A.L. Chin, R. Tong, G. Pickrell, A. Wang, and X. Jia, *Optical Materials Express*, Vol. 7, p. 1813, 2017.
74. J. Zubia and J. Arrue, *Optical Fiber Technology*, Vol. 7, p. 101, 2001.

75. M.A. van Eijkelenborg, M.C.J. Large, A. Argyros, J. Zagari, S. Manos, N.A. Issa, I. Bassett, S. Fleming, R.C. McPhedran, C.M. de Sterke, and N.A. Nicorovici, *Opt. Express*, Vol. 9, p. 319, Sep 2001.
76. T. Golobish, M. Gruda, T. Guliashvili, P. O'Sullivan, A. Scheirer, V. Dan, W.-T. Young, V. Capponi, and P. Chan, Multi-functional hemocompatible porous polymer bead sorbent for removing protein based toxins and potassium from biological fluids, US Patent 11 040 061, assigned to CytoSorbents Corp., June 22, 2021.
77. C. Terai, S. Shimizu, S. Matsushima, I. Tanaka, Y. Okada, T. Miyahara, and S. Nakaji, *Artificial Organs*, Vol. 20, p. 1227, 1996.
78. T. Guliashvili, T. Golobish, M. Gruda, P. O'Sullivan, A. Scheirer, V. Capponi, P. Chan, and W.-T. Young, Use of a hemocompatible porous polymer bead sorbent for removal of endotoxemia-inducing molecules, US Patent 11 065 600, assigned to CytoSorbents Corporation, July 20, 2021.

5

Thermal Insulation

The history of thermal insulation materials is not as long as that of other materials, but the necessity of insulation is as old as building activity (1). Prehistoric people built shelters to protect themselves from the elements, originally using organic materials and later more durable substitutes. However, not only materials found in nature were used, but other materials that were discovered, which were suitable for insulating.

Processing organic materials produced the first insulated panels in the 19th century. Meanwhile an increasing range of artificial materials were developed such as rock wool, fibreglass, foam glass, hollow bricks, and expanded perlite.

The appearance of plastic foams resulted in a huge revolution. Although plastic production was well-known in the 19th century, the first plastic foam was not produced until 1941. Nowadays the most popular insulation materials are plastic foams and mineral wool with only a small amount of natural materials being produced. The overall growth of these products has been substantial due to a wide range of reasons (1).

Poly(styrene) (PS) had been known long before it was used. Its original monomer was the natural styrene that was named after a genus of tropical and Mediterranean trees, called *Styrax* (1). The industrial production of PS started in 1930 (2). The monomer for PS can be commercially manufactured from petroleum.

Poly(styrene) foam was first fabricated in 1931 in the USA. Carl Georg Munters and John Tandberg patented a method for foaming PS (3).

Poly(urethane) (PU) foam was prepared by Otto Bayer by the reaction of glycol and isocyanate. The first commercial insulating panel made from PU foam was produced in 1954 (1).

After the 1950s, several other thermal insulation materials were introduced. Some of them were also plastic foams. Polyester foam (PES) and poly(ethylene) (PE) foam were invented in the 1950s. Phenolic foam and formaldehyde foam appeared in the 1970s and melamine foam arrived in the 1990s (1).

5.1 Prediction Models

Microporous and nanoporous polymeric materials are considered a unique industrial material due to its extremely low thermal conductivity, low density, and high surface area. Therefore, it is important to establish accurate thermal conductivity prediction models. The development of calculation models of the thermal conductivity for these material has been detailed (4).

There are conductive thermal conductivity models and radiative thermal conductivity models.

A lattice-based numerical model was developed to predict the thermal conductivity of sand and modified high thermal backfill material for energy transportation used for underground power cables (5). Both 2D and 3D simulation models were used to analyze and detect differences resulting from a simplification of the model.

The thermal conductivity of the granular mixture is determined numerically considering the volume and the shape of each constituting portion. The numerical method was validated with transient needle measurements and the existing theoretical and semi-empirical models for thermal conductivity prediction. The numerical prediction and the measured values were in agreement to a large extent (5).

The thermal conductivity of porous thermal barrier coatings was evaluated using a five-phase model (6). It was demonstrated that the porosities distributed in the coating strongly affect the thermal conductivity.

The decisive reason for this change in the thermal conductivity can be traced back to a defect morphology and its orientation, depending on the coating deposition technique and process parameters used during deposition.

Here, the two-phase model of Bruggeman (7) was also used as a reference, and a five-phase model was developed to evaluate the thermal conductivity of porous coatings. This approach uses microstructural details of the shape, size, orientation and volumetric fraction of defects of coatings as input parameters. The proposed model can predict thermal conductivity values better than the two-phase model (6).

A mathematical model to accurately calculate the radiative thermal conductivity of microporous or nanoscale porous open cell structures has been presented (8). Due to the limitations of existing radiative thermal conductivity models, a full-scale method based on the Rosseland diffusion equation has been proposed.

By a combination of the full-scale Rosseland diffusion equation and fractal thermal conduction methods, the predicted total thermal conductivity values were well matched with the experimental results for various microscale and nanoscale porous open cell dielectric materials, with an error of less than 15%.

Also, seven influential factors on the thermal conductivity, such as cell size, porosity, cellular pore shape, volume-specific surface area, temperature, refractive index, and extinction index, were investigated (8).

The results showed that the thermal conductivity of porous open cell materials mainly decreased with an increase of the extinction index and/or the porous structure's volume-specific surface area. However, the thermal conductivity increased with an increase of the temperature.

These findings indicated the potential of the full-scale Rosseland diffusion method for use in the design of specific micro/nanoscale porous dielectric structures like polymer foam in the personal energy management device or the silica aerogel in radiative cooling system (8).

5.2 Radiative and Conductive Heat Transfer

A theoretical model that integrates the radiative and conductive heat transfer was presented (9). This model was used to evaluate the thermal energy transport within porous polymer materials.

First, the model was validated by comparing the computed thermal energy flux with the experimental measurements of two porous polymer materials made of wool and polyester. The model was then used to predict the effects of the polymer fiber characteristics (viz., fiber fractional volume, fiber emissivity, fiber radius, and fiber thermal conductivity) on the thermal energy flux within the porous polymer materials (9).

It was found that decreasing the radius of the fiber would significantly reduce the total thermal energy flux through the porous polymer materials. Increasing the emissivity of the fiber or decreasing the thermal conductivity would cause just a slight reduction of the total thermal energy flux.

The fiber fractional volume had a significant influence on the thermal energy flux, and thereby the fiber fractional volume can be optimized in view of improving the thermal insulating performance of the porous polymer materials (9)-

5.3 Studies of Thermal Conductivity

5.3.1 *Macroporous Polymer-Derived SiOC Ceramics*

A three-dimensional reticular macroporous SiOC ceramics structure was made of spherical agglomerates (10). This material has been thermally characterized using a freestanding sensor-based 3ω method.

The freestanding sensor-based 3ω technique has been proposed and developed as a candidate for the conventional 3ω method for the measurement of thermophysical properties (11). The method covers the determination of solids, powders and fluids from cryogenic temperatures to around 400 K (12).

The effective thermal conductivity of the macroporous SiOC ceramics, including the effects of voids, was found to be 0.041 $W m^{-1} K^{-1}$ to 0.078 $W m^{-1} K^{-1}$ at room temperature. This is comparable with an alumina aerogel or a carbon aerogel.

These results suggest that the SiOC ceramics hold a great promise as a thermal insulation material for use at high temperatures. Furthermore, the results reveal that the effective thermal conductivity is limited by the low solid-phase volume fraction for the SiOC series processed at the same conditions. For a SiOC ceramics processed under different pyrolysis temperatures, the contact condition between neighboring particles in the SiOC networks is another key factor that influences the effective thermal conductivity (10).

5.4 Poly(ethylene) Foams

Multifunctional polymer foams are of interest because it is difficult for a single functional foam to adapt to various complicated environments (13).

The preparation of a multifunctional porous ultrahigh-molecular-weight poly(ethylene) (UHMWPE) material with high porosity is still complicated. Thus, it is desirable to develop a simple and effective method that could integrate multiple functions into polymer foam. A multifunctional UHMWPE foam was fabricated via a facile thermally induced phase separation method (13).

The foam has a low density of 8.23 $mg\,cm^{-3}$, a high porosity of 99.2% and a robust superhydrophobicity. The water contact angle is 152°. Due to these properties, the foam exhibits an excellent adsorption capacity for the oil-water separation under different extreme environments. Moreover, the thermal conductivity of the foam goes down to 11.6 $mW\,m^{-1}K^{-1}$, which endows the foam with great thermal insulation performance under various temperature conditions. Thus, this foam can meet the demands of various practical applications (13).

5.5 Rigid Foams

5.5.1 *Aromatic Polymers*

Infrared attenuating agents, such as carbon black powdered amorphous carbon, graphite, and titanium dioxide, have been used as fillers in polymeric foam boards to minimize the thermal conductivity of the material, which, in turn, will maximize the insulating capability for a given thickness (14).

Rigid foam plastic boards are extensively used as thermal insulating materials for a lot of applications. Rigid foam insulating products and processes for making such insulation products were described (14). These foams are formed from a polymer, a blowing agent, and a nanographite.

The nanographite in a foam product acts as a UV stabilizer and gas barrier in the final product. Furthermore, the nanographite reduces the amount of static present during the foaming process due to the increased electric conductivity of the skin of the nanographite polymer foam boards. During the foaming process, the nanographite acts as a nucleating agent and as lubricant.

Preferred thermoplastic polymers are alkenyl aromatic polymers. Suitable alkenyl aromatic polymer materials include alkenyl aromatic homopolymers and copolymers of alkenyl aromatic compounds and copolymerizable ethylenically unsaturated comonomers. The alkenyl aromatic polymer material may further include minor proportions of non-alkenyl aromatic polymers.

Monomers for alkenyl aromatic polymers are collected in Table 5.1. Some of these compounds are shown in Figure 5.1.

Table 5.1 Monomers for alkenyl aromatic polymers (14).

Alkenyl aromatic	Alkenyl aromatic	Alkenyl aromatic
Styrene	α-Methylstyrene	Ethylstyrene
Vinyl toluene	Chlorostyrene	Bromostyrene
Copolymerizable	Copolymerizable	Copolymerizable
Acrylic acid	Methacrylic acid	Ethacrylic acid
Maleic acid	Itaconic acid	Acrylonitrile
Maleic anhydride	Methyl acrylate	Ethyl acrylate
Isobutyl acrylate	*n*-Butyl acrylate	Methyl methacrylate
Vinyl acetate	Butadiene	

Examples of partially halogenated chlorocarbons and mixed chlorofluorocarbons that can be used as blowing agents are shown in Table 5.2. Some of these compounds are shown in Figure 5.2.

The exfoliated graphite advantageously retains its nanostructure during the polymerization process. The expanded or exfoliated graphite is compressed together into flexible thin sheets. The nanographite in the foam comprises a plurality of nanosheets typically in layers, where each layer may contain a few or even a single carbon layer (14). A detailed explanation of the exfoliation of graphite may be found in several monographs (15, 16).

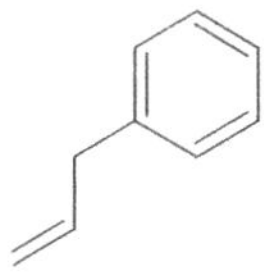

Vinyl toluene

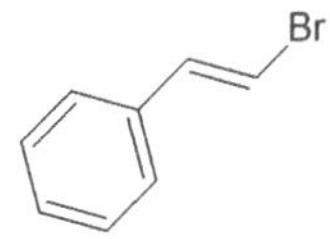

Bromostyrene

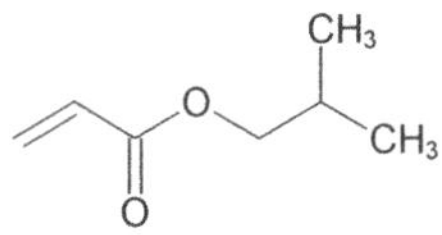

Isobutyl acrylate

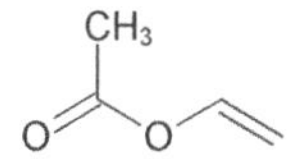

Vinyl acetate

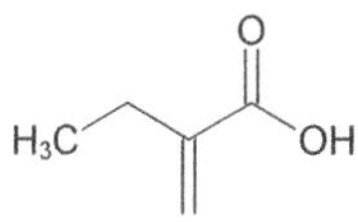

Ethacrylic acid

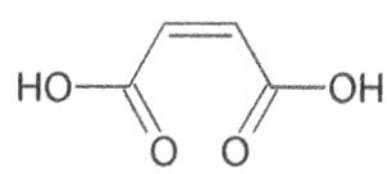

Maleic acid

Figure 5.1 Monomers for alkenyl aromatic polymers (14).

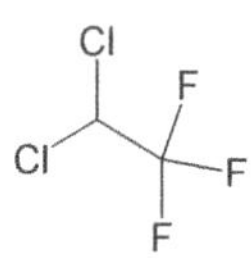

1,1-Dichloro-2,2,2-trifluoroethane

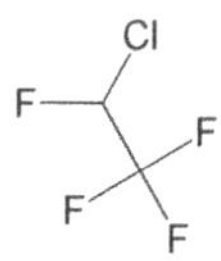

1-Chloro-1,2,2,2-tetrafluoroethane

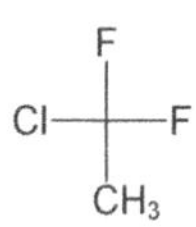

1-Chloro-1,1-difluoroethane

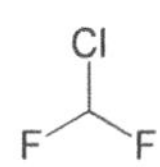

Chlorodifluoromethane

Figure 5.2 Blowing agents.

Table 5.2 Blowing agents (14).

Partially halogentated compound	Shortcut
Methyl chloride	
Methylene chloride	
Chlorodifluoromethane	HCFC-22
Ethyl chloride	
1,1,1-Trichloroethane	
1,1,1-Trifluoroethane	
1,1-Dichloro-1-fluoroethane	HCFC-141b
1-Chloro-1,1-difluoroethane	HCFC-142b
1,1-Dichloro-2,2,2-trifluoroethane	HCFC-123
1-Chloro-1,2,2,2-tetrafluoroethane	HCFC-124
Pentafluoroethane	
Dichloropropane	
Fully halogentated compound	**Shortcut**
Trichloromonofluoromethane	CFC-11
Dichlorodifluoromethane	CFC-12
1,1,2-Trichloro-1,2,2-trifluoroethane	CFC-113
1,2-Dichlorotetrafluoro ethane	CFC-114
Chloroheptafluoropropane	
1,2-Dichlorohexafluoropropane	

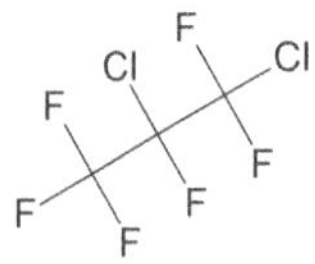

1,2-Dichlorohexafluoropropane

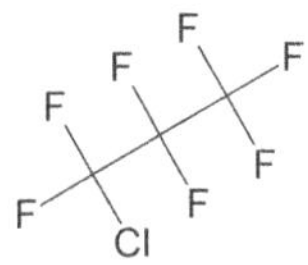

Chloroheptafluoropropane

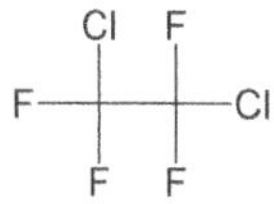

1,2-Dichlorotetrafluoro ethane

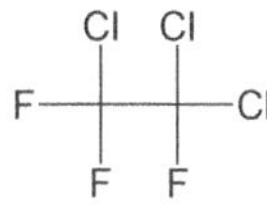

1,1,2-Trichloro-1,2,2-trifluoroethane

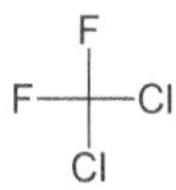

Dichlorodifluoromethane

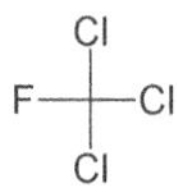

Trichloromonofluoromethane

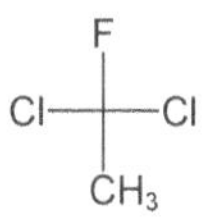

1,1-Dichloro-1-fluoroethane

Pentafluoroethane

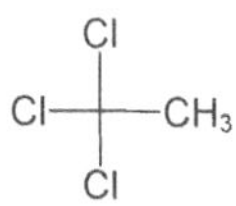

1,1,1-Trichloroethane

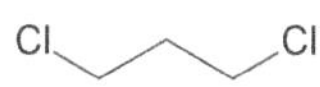

1,3-Dichloropropane

Figure 5.2 (cont) Blowing agents.

A process for making a closed-cell, alkenyl aromatic polymer foam product in which nanoparticle nucleation agents are utilized to control the cell morphology may run as follows (14):

1. Heating an alkenyl aromatic polymer to the temperature above the glass transition temperature of the polymer (for amorphous polymer), or melt point of the polymer (for crystal polymer) to form a polymer melt,
2. Incorporating an appropriate amount of selected nanoparticles (e.g., nanographite) into the polymer melt to alter the polymer property and process behavior, such as rheology and melt strength,
3. Incorporating blowing agents into the polymer melt at elevated pressure,
4. Incorporating other additives, such as flame retardants into the polymer melt, and
5. Extruding and forming a foam product, typically a rigid insulation board, under an atmospheric or sub-atmospheric pressure (partial vacuum) to produce a desired cell morphology, characterized by parameters such as cell size range and distribution, cell orientation and cell wall thickness.

The process conditions for samples in a twin-screw extruder for making foam boards having a width of 16 *in* and a thickness of 1 *in* are shown in Table 5.3.

The addition of nanographite in foaming processing, of preferably 1% to 3% of the polymer, has a profound effect on the thermal resistance.

5.5.2 *PVC*

A rigid poly(vinyl chloride) (PVC)-based foam sheet has been developed that is especially well suited for building construction (17). Also, a system for manufacture of foam sheets rigidized with polymer infiltration has been described (18).

The sheets have a density ranging from 0.12 to 0.66 $g\,cm^{-3}$, which is about 10% to 40% of the solid flexible PVC. Solid flexible PVC typically has a density of 1.1 to 1.35 $g\,cm^{-3}$.

The sheets have a very large number of closed microcells ranging in dimension from 10 μm to 70 μm. The cells have a cell wall of

Table 5.3 Process conditions (14).

Compound in composition	Amount/[%]		
Nanographite	1	–	5
Talc	0.5	–	1.5
Nano-carbon black	0	–	6
Mica	0	–	4
HCFC-142b	11	–	11
CO_2	0	–	0
Working Parameters			
Extruder pressure/[kPa]	13000	–	17000
Die melt temperature/[°C]	117-	–	123
Die pressure/[kPa]	5400-	–	6600
Line speed/[$m\,h^{-1}$]	110-	–	170
Throughput/[$kg\,h^{-1}$]	100	–	100
Die gap/[mm]	0.6	–	0.8
Vacuum/[kPa]	0	–	3.4

a PVC-based polymer. The sheets also have extremely small sized uniformly distributed closed cells of air pockets that enhance the thermal insulation properties.

The method of production of PVC-based sheets uses ultrafine particles of virgin PVC.

Plasticizers are used here. These are chemical species, which dissolve into the PVC composition chains and promote a flexing at the molecular level by a screening mechanism or a hinge mechanism. Suitable plasticizers are collected in Table 5.4. Some compounds are shown in Figure 5.3.

The formation of a foamed product requires a foaming agent that produces fine pores with the mold PVC-based sheet. The foaming agent decomposes at a decomposition temperature producing a large volume of gaseous reaction product. A number of foaming agents for use in foamed PVC production are known. Foaming agents are shown in Table 5.5. These materials are also shown in Figure 5.4.

Here, the preferred foaming agent is azodicarbonamide since it has a decomposition temperature of about 215°C to 219°C. This decomposition temperature is slightly greater than the softening temperature of PVC compositions, which is typically in the range

Table 5.4 Plasticizers for PVC-based sheets (17).

Compound	Compound
Diisodecyl phthalate	Tricresyl phosphate
Dioctyl adipate	Diisodecyl adipate
Dioctyl azelate	Diethylene glycol dibenzoate
Tributyl citrate	Dioctyl sebacate
Trioctyl trimellitate	Tri-2-ethylhexyl phosphate
Dioctyl sebacate	Di(2-ethylhexyl) phosphate
Epoxidized soybean oil	Dioctyl terephthalate

Compound
Tri-2-ethylhexyl trimellitate
Bis(2-ethylhexyl)-1,4-benzenedicarboxylate
2,2,4-Trimethyl 1,3-pentanediol diisobutyrate
Diisononyl 1,2-cyclohexanedicarboxylate

Table 5.5 Foaming agents for PVC-based sheets (17).

Compound	Compound
Azodicarbonamide	Dinitrosopentamethylene tetramine
p-Toluenesulfonyl hydrazide	4,4′-Oxybis(benzene sulfonyl hydrazide)

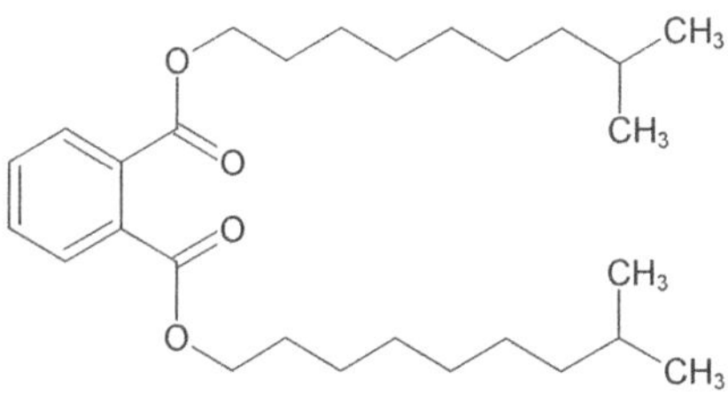

Diisodecyl phthalate

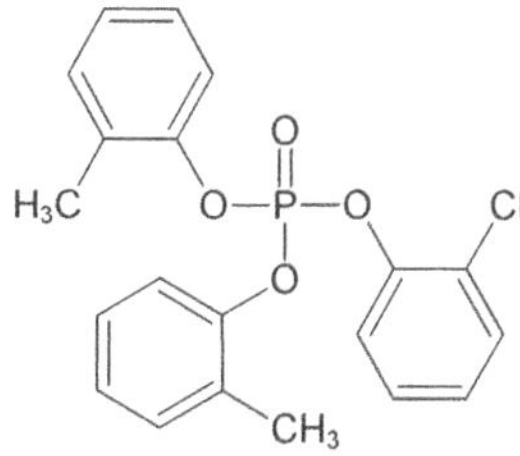

Tricresyl phosphate

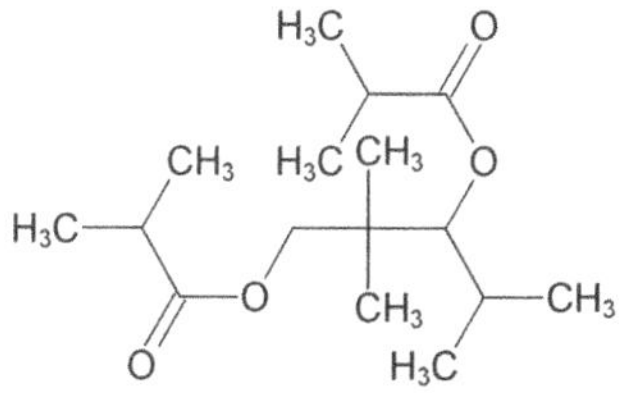

2,2,4-Trimethyl 1,3-pentanediol diisobutyrate

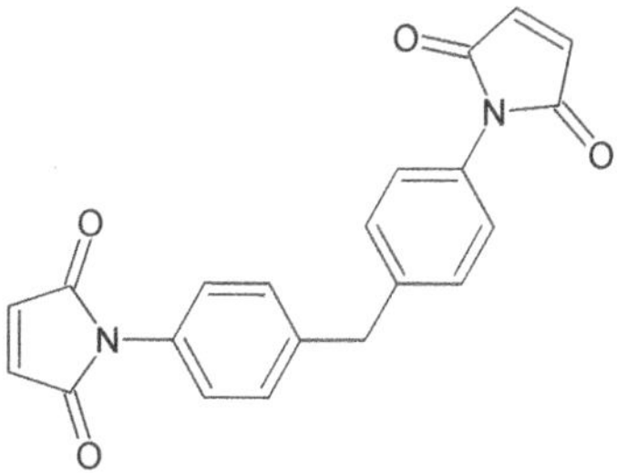

Soybean oil

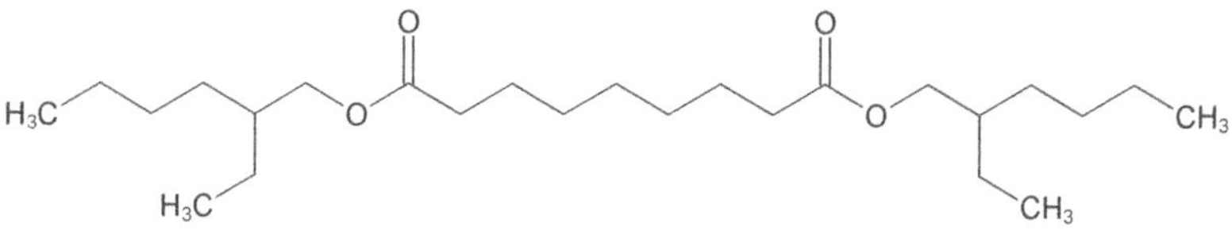

Dioctyl azelate

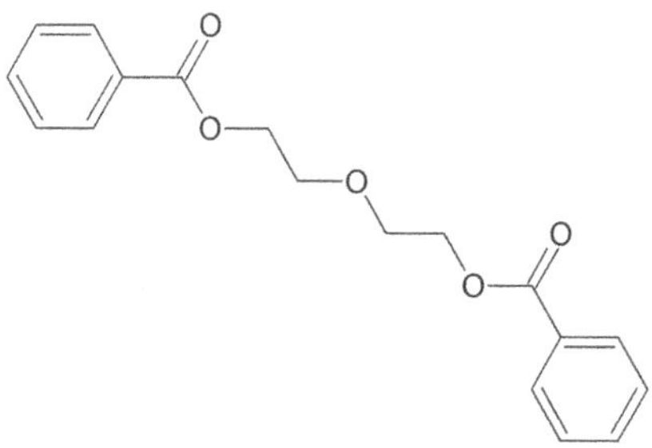

Diethylene glycol dibenzoate

Figure 5.3 Plasticizers for PVC-based sheets (17).

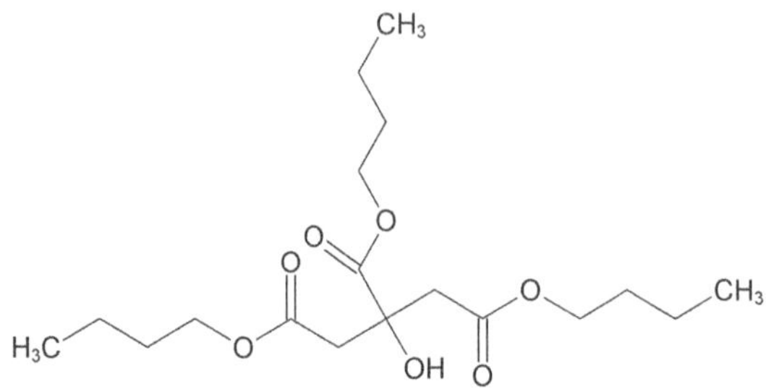

Tributyl citrate

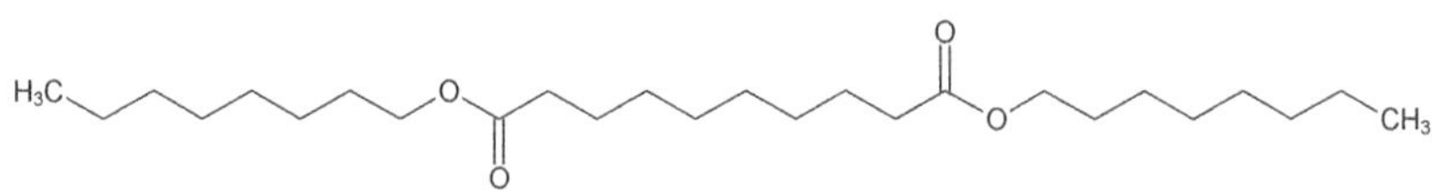

Dioctyl sebacate

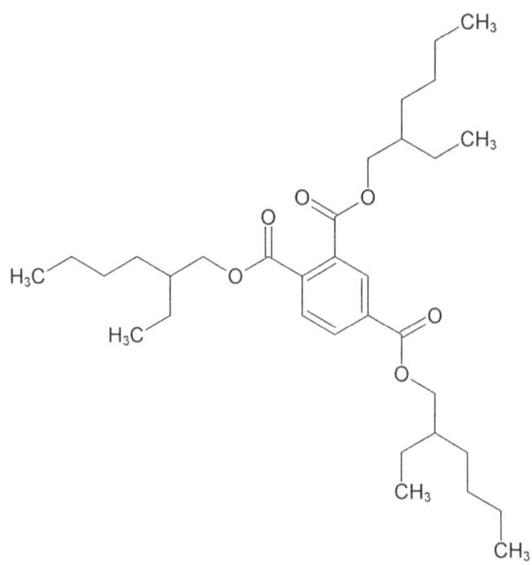

Tri-2-ethylhexyl trimellitate

Figure 5.3 (cont) Plasticizers for PVC-based sheets (17).

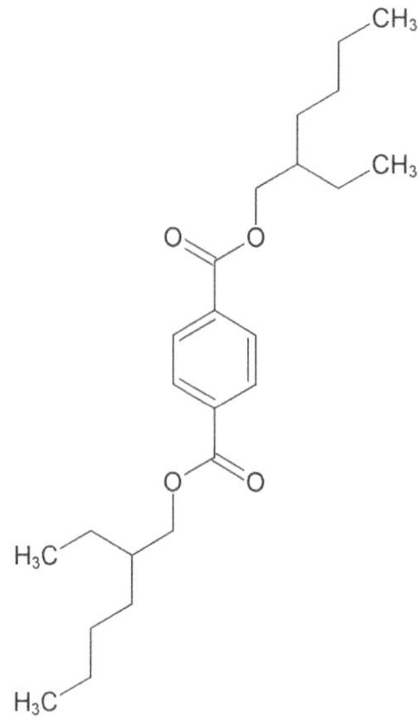

Bis(2-ethylhexyl)-1,4-benzenedicarboxylate

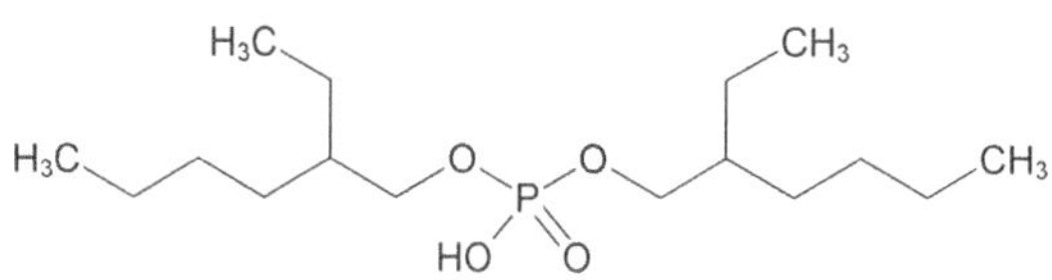

Di(2-ethylhexyl) phosphate

Figure 5.3 (cont) Plasticizers for PVC-based sheets (17).

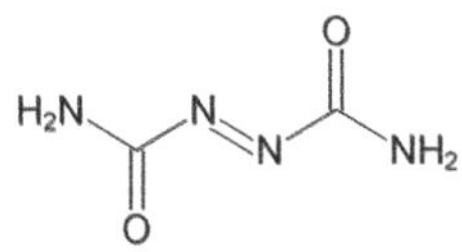

Azodicarbonamide

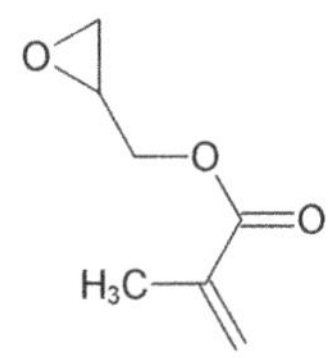

Dinitrosopentamethylene tetramine

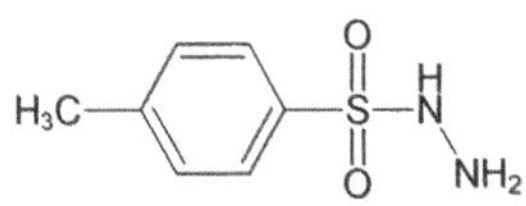

p-Toluenesulfonyl hydrazide

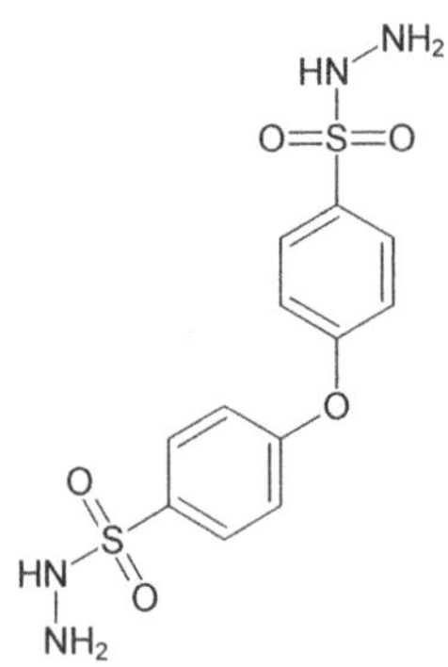

4,4′-Oxybis(benzene sulfonyl hydrazide)

Figure 5.4 Foaming agents for PVC-based sheets.

of 170°C to 190°C. A kicker compound such as ZnO may be used to bring the decomposition temperature of azodicarbonamide to the softening temperature range of the PVC resins.

5.5.3 *Poly(urethane)*

Closed-cell rigid PU foams are useful for insulating appliances and buildings due to their low thermal conductivity and their dimensional stability at low densities. A closed-cell rigid PU foam can be made with a one-shot method and a reaction system that includes a hydrofluoroalkene physical blowing agent and a polyol mixture that contains an aminic polyol (19, 20).

Isocyanates of the reaction system may include any organic poly(isocyanate) that is suitable for the production of rigid PU or urethane-modified poly(isocyanurate) foams. Useful organic poly(isocyanate)s include those having a functionality greater than or equal to 2.0. Also, poly(isocyanate)s can be used that are modified with various groups, such as ester groups, urea groups, biuret groups, allophanate groups, carbodiimide groups, isocyanurate groups, uretdione groups and urethane groups. Examples of isocyanate compounds are collected in Table 5.6 and shown in Figures 5.12 and 5.5.

Table 5.6 Isocyanate monomers (19).

Compound	Compound
Diphenylmethane diisocyanate	Toluene-2,4-diisocyanate
1,5-Naphthalene diisocyanate	1,4-Diisocyanatobenzene
Isophorone diisocyanate	1,6-Diisocyanatohexane
4,4-Diisocyanato-dicyclohexylmethane	

In contrast to perfluoralkenes, hydrogen-containing fluoroalkenes are promising physical blowing agents. These hydrofluoroalkenes may not have a substantial negative affect on atmospheric chemistry. For example, they may, compared to other halogenated blowing agents, contribute little to ozone depletion, and do not substantially contribute to global warming. A hydrofluoroalkene blowing agent may have a boiling point between 10°C and 40°C. Examples of such blowing agents are collected in Table 5.7 and shown in Figure 5.6.

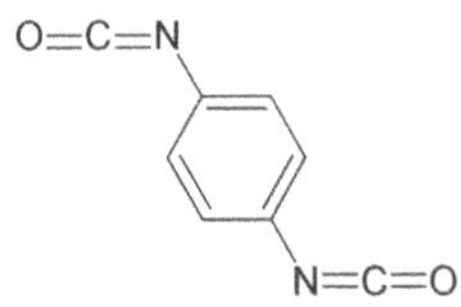

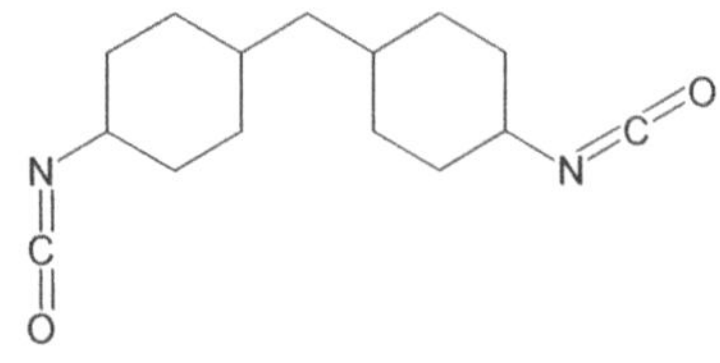

1,4-Diisocyanatobenzene 4,4-Diisocyanato-dicyclohexylmethane

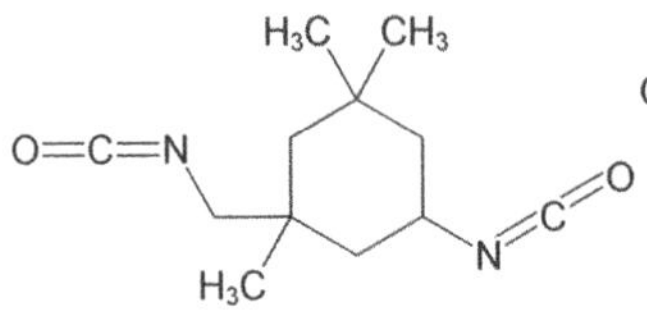

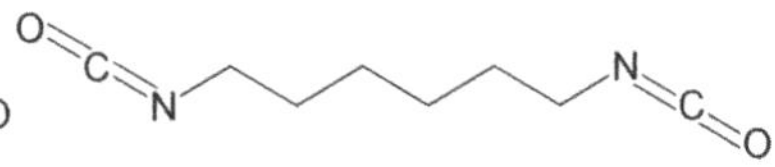

Isophorone diisocyanate 1,6-Diisocyanatohexane

Figure 5.5 Isocyanate monomers.

Table 5.7 Blowing agents (19).

Compound	Boiling point/[°C]
2-Bromopentafluoropropene	25 – 26
1-Bromopentafluoropropene	27 – 28
3-Bromopentafluoropropene	25.5
3,3,4,4,5,5,5-Heptafluoro-1-pentene	32
3-Bromo-1,1,3,3-tetrafluoropropene	33.5
2-Bromo-1,3,3,3-tetrafluoropropene	29 – 32
1-Bromo-2,3,3,3-tetrafluoropropene	32 – 34
1,1,2,3,3,4,4-Heptafluorobut-1-ene	20 – 21
2-Bromo-3,3,3-trifluoropropene	33 – 33.5
1-Bromo-3,3,3-trifluoropropene-1-ene	39 – 39.5
cis-1,1,1,4,4,4-Hexafluoro-2-butene	31 – 36
Hexafluoroisobutene	15
1-Chloro-3,3,3-trifluoropropene	21
2-Chloro-3,3,3-trifluoropropene	14 – 15
1,1,1-Trifluoro-2-butene	17

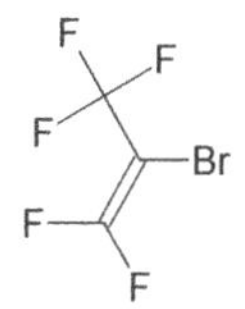

2-Bromopentafluoropropene

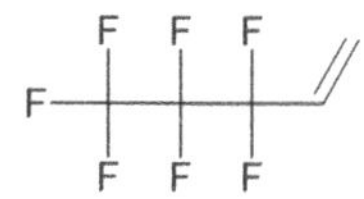

3,3,4,4,5,5,5-Heptafluoro-1-pentene

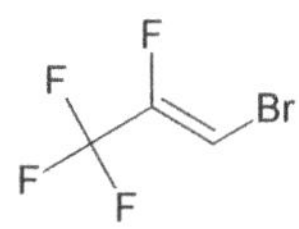

1-Bromo-2,3,3,3-tetrafluoropropene

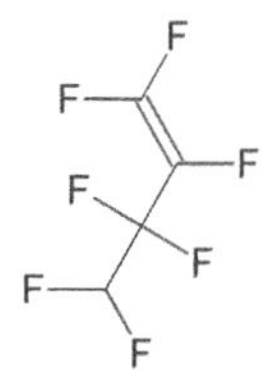

1,1,2,3,3,4,4-Heptafluorobut-1-ene

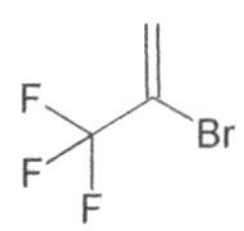

2-Bromo-3,3,3-trifluoropropene

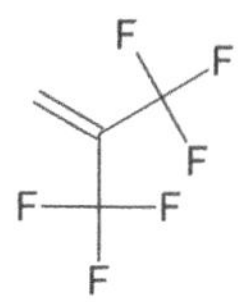

Hexafluoroisobutene

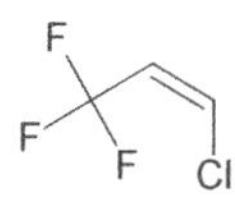

1-Chloro-3,3,3-trifluoropropene

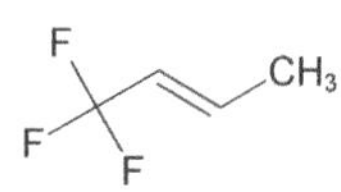

1,1,1-Trifluoro-2-butene

Figure 5.6 Blowing agents (19).

Under certain conditions, the physical blowing agent component may include the hydrofluoroalkene blowing agent and another physical blowing agent. Such suitable physical blowing agents include air, nitrogen, carbon dioxide, hydrofluoroalkanes, alkanes, alkenes, ketones, ethers, and monocarboxylic acid esters boiling below 60°C at 1 *atm*. Examples of saturated hydrofluorocarbons that are suitable co-blowing agents include those that are vaporizable under foaming conditions. These compounds are collected in Table 5.8.

Table 5.8 Co-blowing agents (19).

Compound	Compound
1,1-Difluoroethane	1,1,1,2-Tetrafluoroethane
Pentafluoroethane	1,1,1,3,3-Pentafluoropropane
1,1,1,3,3-Pentaflurobutane	*n*-Butane
n-Pentane	Isopentane
Cyclopentane	1-Pentene
2-Pentene	Acetone
Dimethyl ether	Methyl formate
Ethyl formate	methyl acetate

The blowing agent component may also include chemical blowing agents. For example, water or carboxylic acids may react with isocyanate under foam-forming conditions to liberate CO_2.

Suitable catalysts for curing can be urethane catalysts, isocyanurate catalysts, and combinations of these catalysts. Catalysts are shown in Table 5.9 and in Figure 5.7.

In addition, PU foam-forming compositions to get rigid PU foams have been described (21). Such composites can be used as floor insulating elements for refrigerated tractor trailers.

These rigid PU foams are produced from an isocyanate-reactive composition that contains (21):

1. A polyol blend,
2. A hydrochlorofluoroolefin, and
3. A tertiary amine catalyst.

Tertiary amine catalysts are known as *blow catalysts* since they have a greater effect on the water-poly(isocyanate) blowing reaction (21).

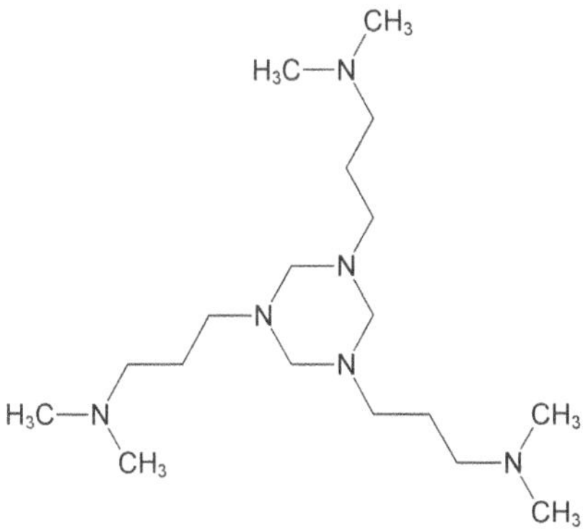

N,N′,N″-Dimethylamino-propylhexahydrotriazine

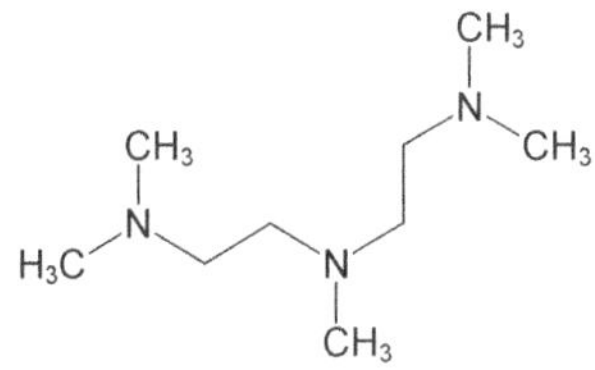

N,N,N′,N″,N″-pentamethyl-diethylene triamine

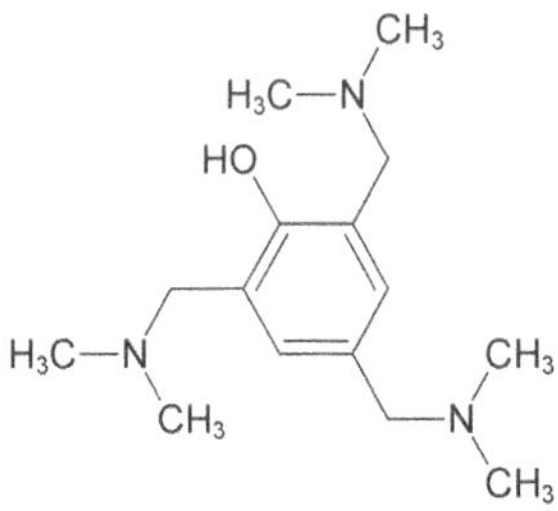

2,4,6-Tris-(dimethylaminomethyl) phenol

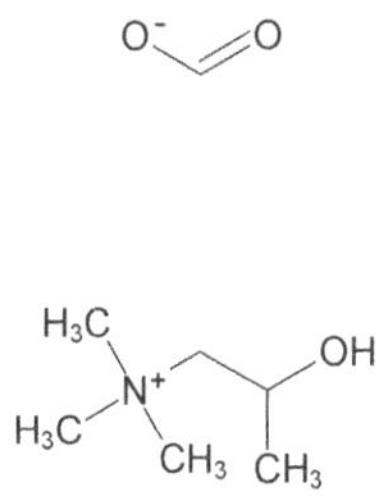

2-Hydroxpropyl trimethyl-ammonium formate

Figure 5.7 Catalysts.

Table 5.9 Catalysts (19).

Tertiary amine catalysts
Dimethylcyclohexylamine Benzyl dimethylamine *N,N,N',N'',N''*-pentamethyldiethylene triamine 2,4,6-Tris-(dimethylaminomethyl)-phenol Triethylene diamine *N,N*-Dimethyl ethanolamine
Quaternary ammonium salt catalysts
2-Hydroxpropyl trimethylammonium formate *N,N',N''*-Dimethylaminopropylhexahydrotriazine
Organometallic catalysts
Potassium octoate Potassium acetate Dibutyltin dilaurate Dibutlytin diacetate

5.6 Microporous Foams

5.6.1 Microporous Poly(styrene)

Microporous PS spheres could be prepared by a Pickering emulsion polymerization method (22–24).

The emulsions were stabilized by amino-functionalized silica nanoparticles and the mesoporous structure was induced by *i*-octane. In order to verify that the prepared microporous PS spheres have a porous structure, the morphology and pore size were studied by scanning electron microscopy and BET. Also, near-infrared light reflectivity and visible light transmittance of the coatings with microspheres in an epoxy soybean oil acrylic resin coating were measured.

The thermal insulation properties of microporous PS spheres filled coating were systematically studied by a thermal conductivity test and a heat transfer rate test (24). Compared with traditional hollow glass microbeads, the coating prepared by adding microporous PS spheres has higher near-infrared reflectivity, better visible light transmittance and a comparable thermal insulation ability.

The results showed that the introduction of microporous PS spheres will result in a coating with a low thermal conductivity and a high light- and heat-shielding performance, thereby improving the thermal insulation performance of the coating.

So, the microporous PS spheres serve as a functional filler for coatings for thermal insulation applications (24).

5.6.2 Conjugate Microporous Foams

Thermal insulation materials are widely used, such as building materials, cold storage, and air-conditioning equipment, and even stuffing in clothes (25).

The thermal conductivity is one of the main indicators to present the performance of thermal insulation materials, i.e., the thermal conductivity is below 120 $mW\,m^{-1}K^{-1}$ (26). The thermal conductivity depends on the composition, internal structure, bulk density, as well as the average temperature and the humidity of the environment of the application.

1,1,2,2-Tetrakis (4-bromophenyl) ethene and 1,3,6,8-tetrabromopyrene were used to synthesize conjugated microporous polymer (CMP) foams These foams possess planar constructions and rich aromatic rings as ligands.

The two ligands may generate the CMP foams with a superior hydrophobicity and micropore structure, which could adjust their heat insulation performance. The pore size of the CMP could be accurately controlled by the functional group structure of building units.

The procedure for the synthesis of CMP-ED was reported as follows (26):

Preparation 5–1: 1,1,2,2-Tetrakis (4-bromophenyl) ethene (648 *mg*), 1,4-diethynylbenzene (504.6 *mg*), CuI (66.67 *mg*), and tetrakis-(triphenylphosphine) palladium (133.33 *mg*) were added into a three-mouth flat-bottom glass test tube. The degassing operation was performed through vacuuming the tube three times, followed by nitrogen backfill. Then, 8 *ml* toluene and 8 *ml* triethylamine were added into the tube and the mixture were heated up to 80°C under magnetic stirring with nitrogen protection. After the solution became uniform mixture, the magneton was removed and the reaction lasted another 72 *h*.

The product was obtained as the temperature cooled down naturally and washed by dichloromethane (3 × 10 *ml*), acetone (3 × 10 *ml*), deionized water (3 × 10 *ml*) and methanol (3 × 10 *ml*) to remove the remaining monomers and catalyst. Finally, the materials were further purified by Soxhlet extraction for 3 *d* and treated in an vacuum oven at 60°C for 12 *h*.

The procedure for CMP-PT synthesis was similar to CMP-ED. The only difference was that the 1,1,2,2-tetrakis (4-bromophenyl) ethene and 1,4-diethynylbenzene were replaced by 1,3,6,8-tetrabromo-pyrene and 1,3,5-triethynyl-benzene. The compounds used here are shown in Figure 5.8.

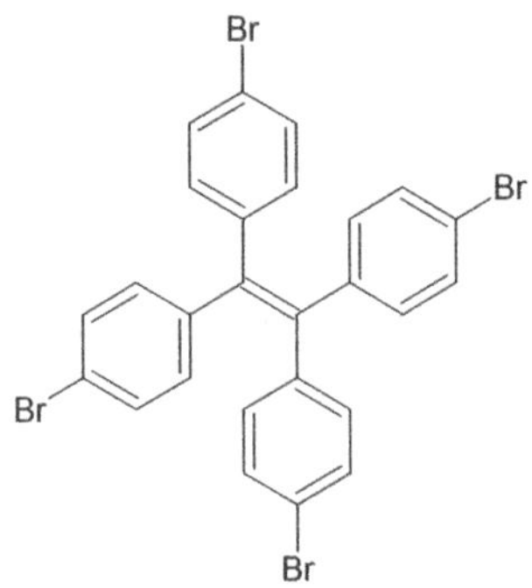

1,1,2,2-Tetrakis (4-bromophenyl) ethene

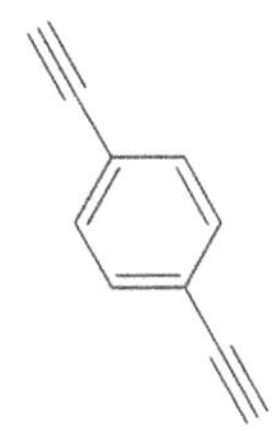

1,4-Diethynylbenzene

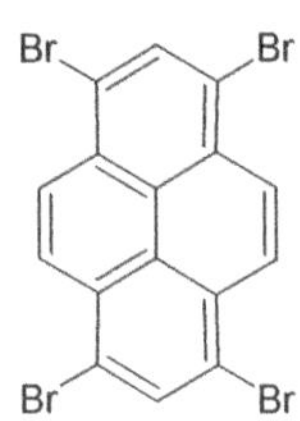

1,3,6,8-tetrabromo-pyrene

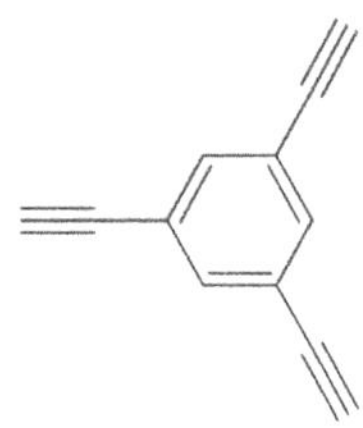

1,3,5-Triethynyl-benzene

Figure 5.8 Monomers for conjugated microporous polymers.

5.7 Resilient Porous Polymer Foams

The requirement of energy efficiency makes demands for materials with superior thermal insulation properties.

Inorganic aerogels are excellent thermal insulators, but are difficult to produce on a large-scale, are mechanically brittle, and their structural properties strongly depend on their density.

The scalable generation of low-density, hierarchically porous, poly(propylene) (PP) foams was described. The foaming process runs as follows (27):

Preparation 5–2: The melt strength of PP was improved by blending 3% poly(tetrafluoroethylene) using a previously described method (28, 29).

Foams of PP/ PTFE were prepared using a 50 *t* injection molding machine (Arburg 270C). A gas injection system (Trexel MuCell) was used to deliver 10% CO_2 into the injection molding machine. The mold cavity temperature was maintained at 85°C. A full shot of the polymer and CO_2 was injected into the mold cavity at an injection flow rate of 100 $cm^3 s^{-1}$, up to a cavity-packing pressure of 18 *MPa*. The polymer/gas mixture was allowed to dwell in the cavity for 40 *s*.

The mold was then opened at a mold-opening rate of 20 $mm\,s^{-1}$. The rapid depressurization from mold-opening caused cell nucleation and growth. The mold-opening distance was varied between 1.5 and 50 *mm*. Samples were extracted from the core of the injection molded foam for characterization. When the mold-opening distance was 1.5 *mm*, the samples extracted from the core of the injection molded foam were subjected to a subsequent foaming process in a high-pressure/temperature-regulated autoclave using CO_2 at various temperatures and pressures for a saturation time of 1.5 *h*. The foams were collected after rapid depressurization of the chamber for characterization.

Industrial-scale foam-processing equipment was used with a thermal conductivity lower than that of commercially available high-performance thermal insulators, such as superinsulating Styrofoam. The reduction of the thermal conductivity is attributed to the restriction of air flow caused by the porous nanostructure in the cell walls of the foam.

In contrast to inorganic aerogels, the mechanical properties of the foams are less sensitive to density, suggesting efficient load transfer through the skeletal structure. The scalable fabrication of hierarchically porous polymer foams opens up new perspectives for the scalable design and development of novel superinsulating materials (27).

5.8 Electrically Conductive Networks

Lightweight, biodegradable, thermally insulating, and electrically conductive materials are important to achieve a sustainable development (30). The fabrication of such multifunctional materials is very challenging.

5.8.1 *Poly(lactic acid)*

A facile, and ecofriendly method for the large-scale fabrication of ultralow-threshold and biodegradable porous poly(lactic acid) (PLA)/multiwalled carbon nanotube (MWCNT) was reported (30). These materials can be used for high-performance thermal insulation and electromagnetic interference (EMI) shielding applications.

Due to the unique structure of the microporous PLA matrix embedded by conductive 3D MWCNT networks, a lightweight porous PLA/MWCNT with a density of 0.045 $g\,cm^{-3}$ possesses a percolation threshold of 0.00094% by volume, which is the minimum value reported so far. Furthermore, the material exhibits excellent thermal insulation performance with a thermal conductivity of 27.5 $mW\,m^{-1}K^{-1}$, which is much lower than the best value of common thermal insulation materials. Moreover, it also shows outstanding EMI shielding performance characterized by its high shielding effectiveness values and absorption-dominated shielding feature.

More importantly, its specific EMI shielding effectiveness is as high as 1010 $dB\,cm^3g^{-1}$, which is superior to that of other shielding materials reported so far. Thus, this novel multifunctional material and its general fabrication methodology provide a promising way to meet the growing demand for high-performance multifunctional materials in sustainable development (30).

5.8.2 *Natural Rubber*

Natural rubber/carbon nanotubes composite foams with high electrical conductivity and excellent EMI performance were developed through a multi-step process consisting of (31):

1. Carbon nanotubes (CNTs) assembled on natural rubber latex particles,
2. Pre-crosslinking of natural rubber,

3. Supercritical carbon dioxide foaming of pre-crosslinked composite samples, and
4. Post-crosslinking of foamed composite samples.

Preparation 5–3: A certain amount of CNT and cetyl trimethylammonium bromide (CTAB) (c.f. Figure 5.9) as a surfactant, was obtained with a weight ratio of carbon nanotube to CTAB of 1:4. These materials were uniformly dispersed into water by using an ultrasound probe (UP200S Hielscher sonifier, temperature: 0°C, cycle: 0.5, amplitude: 80%) for 30 *min*. A natural rubber latex was added into the obtained solution containing 0.5 $g\,l^{-1}$ CNT by ultrasound for 30 *min*. Then, dicumyl peroxide (c.f. Figure 5.9) was added and stirred vigorously for a further 10 *min* at 50°C. Afterwards, the obtained dispersion was filtered, washed and dried. Several rubber composites were prepared, whose formulations are reported in Table 5.10.

In order to produce pre-crosslinked rubber sheet specimens, the various formulations were compression molded at 120°C and 10 *bar* for 3 *min* and then cooled at room temperature. With this procedure, it was possible to preserve the segregated morphology originated by the tailored assembling of CNTs around the natural rubber latex particles. Several composites with segregated network were denoted as N-NR/CNTx, where x was the CNTs concentration expressed as weight percentage of CNTs with respect to the natural rubber. N-NR/CNTx sheet specimens were placed into a high pressure vessel and saturated with $ScCO_2$ at 40 *bar* and 35°C for three times: 30 *min*, 60 *min*, and 120 *min*. After the saturation time, the high-pressure CO_2 was released quickly to atmospheric pressure and so the samples were foamed. Finally, the foamed specimens were quickly (in the range of a few seconds) placed into an oven at 170°C for 90 *s* in order to further crosslink the natural rubber.

In addition, the composite with 3.31% of CNTs was further mixed in a Haake Minilab twin-screw extruder (Thermo Electron Corp., Hamburg, Germany) at room temperature for 10 *min*. Then, this composite was pre-crosslinked, foamed and post-crosslinked following the same procedures previously described for samples with segregated network.

The formulations are detailed in Table 5.10 and the properties of these composite foams are shown in Table 5.11.

A closed-cell porous structure and a segregated CNTs network were observed in the fabricated foams. Due to this morphology, these composite foams exhibit low density, good mechanical properties, and a high electrical conductivity. The density of F-NR/CNTs composite foams is shown in Figure 5.10.

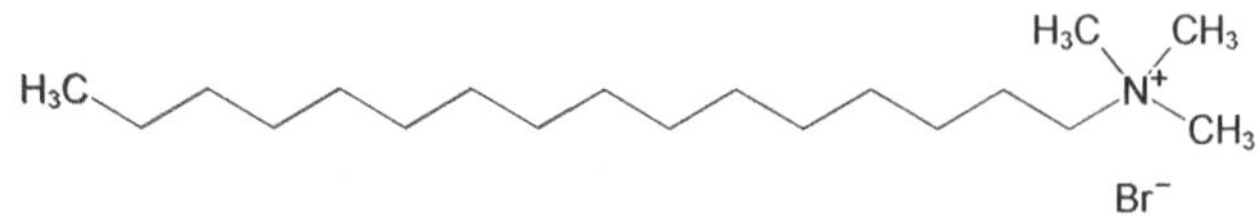

Cetyl trimethylammonium bromide

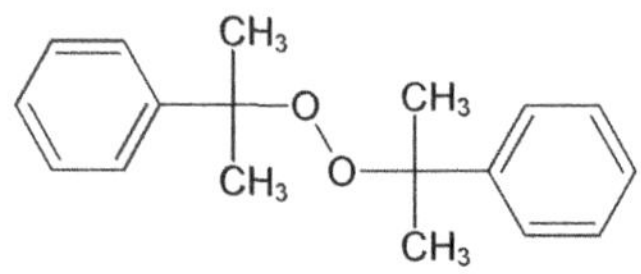

Dicumyl peroxide

Figure 5.9 Surfactant and radical initiator.

Table 5.10 Compositions of N-NR/CNTs composites (31).

Sample	NRL/[*g*]	CNT/[*g*]	CTAB/[*g*]	DCP/[*g*]
NR	8.3	0	0	0.25
N-NR/CNT0.82	8.3	0.05	0.2	0.25
N-NR/CNT1.68	8.3	0.1	0.4	0.25
N-NR/CNT3.31	8.3	0.2	0.8	0.25
N-NR/CNT4.87	8.3	0.3	1.2	0.25
N-NR/CNT6.40	8.3	0.4	1.6	0.25

NRL:	Natural rubber latex
CNT:	Carbon nanotube
CTAB:	Cetyl trimethylammonium bromide
DCP:	Dicumyl peroxide

Table 5.11 Properties of the N-NR/CNTs composites (31).

Sample	Saturation time/[*min*]	Mean cell diameter/[μm]	Cells per cm^3
NR foam	60	13.46	2.11×10^8
F-NR/CNT1.68	60	23.11	1.04×10^8
F-NR/CNT3.31	30	5.51	4.85×10^8
F-NR/CNT3.31	60	8.57	3.58×10^8
F-NR/CNT3.31	120	6.32	5.45×10^8
F-NR/CNT4.87	60	3.92	1.77×10^8

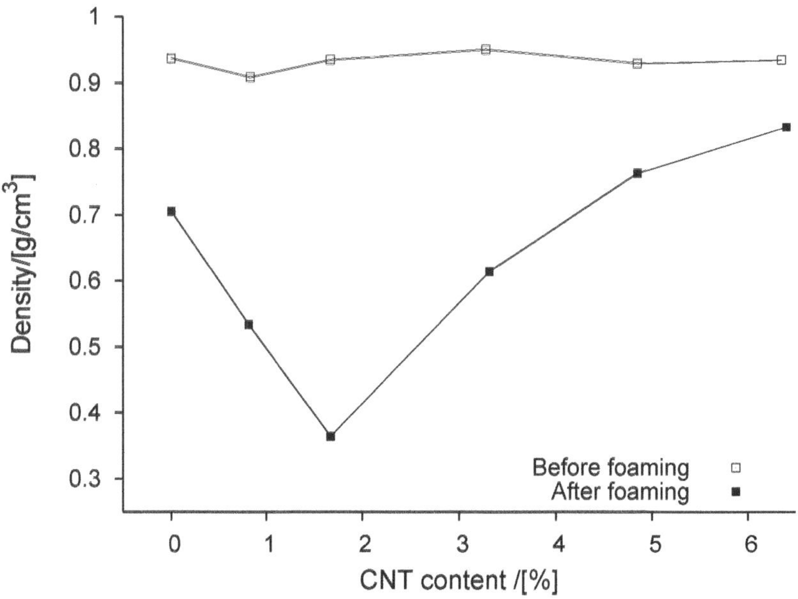

Figure 5.10 Density of F-NR/CNTs composite foams (31).

Owing to the multiple radiation reflections and scattering between the cell-matrix interfaces, the composite foams showed an excellent specific shielding effectiveness of 312.69 $dB\,cm^3 g^{-1}$ for a foam with 6.4% of CNTs. This is significantly higher than those already published for rubber composites containing comparable filler content. Furthermore, the analysis of the EMI shielding effectiveness highlights that the absorption efficiency is more significant than the reflection efficiency. This implies that most of the incident electromagnetic radiation is dissipated in the form of heat (31).

5.9 Electroconducting Polymer Coatings

Composite systems containing electroconducting polymer coatings, such as poly(aniline) and poly(pyrrole) were applied to porous films of semicrystalline polymers, i.e., PE, acPP, poly(vinylidene fluoride) (32).

The porous supports were obtained in a process based on polymer melt extrusion with subsequent annealing, uniaxial extensions, and thermal stabilization. Conducting coatings were formed by the oxidative polymerization of the monomers directly onto the porous supports.

The structure (overall porosity, permeability, pore sizes, factor of orientation) and morphology (specific surface and character of the film surface) of the supports were characterized by sorptometry, filtration porosimetry, atomic force microscopy, and X-ray scattering techniques.

The overall porosity P can be calculated as:

$$P = \frac{\rho - \rho_P}{\rho} \times 100. \tag{5.1}$$

Here, ρ is the density of the dense films and ρ_P is the density of the porous films, which can be measured gravimetrically.

Some properties of the porous structure of the supports is collected in Table 5.12.

Table 5.12 Properties of the porous structure of the supports (32).

Support material	Thickness /[μm]	Specific surface/[$m^2 g^{-1}$]	Maximum channel size/[nm]
PE	11	54	640
PP	21	104	260
PVDF	20	38	<60

It was observed that the porous supports have a strongly developed relief surface, which is formed in the pore formation process (32).

The porous supports have an oriented structure, and the surface of the composites is defined by the morphology inherent in the conducting component (32).

5.10 Foam Insulation Structure

A foam insulation structure has been described that consists of (33):

1. A polymer foam layer having opposing major surfaces and gas-filled cells that contain a physical blowing agent, and

2. A multi-layer sheet affixed to at least one of said opposing major surfaces of the polymer foam layer.

Here, the multi-layer sheet includes:

1. A first noncellular layer containing at least 50% of one or more thermoplastic resins, wherein the one or more thermoplastic resins is not a poly(lactide) resin and has a Vicat softening temperature (ASTM D1525 (34)) of at least 70°C, and
2. A second noncellular layer containing at least 50% of a poly(lactide) resin, wherein the first noncellular layer of the multi-layer sheet is sealingly affixed to the polymer foam layer and the second noncellular layer of the multi-layer sheet is sealingly affixed to the first noncellular layer of the multi-layer sheet.

It has been found that a poly(lactide) resin is a highly effective barrier to the permeation of atmospheric gases and certain blowing agents from the insulation foam.

The foam insulation structure therefore experiences a surprisingly slow loss of thermal insulation efficiency over time due to diffusion of atmospheric gases and blowing agents through the poly(lactide) layers of the multi-layer sheet

Often, the insulation foam layer is produced in a so-called pour-in-place process in which liquid foam precursors are applied to and react at the surface of the facing layers to form the foam. The foam so produced adheres to the facing layers to form the foam insulation structure.

The foam precursors in such processes typically include at least one poly(isocyanate), water, a physical blowing agent, and most typically one or more polyols. The foaming reaction is exothermic and can produce local temperatures of 60°C to 160°C, or even higher.

The physical blowing agent is one or more compounds having a boiling temperature of −10°C to 100°C. The physical blowing agent volatilizes during the production of the polymer foam to produce a gas that expands the polymer. Physical blowing agents are collected in Table 5.13. Some compounds are shown in Figure 5.11.

Unfortunately, the exposure to such high temperatures and to the reactive precursors can damage the polymeric liner materials.

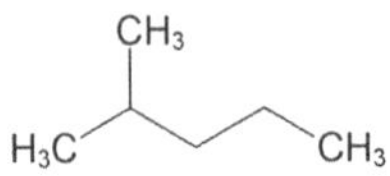

i-Hexane

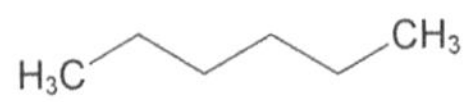

n-Hexane

Cyclobutane

Cycloheptane

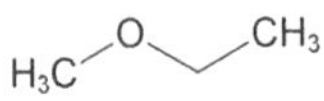

Ethyl methyl ether

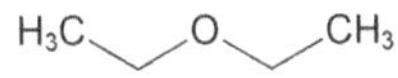

Diethyl ether

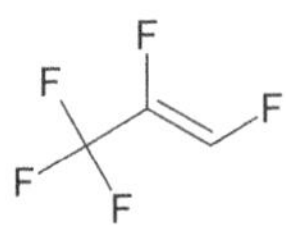

1,2,3,3,3-Pentafluoropropene

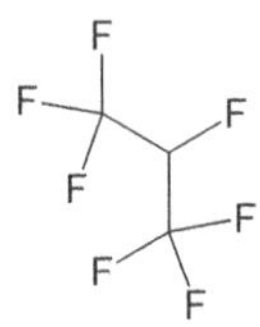

1,1,1,2,3,3,3-Heptafluoropropane

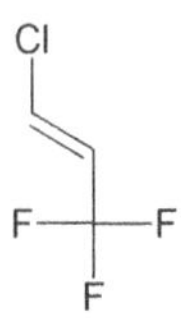

1-Chloro-3,3,3-trifluoropropene

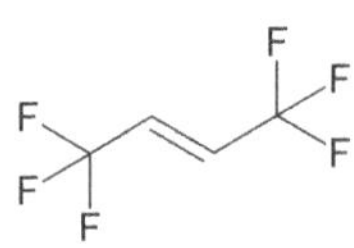

1,1,1,4,4,4-Hexafluorobut-2-ene

Figure 5.11 Physical blowing agents.

Table 5.13 Physical blowing agents (33).

Compound	Compound
1,1,1,3,3-Pentafluoropropane	1,1,1,3,3-Pentafluorobutane
1,1,1,2,3,3,3-Heptafluoropropane	1,1,1,2-Tetrafluoroethane
Trifluoropropene	1,3,3,3-Tetrafluoropropene
1,1,3,3-Tetrafluoropropene	2,2,3,3-Tetrafluoropropene
1,2,3,3,3-Pentafluoropropene	1,1,1-Trifluoropropene
1,1,1,3,3-Pentafluoropropene	1,1,2,3,3-Pentafluoropropene
(Z)-1,1,1,2,3-Pentafluoropropene	1-Chloro-3,3,3-trifluoropropene
1,1,1,4,4,4-Hexafluorobut-2-ene	Dimethyl ether
Ethyl methyl ether	Diethyl ether
n-Butane	*i*-Butane
n-Pentane	*i*-Pentane
n-Hexane	*i*-Hexane Cyclobutane
Cyclopentane	Cyclohexane
Cycloheptane	Cyclooctane

This restricts the range of liner materials that can be used in these *pour-in-place* processes (33).

The polymer foam may include one or more thermoplastic polymers and/or one or more thermoset polymers. Thermoplastic polymers are PS, styrene copolymers such as styrene-acrylonitrile copolymers and styrene-acrylic acid copolymers, PE, and blends of any two or more of the foregoing materials. Thermoset polymers are, isocyanate-based polymers such as PUs, poly(urea)s, PU-ureas, poly(isocyanurate)s, poly(urethane-isocyanurate)s, poly(urea-isocyanurate)s, and poly(urethane-urea-isocyanurate)s, which are produced by the polymerization of a poly(isocyanate) (33).

5.11 Passive Cooling

Porous polymer materials achieve thermal insulation and solar reflection property at the same time because of their own porous structure. This makes the passive cooling property more superior, so as to reduce the indoor temperature more effectively and achieve the purpose of energy saving. A pore structure was constructed in PVC matrix by chemical etching with calcium carbonate ($CaCO_3$)

particles as pore template (35). Then, the effects of porosity on the structure and properties of PVC porous films were studied.

The thermal insulation performance of PVC porous films improves with the increase of porosity, which is better than that of PVC films and PVC/$CaCO_3$ composite films. The thermal conductivity of sample with the porosity of 67% by volume is 0.0868 $W m^{-1} K^{-1}$. The total solar reflectance of a sample with 67 % by volume is 96.34%, which is 454.9% higher than that of PVC films (35).

Although the thermal emissivity from 3 to 5 μm of PVC porous films is decreased, it still remains at a high level of about 0.83 in the whole infrared region. In a temperature test, the final temperature of sample with the porosity of about 67% by volume is only 25.2°C, 23.8°C lower than that of PVC films, and 1.2°C higher than the room temperature. The cooling property of PVC/TiO_2 composite films with 10% by volume TiO_2 was also measured for comparison, and the final temperature is 29.1°C, which is between that of samples with porosity of about 20% by volume and 33% by volume. The PVC porous material prepared in this study has the advantages of low thermal conductivity, good solar reflectance and simple preparation, which has potential application prospects in the field of passive cooling materials (35).

5.11.1 *Radiative Cooling*

Radiative cooling is a passive technology that lowers the surface temperatures. It has shown great potential in high temperate regions (36).

A highly reflective polymeric coating has been proposed (36). Here, $BaSO_4$ particles are dispersed in a P(VdF-HFP) matrix for radiative cooling. Through the strong Mie scattering of sunlight and intrinsic bond vibration, the average substrate-independent solar reflectance and infrared emittance within the 8 μm to 13 μm atmospheric window could reach 97% and 94.2%, respectively.

Under direct exposure to the sky, surfaces could maintain close to ambient temperature even when the solar intensity was as high as 1000 $W m^{-2}$, while separately achieving 5.7°C below ambient temperature during night-time with an effective cooling power of 54.4 $W m^{-2}$.

With a scalable fabrication process, this cost-effective single-layer coating can be easily applied to diverse substrates, which is suitable for real-world applications in the tropics (36).

Passive daytime radiative cooling materials significantly alleviate the current energy shortage and global warming without consuming any energy (37). However, conventional inorganic nanoparticles can hardly achieve excellent cooling effect during daytime due to their low reflectivity.

Broadband super white nanoparticle-based radiators can demonstrate an excellent solar reflectivity under direct sunlight and low absorption. So, introduced TiO_2 nanoparticles were used with the advantages of a high refractive index, high infrared emissivity and low solar absorption into the porous cellulose-based film via a facile phase separation strategy. The cavities were distributed around 5 μm in size due to the limited evaporation of solvents, while the introduction of TiO_2 nanoparticles effectively enhanced the solar reflectance to 0.97.

Moreover, the scattering model demonstrated that the high solar reflectivity was mainly produced by the randomly agglomerate effect of nanoparticles and these nanoclusters also contribute to a high mid-infrared emissivity. As a result, the photonic film presented about 10°C cooling capacity even under the high density solar radiation (~897 $W m^{-2}$) according to an outdoor measurement.

So, the random nanoclusters filling strategy avoids the complicated process of specific size particles and has significant potential to propel the polymer-based cooling technology to satisfy the needs of large-scale thermal management (37).

5.11.2 *Passive Building Cooling*

A review of the applications, modeling and experiments of porous materials in building energy technologies has been presented (38).

Passive building cooling without any electricity input is highly desirable in pursuing low energy consumption and environmental protection (39). However, widespread adoption of existing techniques is restrained by the complex system design or low cooling power.

5.11.2.1 *Bilayer Film*

An efficient passive cooling approach with a bilayer porous polymer film was proposed, which comprises a hygroscopic hydrogel and a hydrophobic top layer with hierarchical pores. The hydrogel implements evaporative cooling in the daytime and regenerates itself at night.

Further, the top layer protects and radiatively cools the hydrogel, which enhances the cooling power during day and helps the hydrogel regeneration at night. With this synergistic effect, the bilayer film attains a remarkable subambient temperature drops of ca. 7°C and an effective cooling power of ~150 $W\,m^{-2}$ under direct sunlight, showing great potential for low-cost, efficient and scalable passive building cooling (39).

5.11.2.2 *Poly(ethylene) Aerogel*

Air-conditioning in buildings is energy intensive and takes up large proportion of global energy consumption. To pursue low energy expenditure, herein, a radiant air-conditioning strategy was proposed (40).

Here infrared transparent and low thermal conductive porous PE aerogels are used (40).

The PE aerogel was was fabricated using a thermal phase separation followed by a solvent extraction and critical point drying.

The aerogel allows a direct heat exchange between the cooling panel and the human body via radiation and avoids cooling the entire space. Compared with traditional space air-conditioning, a 6-mm PE aerogel assisted radiant air-conditioning can reduced the electric power by 31% under the same cooling and ambient temperature.

Moreover, high solar reflectance and low thermal conductivity of the porous PE hold a great promise of the strategy in working outdoors. it could be demonstrated that 66% of energy saving can occur, compared with a direct space cooling process in a typical summer day in Wuhan. So, the results of this study show a great potential for the cooling method in building cooling both in closed spaces and open conditions (40).

5.12 Sulfur-Containing Polymers

The removal of mercury in the gas phase is vital in gas processing industries. The preparation of porous sulfur copolymers for capturing elemental mercury in the gas phase was reported (41).

Porous sulfur copolymers were prepared using porogen and an inverse vulcanization technique (42). Sulfur powder was mixed with a water-soluble porogen (NaCl) and crosslinked with (divinylbenzene) at 165°C. A porous structure was later obtained by dissolving the porogen in water. The porosity of the copolymer was controlled by varying the amount of salt and crosslinker. The effect of pore volume and temperature on the adsorption of mercury from gas phase has been estimated.

Samples with larger pore volumes showed a significant increase in adsorption capacity, reaching approximately 151 $\mu g\, g^{-1}$ at 60°C. A high sulfur content and the porous structure enhanced the adsorption capacity. Furthermore, the thermal conductivities of the porous copolymers were studied at a range temperature.

Copolymers with relatively high crosslinking densities showed the very low thermal conductivity of approximately 0.048 $W\, m^{-1} K^{-1}$ at 25°C, which is much lower than that of pristine sulfur.

The abundant availability of sulfur and a simple method of preparation could make porous sulfur copolymers potential candidates for mercury capture and also for thermal insulation (41).

5.13 Nanocellular Polymers

5.13.1 *Poly(methyl methacrylate) Thermoplastic Poly(urethane) Composites*

Nanocellular polymers have been considered promising multifunctional materials with unique properties. The key challenge is to fabricate nanocellular polymers with adequate cell sizes and densities. A method to prepare nanocellular poly(methyl methacrylate) (PMMA)/thermoplastic poly(urethane) (TPU) with tunable structures has been reported (43).

The nanocellular PMMA/TPU shows a ductile fracture behavior, while the microcellular PMMA has a brittle fracture behavior. The nanocellular PMMA/TPU presents a tensile toughness and an impact toughness which are respectively 180-fold and 762% higher than the microcellular PMMA.

The nanocellular structure combined with the nanostructured TPU greatly promotes the movement, shear slip and nano fibrillation of the polymer chains, thus leading to this remarkably enhanced toughness. The nanocellular PMMA/TPU with an average cell size of 205 *nm* and an expansion ratio of 8.0 displays a thermal conductivity of 24.8 $mWm^{-1}K^{-1}$ and a high compressive strength. So, a superinsulating performance could be achieved with porous polymers. The reduced air's thermal conductivity due to the Knudsen effect combined with the reduced polymer's thermal conduction due to the large expansion ratio are responsible for this superinsulating performance.

So, this multifunctional nanocellular material offers a promising solution for advanced thermal insulation applications (43).

5.13.2 *Poly(methyl methacrylate) Multiwalled Carbon Nanotube Composites*

The foaming behavior of PMMA/MWCNTs nanocomposites and thermally insulating, electrical, and mechanical properties of the nanocomposite foams were investigated (44).

First, PMMA/MWCNT nanocomposites containing various amounts of MWCNTs were prepared by combining solution and melt blending methods. Then these materials were foamed using CO_2. The foaming temperature and MWCNT content were varied for regulating the structure of the PMMA/MWCNT nanocomposite foams.

The measurements of the electrical conductivity showed that the MWCNTs have little effect on the electrical conductivity of the foams with a large expansion ratio (44).

Also, the thermal conductivities of both solid and foamed PMMA/MWCNT nanocomposites were measured to evaluate their thermally insulating properties. The gas conduction, solid conduction, and thermal radiation of the foams were calculated for clarifying the effects of cellular structure and MWCNT content on thermal insulation properties (44).

The results demonstrate that MWCNTs endow foams with enhanced thermal insulation performance by blocking thermal radiation. In addition, compressive testing showed that MWCNTs improve the compressive strength and the rigidity of the foams.

So, this research is essential for optimizing environmentally friendly thermal insulation nanocomposite foams with enhanced thermal insulation and compressive mechanical properties (44).

5.14 Household Applications

Household appliances such as dishwashers, clothes driers, washing machines, vacuum cleaners, etc., often are manufactured with thermal insulation and are also insulated to reduce noise and vibration (45),

Here, the acoustic insulating material is commonly a fiber insulation batt, a bitumen or asphalt mastic, or a high-density polymer. These materials provide acoustic insulation, but are poor thermal insulators. Accordingly, it has been necessary to incorporate multiple layers of insulating material to obtain both acoustic and thermal insulation. This leads to undesired increases in the weight of the products and also to an increase of the production costs.

An insulation structure has been described with at least one layer of a viscoelastic *PU* foam, The viscoelastic *PU* has the following properties (45):

1. It has a resiliency value of at most 15% as measured according to ASTM 3574 (46),
2. It has a recovery time of at least 3 *s* as measured according to ASTM D3574 Test, and
3. The viscoelastic foam is made in the reaction of an aromatic poly(isocyanate) with a mixture of isocyanate-reactive materials that includes at least 20%, based on the combined weight of all isocyanate-reactive materials in the mixture, of at least one polyol having a molecular weight of at least 750, a hydroxyl equivalent weight of 225 to 450 and 2 to 4 hydroxyl groups per molecule, and water in an amount of at least 0.2 parts per 100 parts by weight of the mixture of isocyanate-reactive materials, wherein the isocyanate index is 60 to 100.

The aromatic poly(isocyanate)s are compounds having an average of at least 2 aromatically bound isocyanate groups per molecule. Examples of useful aromatic poly(isocyanate)s are collected in Table 5.14. Some compounds are shown in Figure 5.12.

Table 5.14 Monomers for aromatic poly(isocyanate)s (45).

Monomer
m-Phenylene diisocyanate
Toluene-2,4-diisocyanate
Toluene-2,6-diisocyanate
1,5-Naphthalene diisocyanate
1,3-Bis(isocyanatomethyl)cyclohexane
1,4-Bis(isocyanatomethyl)cyclohexane
Methoxyphenyl-2,4-diisocyanate
Diphenylmethane-4,4′-diisocyanate
Diphenylmethane-2,4′-diisocyanate
4,4′-Biphenylene diisocyanate
3,3′-Dimethoxy-4,4′-biphenyl diisocyanate
3,3′-Dimethyl-4-4′-biphenyl diisocyanate
3,3′-Dimethyldiphenyl methane-4,4′-diisocyanate
4,4′,4″-Triphenyl methane triisocyanate
Toluene-2,4,6-triisocyanate
4,4′-Dimethyldiphenylmethane-2,2′,5,5′-tetraisocyanate

Modified aromatic poly(isocyanate)s that contain urethane, urea, biuret, carbodiimide, uretoneimine, allophanate, isocyanurate or other groups formed by reaction of an isocyanate group with itself or an isocyanate-reactive compound, such as an alcohol, amine or water, are also useful.

Preferably, the poly(isocyanate) can be diphenylmethane-4,4′-diisocyanate, diphenylmethane-2,4′-diisocyanate, toluene-2,4-diisocyanate, toluene-2,6-diisocyanate or mixtures thereof.

Suitable catalysts include, for example, including tertiary amines, cyclic amidines, tertiary phosphines, various metal chelates, acid metal salts, strong bases, various metal alcoholates and phenolates and metal salts of organic acids. Examples of metal-containing catalysts are bismuth, cobalt and zinc salts. Catalysts of most importance are tertiary amine catalysts, cyclic amidines and tin catalysts.

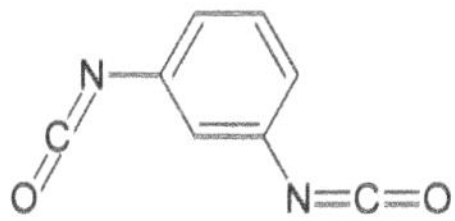

m-Phenylene diisocyanate

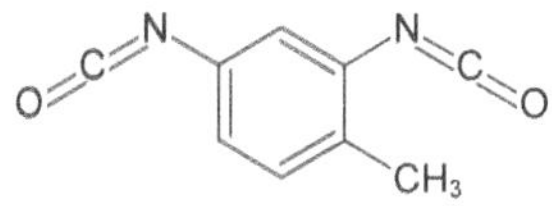

Toluene-2,4-diisocyanate

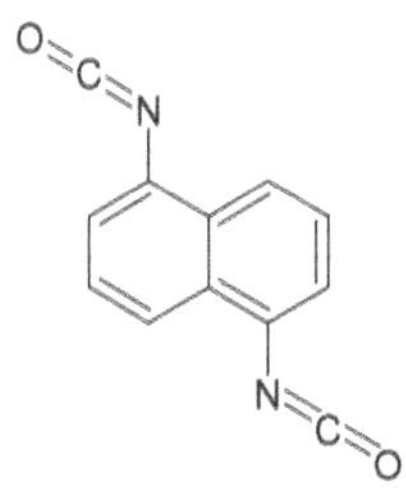

1,5-Naphthalene diisocyanate

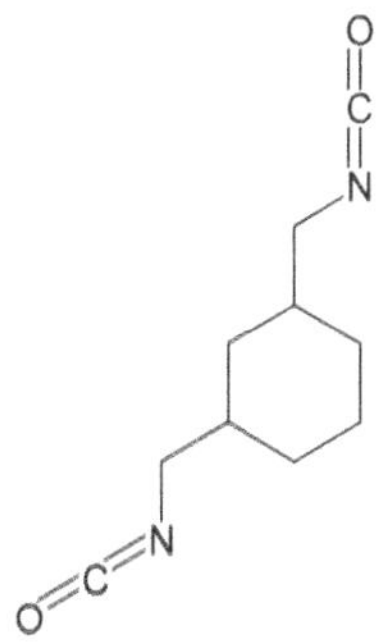

1,3-Bis(isocyanatomethyl) cyclohexane

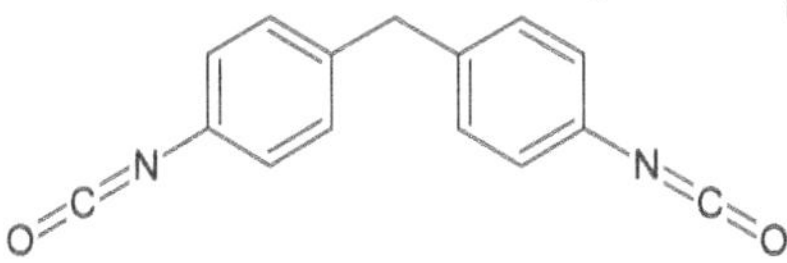

Diphenylmethane-4,4′-diisocyanate

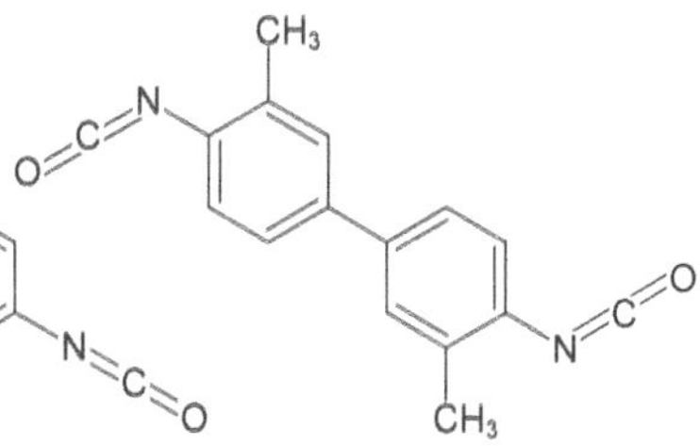

3,3′-Dimethyl-4-4′-biphenyl diisocyanate

Figure 5.12 Monomers for aromatic poly(isocyanate)s.

Examples of tertiary amine catalysts are collected in Table 5.15. Some of these catalysts are shown in Figure 5.13.

Table 5.15 Tertiary amine catalysts (45).

Catalyst
Trimethylamine
Triethylamine
N-Methylmorpholine
N-Ethylmorpholine
N,N-Dimethylbenzyl amine
N,N-Dimethyl ethanolamine
N,N,N′,N′-Tetramethyl-1,4-butanediamine
N,N-Dimethylpiperazine
1,4-Diazobicyclo-2,2,2-octane
Bis(2-dimethylaminoethyl)ether
Triethylene diamine
Dimethylaminopropyl amine

Examples of tin catalysts are stannic chloride, stannous chloride, stannous octoate, stannous oleate, dimethyltin dilaurate, dibutyltin dilaurate, tin ricinoleate and other tin compounds, e.g., tin mercaptides, tin thioglycolates. Some of these tin catalysts are shown in Figure 5.14.

The catalysts are typically used in small amounts; for example, each catalyst being employed from about 0.0015% to about 5%.

The viscoelastic foam is prepared by forming a reaction mixture containing the mixture of isocyanate-reactive materials and the poly(isocyanate)s, then curing the mixture. Water and various other isocyanate-reactive materials can be mixed together before they are combined with the poly(isocyanate).

Alternatively, they can be combined with the poly(isocyanate) individually, i.e., as separate streams, or they can be formed into one or more submixtures that are then combined with the poly(isocyanate).

Due to the speed of the curing reaction, it is preferred to combine the water and other isocyanate-reactive compounds with the poly(isocyanate) simultaneously or nearly simultaneously (such as within 5 seconds). The reaction usually proceeds spontaneously at room temperature, i.e., 22°C.

Elevated temperatures can be used to speed the cure or drive it

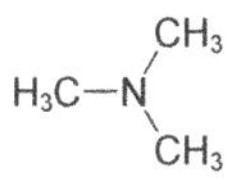

Trimethylamine

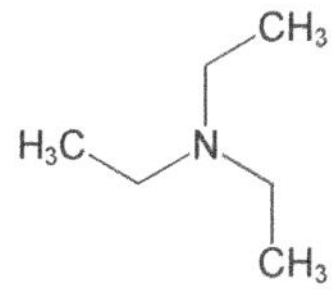

Triethylamine

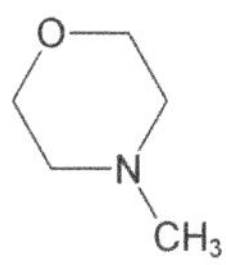

N-Methylmorpholine

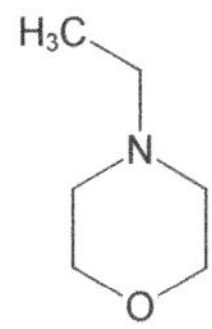

N-Ethylmorpholine

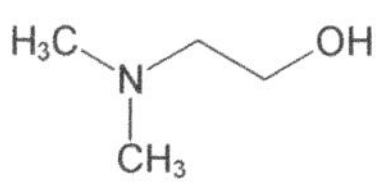

N,*N*-Dimethyl ethanolamine

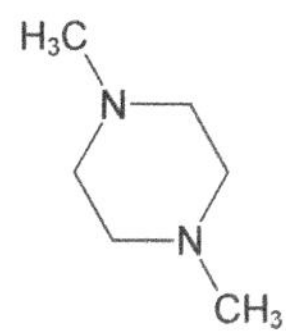

N,*N*-Dimethylpiperazine

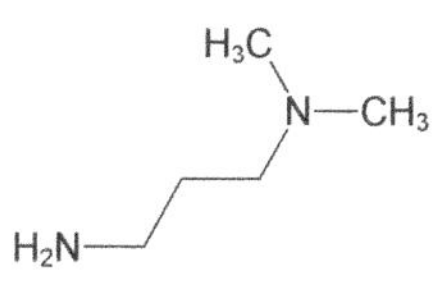

Dimethylaminopropyl amine

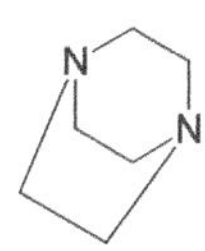

1,4-Diazobicyclo-2,2,2-octane

Figure 5.13 Tertiary amine catalysts.

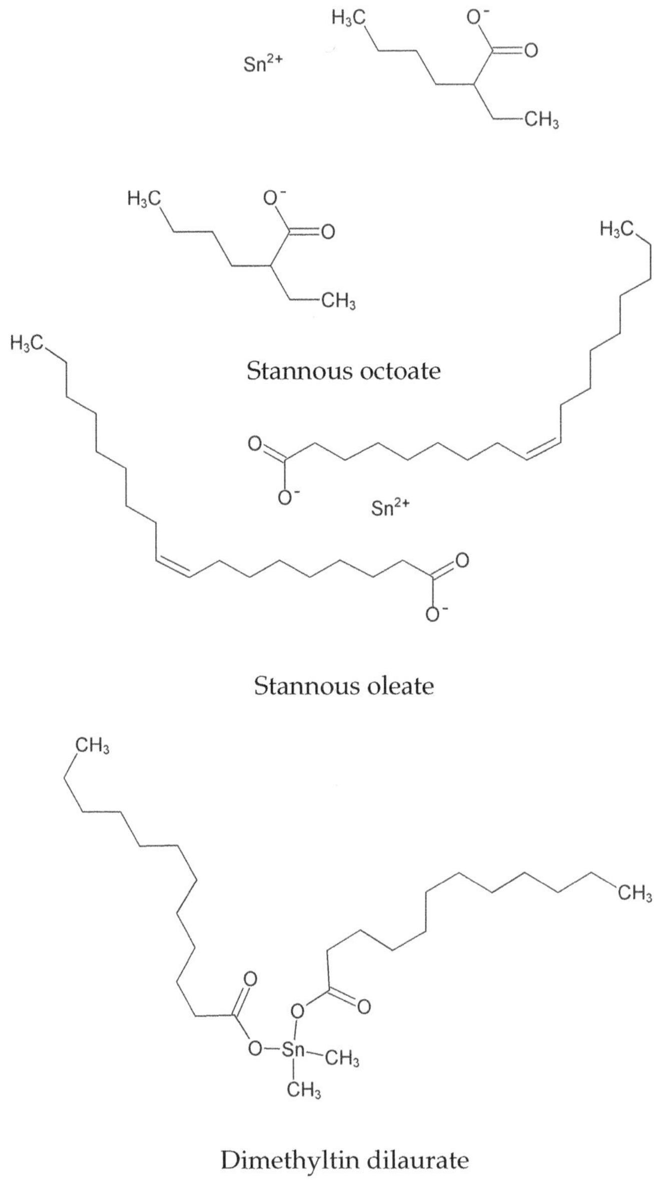

Figure 5.14 Tin catalysts.

toward completion. This can be done by heating some or all of the ingredients prior to combining them, by applying heat to the reaction mixture as it cures, or some combination of each. The curing is continued until the reaction mixture has expanded and cured sufficiently to form a stable foam.

The layer of viscoelastic foam in the insulation structure can be 1 *mm* to 100 *mm* thick. A greater thickness corresponds to greater acoustic and thermal insulation, but the weight increases with increasing thickness.

Viscoelastic foams can be prepared from formulations shown in Table 5.16.

Table 5.16 Formulations for viscoelastic foams (45).

Component	No. 1	No. 2	No. 3
	Parts per weight		
Monoethylene glycol	0.585	0.6	0.573
Glycerine	0.195	0.2	0.191
Diethanolamine	0.78	0.8	0.763
Cell Opener (ethylene oxide propylene oxide copolymer)	3.9	0	3.817
Bis(2-dimethylaminoethyl)ether catalyst	0.146	0.150	0.219
Triethylene diamine catalyst	1.949	4.200	3.053
Tin catalyst	0.195	0.33	0.315
Water	0.224	0.500	0
Trifunctional poly(propylene oxide)	75.458	77.420	70.076
Trifunctional ethylene oxide-capped poly(propylene oxide)	9.747	10	9.542
Trifunctional, poly(propylene oxide)	4.873	5	4.771
Ortegol 501 surfactant	1.95	0	1.91
Tegostab B8715 LF2 surfactant	0	0.8	0
Zeolite	0	0	4.77
Volume density/[$kg\,cm^{-3}$]	500	400	>800

Test specimens were prepared as follows (45):

Preparation 5–4: All the components except the isocyanate were mixed to form a curative. The curative and isocyanate were processed through a spray robot to form a reaction mixture that is sprayed onto steel plates

(500×500×0.5 *mm*, weighing 650 *g*) and cured on the steel plates. The weights of the coated plates and the foam thickness are measured in each case.

The viscoelastic PU foam has a lambda value of less than 0.07 $W m^{-1} K^{-1}$. The very low lambda values indicate that the viscoelastic foams provide an excellent thermal resistance (45).

5.14.1 *Refrigerator*

Insulated walls of a refrigerator have an outer case made of a steel plate, an inner case molded from acrylonitrile-butadiene-styrene or high impact poly(styrene), and a thermal insulation formed of hard PU foam and interposed there between. The hard PU foam provides a refrigerator with a thermal insulation property and rigidity.

In particular, the hard PU foam used for refrigerators is formed from closed cells and low thermal conductivity and low density of a foaming gas filled in the cells provide thermal insulation property to the refrigerators. Also, the refrigerator may have rigidity due to a chemically crosslinked structure formed of a large number of functional groups of polyol and isocyanate.

A composition for preparing a PU is shown in Table 5.17.

Table 5.17 Poly(urethane) composition (47).

Compound	Amount/[*g*]
Polyol	100
Surfactant	2.5
Blowing catalyst	0.45
Gelling catalyst	2.10
Trimerization catalyst	0.7
Water	1.90
Cyclopentane	16.5
Poly(methylene diphenyl diisocyanate)	149.0
Cellulose microfiber	0.27

A cellulose microfiber having a length of 50 μm to 70 μm was used. Also, several other examples, in total 14, have been presented (47).

5.15 Fluid Storage Tank

Sealed and thermally insulating membrane tanks are used for the storage of liquefied natural gas. These tanks may be installed on the land or on a floating structure. In the case of a floating structure, the tank may be intended for the transportation of liquefied natural gas or for receiving liquefied natural gas used as fuel for the propulsion of the floating structure.

A method for producing a sealed and thermally insulating wall for a fluid storage tank has been presented. The method consists of the following steps (48):

1. Attaching a plurality of anchoring elements to a support structure,
2. Installing modular formwork elements on the support structure, the modular formwork elements having a shape that protrudes relative to the support structure and that defines, with the support structure and the plurality of anchoring parts, compartments having an open side opposite the support structure,
3. Spraying insulating foam into said compartments through the open side so as to form a plurality of insulating sectors made from sprayed insulating foam,
4. Arranging insulating junction elements in a stressed position in which they are stressed between said insulating sectors and capable of expanding when said insulating sectors contract, so as to ensure continuity of the thermal insulation, and
5. Attaching a sealing membrane to said anchoring elements.

This method for producing a wall benefits from the advantages of the use of a formwork, since it is simple to carry out and enables the layer of thermal insulation to be divided up so as to limit the mechanical stresses due to the temperature differences between its external and internal surfaces, while ensuring continuity of the thermal insulation when the wall is subjected to low temperatures.

An insulating junction element contains a profiled element having two resilient flanges which, in a stressed position between the insulating sectors, are stressed toward one another and exert a reactive force tending to separate them from one another. The profiled element with two resilient flanges is produced from a foam made of a polymer selected from PU, melamine (1,3,5-triazine-2,4,6-triamine), PE, PP, PS, and silicone. An insulating junction element comprises a strip made of a compressible material selected from among glass wool, polyester wadding, and foams of PU, melamine, PE, PP, PS, or silicone. The method comprises a step of trimming the insulating sectors.

The thermally insulating layer of the block is made of polymer foam with a density of more than 100 $kg\,m^{-3}$, or of wood (48).

5.16 Thermal Insulation for High Explosives

Warheads and guided weapons filled with high explosives are continuously and periodically exposed to changes in external temperature, severe environments such as fires, and external environmental factors such as impacts during the storage, transportation and operation thereof (49).

In particular, changes in the external temperature cause changes in high explosives placed in warheads and guided weapons, thus reducing their storage life. When high explosives are exposed to flame in a situation, such as a fire, spontaneous ignition occurs due to rapid or slow heating, which, in the worst case, undesirably results in an explosion.

A thermal insulation foam for high explosives has been developed. This can be applied to the inner surface of warheads and guided weapons filled with high explosives, thus maximizing the storability and survivability of warheads and guided weapons despite changes in temperature and external environmental factors such as impacts.

The thermal insulation foam can be manufactured by mixing thermally expanded microspheres, a polysiloxane-based prepolymer, isophorone diisocyanate, and an additive such as a diol chain extender, a plasticizer, a urethane reaction catalyst or a curing catalyst.

The curing catalyst may be a ferric acetylacetonate, a tin-based curing catalyst such as dibutyltin dilaurate, triphenyl bismuth, and 3,5-dinitrosalicylic acid. These compounds are shown in Figure 5.15.

The thermal insulation foam has a low surface energy compared

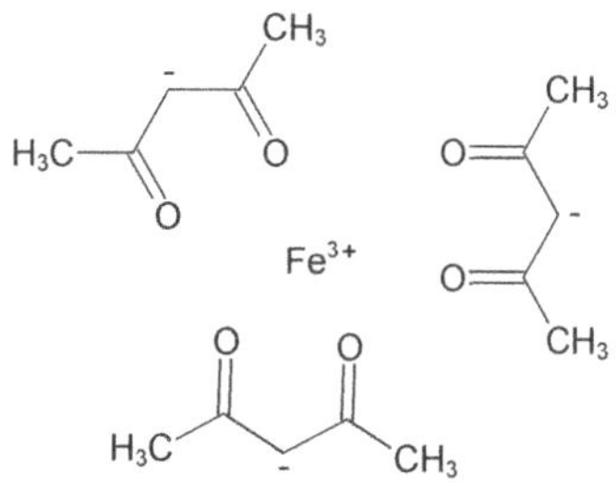

Ferric acetylacetonate

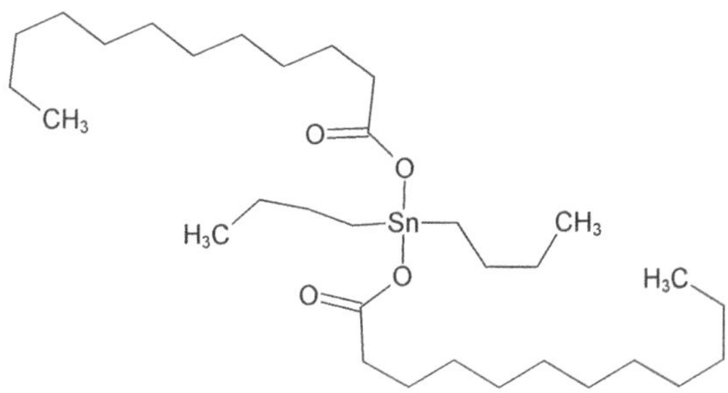

Dibutyltin dilaurate

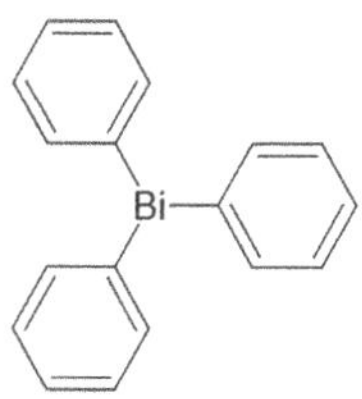

Triphenyl bismuth

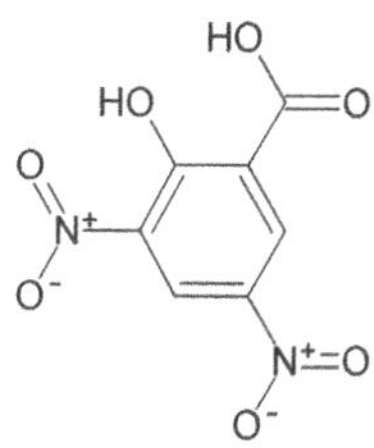

3,5-Dinitrosalicylic acid

Figure 5.15 Curing catalysts.

to the hydroxyl-terminated polybutadiene-based PU, thus exhibiting low interfacial adhesion with the high explosives (49).

5.17 Aerogels

5.17.1 Polysaccharide-Based Aerogels

To reduce energy losses due to insufficient thermal insulation is one of the current hot topics. Various commercial porous materials are used with the best conductivity around 0.03–0.04 $W\,m^{-1}K^{-1}$ (50).

Aerogels are the only known materials with intrinsic thermal superinsulating properties, i.e., with a thermal conductivity below that of air in ambient conditions, 0.025 $W\,m^{-1}K^{-1}$.

The classical thermal superinsulating aerogels are based on silica and some synthetic polymers, with conductivity 0.014–0.018 $W\,m^{-1}K^{-1}$. Aerogels based on natural polymers are new materials created at the beginning of the 21st century (50).

The properties of polysaccharide-based aerogels have been detailed (50),

5.17.1.1 Xerogels

Scalability is a common challenge in the structuring of nanoscale particle dispersions, particularly in the drying of these dispersions for producing functional, porous structures such as aerogels (51). The aerogel production relies on supercritical drying, which exhibits only a poor scalability. A solution to this scalability limitation is the use of evaporative drying under ambient pressure. However, the evaporative drying of wet gels comprising nanoscale particles is accompanied by a strong capillary force.

So, it is challenging to produce evaporative-dried gels or *xerogels* that possess the specific structural profiles of aerogels such as mesoscale pores, high porosity, and high specific surface area.

Xerogels can be prepared by using common oven dryers or even by air drying, such that xerogels have potentially no limitation in dimension and are scalable (52–54).

The representative material of aerogels is silicon dioxide. Silica aerogels are composed of interconnected spherical silica particles, creating a skeletal structure that is too brittle to withstand the capillary force in evaporative drying, resulting in crack formation or collapse of the xerogel structure.

A structure of mesoporous xerogels with high porosity of around 80% and high specific surface area of greater than 400 m^2g^{-1} could be achieved by exploiting cellulose nanofibers (CNFs) as the building blocks with tunable interparticle interactions. CNFs are sustainable, wood-derived materials with high strength. In this study, the few-nanometer-wide CNFs bearing carboxy groups were structured into a stable network via ionic inter-CNF interaction. The outline of the resulting xerogels was then tailored into a regular, millimeter-thick, board-like structure.

Several characterization techniques highlighted the multifunctionality of the CNF xerogels combining outstanding strength (compression E = 170 *MPa*, s = 10 *MPa*; tension E = 290 *MPa*, s = 8 *MPa*), moderate light permeability, thermal insulation (0.06–0.07 $W\,m^{-1}K^{-1}$), and flame self-extinction. As a potential application of the xerogels, daylighting yet insulating, load-bearing wall members can be thus proposed (51).

5.17.2 *Silica Aerogels*

A hydrophobic silica-based aerogel (FHSA) was prepared using a facile sol-gel process and an ambient pressure drying method (55).

Here, methyl trimethoxysilane was used as the precursor and acid and base as the catalyst. Functional precursors such as dimethyl dimethoxysilane and 3-aminopropyl triethoxysilane were added as coprecursors to impart to the aerogels not only mechanical properties but also functionalization, such as CO_2 capture and air purification. The above-mentioned compounds are shown in Figure 5.16. Moreover, the incorporation of polymeric chains into silica aerogels was another strategy to produce flexible silica-based aerogels.

In detail, the synthesis runs as follows (55):

Preparation 5–5: Terephthalaldehyde, 3-aminopropyl triethoxysilane, methyl trimethoxysilane, and dimethyl dimethoxysilane in a fixed molar ratio of 1:2:2:2 were involved in the reaction as the silicon sources.

First, 0.284 *g* of TPAL, 1 *ml* of 3-aminopropyl triethoxysilane, 0.4 *ml* of methyl trimethoxysilane, and 0.4 *ml* of dimethyl dimethoxysilane were added into a 50 *ml* glass beaker, which contained 18 *ml* of EtOH as the

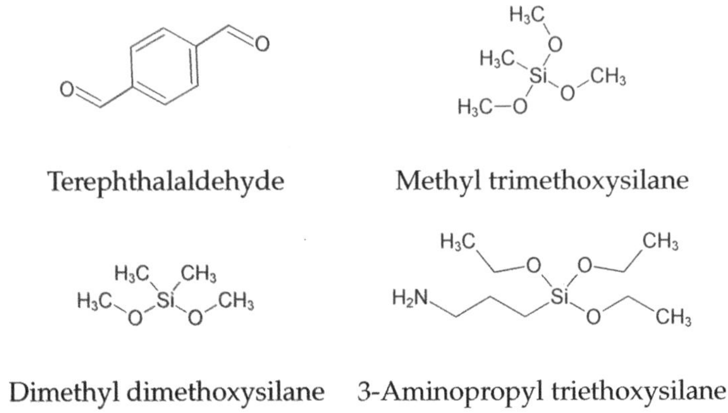

Figure 5.16 Silicone monomers.

solvent. Except as noted, the mentioned solution is prepared under a constant magnetic stirring speed of 700-800 r/min. After stirring for 15 *min* at a room temperature of 25°C, the mixture became a clear solution. Then, 1 *ml* of deionized water was slowly added under stirring conditions to induce the hydrolysis of the silicon sources.

Then, 0.2 *ml* of HAc was added drop by drop to facilitate the gelation procedure. After further stirring for 15 *min*, the solution was transferred into a 20 *ml* cylindrical plastic mold. The mold was sealed and put into an oven at 60°C to experience the sol-gel process. The gelation was formed after 6 *h* and aged for another 6 *h* in the oven. Then, the gel was taken out from the mold and washed with EtOH at 60°C three times every 12 *h* to remove the water and other unreacted solvents in the pores. Next, the EtOH was replaced using the n-hexane, whose surface free energy is lower, with the above-mentioned similar operation, except at room temperature. Finally, the alcogel was dried via ambient pressure drying at a medium temperature of 40°C for 24 *h* to obtain the FHSA.

The FHSA was treated at different temperatures varying from −196 to 450°C to evaluate its thermal and mechanical performances. The evolutions of the physical property, hydrophobicity, microstructure, pore structure, and chemical structure of the FHSA with the various treatment temperatures were investigated comprehensively.

The structure of the FHSA did not show an obvious change after treatment in the liquid nitrogen. The bulk density of the FHSA increased from 0.047 to 0.077 $g\,cm^{-3}$ when the thermal treatment temperature increased from 25°C to 450°C. The specific surface area and the pore volume of the FHSA increased with the treatment temperature owing to the decomposition of the organic moieties. Fourier transform infrared spectroscopy spectra showed that the methyl groups in the FHSA had an excellent thermostability up to 400°C. The water contact angles of the FHSA after treatment at various temperatures are shown in Table 5.18.

Table 5.18 Water contact angles and pore volume (55).

Temperature/[°C]	Contact angle/[°]	Pore volume/[$cm^3 g^{-1}$]
−196	131	0.053
25	151	
200	162	0.143
300	150	0.151
350	132	0.206
400	119	0.216
450	34	0.267

The thermal conductivity of the FHSA at a low temperature of −10°C was 0.022 $W m^{-1} K^{-1}$. The reversible deformation rate of the FHSA was more than 80% within 100 compression cycles. After a treatment in liquid nitrogen, the reversible deformation rate of the FHSA remained at 50% (55).

The synthesis method of the FHSA is simple, the resulting FHSA showed good performance both in thermostability and flexibility, and it is promisingly applied for thermal insulation and sealing in ultralow-temperature environments (55).

Energy consumption is a worldwide topic and it is an urgent requirement for the development of energy saving materials (56). Thermal insulation materials with a low thermal conductivity show a significant potential to reduce energy waste.

Many porous materials are used as thermally insulating materials to reduce heat loss. Most of these materials have a low heat stability, inferior mechanical properties, and inflammability (57). Here, an efficient thermally insulating and flame retardant SiO_2/poly(imide) (PI) nanocomposite aerogel with hierarchically porous structures was fabricated by a *co-gel* strategy.

The co-gel process is to *in-situ* generate a homogeneous nanocomposite gel using water-soluble poly(imide) and silica precursors. Subsequently, the complex gel underwent freeze-drying and the imidization process. The target nanocomposite aerogel with a binary network structure in which the poly(imide) and silica particles were interspersed was acquired. The so prepared SiO_2/PI nanocomposite aerogels also exhibited highly hydrophobic properties (water contact angle of 120–127°), allowing it to resist moisture, which is an important property of thermal insulation materials. The so prepared SiO_2/PI nanocomposite aerogel possessed excellent mechanical property with compressive moduli of 1.96 MPa and specific modulus of 52.7 m^2s^{-2}.

In comparison to most polymer-based aerogel and aerogel-like materials, the so prepared SiO_2/PI nanocomposite aerogels exhibit an excellent flame resistance with a low fire growth rate of 0.2–0.6 $W s^{-1}$ and excellent thermal insulation properties with low thermal conductivities that varied from 31.1 $mW m^{-1}K^{-1}$ to 58.5 $mW m^{-1}K^{-1}$ between 25°C and 300°C.

So, this nanocomposite aerogel has very broad application prospects in the areas of thermal insulation in which fire safety is required (57).

5.17.3 *Aerogel Fibers*

An aerogel fiber, with the characteristics of ultralow density, ultrahigh porosity, and high specific surface area, is the most potential candidate for manufacturing a wearable thermal insulation material (58). However, aerogel fibers generally show weak mechanical properties and complex preparation processes.

By first preparing a cellulose acetate (CA)/poly(acrylic acid) (PAA) hollow fiber using a coaxial wet-spinning technique followed by injecting the silk fibroin (SF) solution into the hollow fiber, the CA/PAA-wrapped SF aerogel fibers toward textile thermal insulation were successfully constructed after freeze-drying. The sheath (CA/PAA hollow fiber) possesses a multiscale porous structure, including micropores (11.37 ± 4.01 μm), sub-micron pores (217.47 ± 46.16 nm), as well as nanopores on the inner (44.00 ± 21.65 nm) and outer (36.43 ± 17.55 nm) surfaces, which is crucial to the formation of a SF aerogel core. Furthermore, the porous CA/PAA-wrapped SF aerogel fibers have many advantages, such as low density (0.21 $g cm^{-3}$), high porosity (86%), high strength at break (2.6 ± 0.4 MPa), as well as potential continuous and large-scale production.

The delicate structure of multiscale porous sheath and ultralow-density SF aerogel core synergistically inhibit air circulation and limit convective heat transfer. Meanwhile, the high porosity of aerogel fibers weakens heat transfer and the SF aerogel cellular walls prevent infrared radiation.

The results showed that the mat composed of these aerogel fibers exhibits excellent thermal insulating properties with a wide working temperature from −20 to 100°C. Therefore, this SF-based aerogel fiber can be considered as a practical option for high performance thermal insulation (58).

Strong, continuous, and highly porous coaxial fibers with a CNF aerogel core and a cellulose-rich sheath were fabricated by wet-spinning hollow fibers and infusing them with an aerogel precursor for high-performance thermal insulators (59).

The sheath contained multiscale pores, including microvoids with 14.5 μm and sub-micron pores of 133 nm in bulk, as well as ca. 25 nm to 26 nm surface nanopores, to function as a template and protective sheath for the microporous CNF aerogel core. The porous coaxial fibers showed many desirable qualities, including a low density of 0.2 $g\,cm^{-3}$, a high porosity of 85%, a high specific tensile strength (23.5±2.5 $MPa\,g\,cm^{-3}$), wide working temperatures of 20°C to 150°C, a continuous and large-scale producibility, as well as a good biodegradability.

The unique combination of multiscale porous sheath and ultralow density aerogel core synergistically minimizes heat conductivity by all three mechanisms, i.e., air circulation is restrained to limit convective heat transfer, while the poorly conducting cellulose permits little conductive heat transfer and the highly crystalline aerogel cellular walls prohibit infrared radiation, effectively suppressing radiative heat transfer under extreme temperatures (59).

References

1. D. Bozsaky, *Periodica Polytechnica Architecture*, Vol. 41, p. 49, 2010.
2. BASF, AG, *Plastics*, Vol. 7, p. 2, 2005.
3. C.G. Munters and J.G. Tandberg, Heat insulation, US Patent 2 023 204, assigned to The authors, December 3, 1935.
4. H. Yu, H. Zhang, J. Zhao, J. Liu, X. Xia, and X. Wu, *Frontiers of Physics*, Vol. 17, p. 23202, 2022.

5. Z.H. Rizvi, D. Shrestha, A.S. Sattari, and F. Wuttke, *Heat and Mass Transfer*, Vol. 54, p. 483, 2018.
6. R.S. Ghai, K. Chen, and N. Baddour, *Coatings*, Vol. 9, p. 101, February 2019.
7. D.A.G. Bruggeman, *Annalen der Physik*, Vol. 416, p. 636, 1935.
8. H. Yu, H. Zhang, P. Buahom, J. Liu, X. Xia, and C.B. Park, *Energy*, Vol. 233, p. 121140, 2021.
9. H. Wu, J. Fan, and N. Du, *Journal of Applied Polymer Science*, Vol. 106, p. 576, 2007.
10. L. Qiu, Y.M. Li, X.H. Zheng, J. Zhu, D.W. Tang, J.Q. Wu, and C.H. Xu, *International Journal of Thermophysics*, Vol. 35, p. 76, 2014.
11. L. Qiu, D.W. Tang, X.H. Zheng, and P. Su, G, *Review of Scientific Instruments*, Vol. 82, p. 045106, 2011.
12. L. Qiu, X. Zheng, G. Su, and D. Tang, *International Journal of Thermophysics*, Vol. 34, p. 2261, 2013.
13. M. Su, Y. Pan, G. Zheng, C. Liu, C. Shen, and X. Liu, *Polymer*, Vol. 218, p. 123528, 2021.
14. R. Breindel, R. Loh, J.P. Rynd, Y. Delaviz, and M.E. Polasky, Mercury capture using functionalized porous organic polymer with hierarchical porosity, US Patent 10 184 037, assigned to Owens Corning Intellectual Capital LLC, January 22, 2019.
15. C.L. Mantell, *Carbon and Graphite Handbook*, Interscience, New York, 1968.
16. H. Zabel and S.A. Solin, eds., *Graphite Intercalation Compounds I: Structure and Dynamics*, Springer Series in Materials Science, Springer Berlin Heidelberg, 2013.
17. I. Michael A. Pero, System for manufacture of foam sheets rigidized with polymer infiltration, US Patent 9 694 897, assigned to Michael A. Pero, III, July 04, 2017.
18. I. Michael A. Pero, System for manufacture of foam sheet rigidized with polymer infiltration, US Patent 10 053 206, assigned to Michael A. Pero, III, August 21, 2018.
19. S.N. Singh, J. Wu, and A.J. Hamilton, Rigid polyurethane foam and system and method for making the same, US Patent 9 238 722, assigned to Huntsman International LLC, January 19, 2016.
20. S.N. Singh, J. Wu, and A.J. Hamilton, Rigid polyurethane foam and system and method for making the same, US Patent 9 587 080, assigned to Huntsman International LLC, March 07, 2017.
21. S.G. Rider, B.W. Parks, and C.C. Miller, Rigid polyurethane foams suitable for use as panel insulation, US Patent 10 851 196, assigned to Covestro LLC, December 01, 2020.
22. S.U. Pickering, *J. Chem. Soc., Trans.*, Vol. 91, p. 2001, 1907.

23. Wikipedia contributors, Pickering emulsion — Wikipedia, the free encyclopedia, https://en.wikipedia.org/w/index.php?title=Pickering_emulsion&oldid=1021378483, 2021. [Online; accessed 25-August-2021].
24. G. Sun, L. Yang, and R. Liu, *Progress in Organic Coatings*, Vol. 151, p. 106023, 2021.
25. G. Hayase, *Bulletin of the Chemical Society of Japan*, Vol. 94, p. 70, 2021.
26. N. Feng, S. Wu, D. Song, Y. Li, N. Lu, L. Sun, T. Yu, A. Li, and W. Deng, *RSC Advances*, Vol. 11, p. 13957, 2021.
27. A. Rizvi, R.K.M. Chu, and C.B. Park, *ACS Applied Materials & Interfaces*, Vol. 10, p. 38410, 2018.
28. A. Rizvi, A. Tabatabaei, M.R. Barzegari, S.H. Mahmood, and C.B. Park, *Polymer*, Vol. 54, p. 4645, 2013.
29. A. Rizvi, R.K.M. Chu, J.H. Lee, and C.B. Park, *ACS Applied Materials & Interfaces*, Vol. 6, p. 21131, 2014.
30. G. Wang, L. Wang, L.H. Mark, V. Shaayegan, G. Wang, H. Li, G. Zhao, and C.B. Park, *ACS Applied Materials & Interfaces*, Vol. 10, p. 1195, 2018.
31. Y. Zhan, M. Oliviero, J. Wang, A. Sorrentino, G.G. Buonocore, L. Sorrentino, M. Lavorgna, H. Xia, and S. Iannace, *Nanoscale*, Vol. 11, p. 1011, 2019.
32. G.K. Elyashevich, D.I. Gerasimov, I.S. Kuryndin, V.K. Lavrentyev, E.Y. Rosova, and M.E. Vylegzhanina, *Coatings*, Vol. 12, p. 51, January 2022.
33. N. Hossieny, O.A. Owusu, M.A.W. Natal, and D. Venkatraman, Polymer foam insulation structure having a facing of a multi-layer sheet that contains a heat resistant polymer layer and a polylactide resin layer, US Patent 10 807 344, assigned to NatureWorks LLC, October 20, 2020.
34. Standard test method for Vicat softening temperature of plastics, ASTM Standard D1525, ASTM International, West Conshohocken, PA, 2017. https://www.astm.org/Standards/D1525.htm.
35. Y. Tao, Z. Mao, Z. Yang, and J. Zhang, *Energy and Buildings*, Vol. 225, p. 110361, 2020.
36. D. Han, J. Fei, J. Mandal, Z. Liu, M.P. Wan, H. Li, A.P. Raman, and B.F. Ng, *Available at SSRN 3850366*, pp. 1–30, 2021.
37. X. Chen, M. He, S. Feng, Z. Xu, H. Peng, S. Shi, C. Liu, and Y. Zhou, *Optical Materials*, Vol. 120, p. 111431, 2021.
38. S. Rashidi, J.A. Esfahani, and N. Karimi, *Renewable and Sustainable Energy Reviews*, Vol. 91, p. 229, 2018.
39. C. Feng, P. Yang, H. Liu, M. Mao, Y. Liu, T. Xue, J. Fu, T. Cheng, X. Hu, H.J. Fan, and K. Liu, *Nano Energy*, Vol. 85, p. 105971, 2021.
40. R. He, Y. Liao, J. Huang, T. Cheng, X. Zhang, P. Yang, H. Liu, and K. Liu, *Materials Today Energy*, Vol. 21, p. 100800, 2021.

41. A.M. Abraham, S.V. Kumar, and S.M. Alhassan, *Chemical Engineering Journal*, Vol. 332, p. 1, 2018.
42. J.M. Chalker, M.J.H. Worthington, N.A. Lundquist, and L.J. Esdaile, Synthesis and applications of polymers made by inverse vulcanization in X. Jiang, ed., *Sulfur Chemistry*, Topics in Current Chemistry, chapter 4, pp. 125–151. Springer, Cham, 2019.
43. G. Wang, J. Zhao, L.H. Mark, G. Wang, K. Yu, C. Wang, C.B. Park, and G. Zhao, *Chemical Engineering Journal*, Vol. 325, p. 632, 2017.
44. T. Li, G. Zhao, G. Wang, L. Zhang, and J. Hou, *Macromolecular Materials and Engineering*, Vol. 304, p. 1800789, 2019.
45. G. Lista, S. Scussolin, and G. Casagrande, Appliances having sound and thermal insulation based on viscoelastic polyurethane, US Patent 10 537 224, assigned to Dow Global Technologies LLC, January 21, 2020.
46. Standard test methods for flexible cellular materials-slab, bonded, and molded urethane foams, ASTM Standard D3574, ASTM International, West Conshohocken, PA, 2017. https://www.astm.org/Standards/D3574.htm.
47. J.R. Choi, I.Y. Kim, J.-K. Kim, W. Jang, and Y.S. Jung, Polyurethane foam, refrigerator including the same, and method of manufacturing polyurethane foam, US Patent 10 351 659, assigned to Samsung Electronics Co. Ltd., July 16, 2019.
48. L. Ducoup, G.L. Roux, V. Longuet, and J. Pelle, Method for producing a sealed and thermally insulating barrier for a storage tank, US Patent 10 317 012, assigned to Gaztransport et Technigaz SARL, June 11, 2019.
49. J.-S. Kim and Y.-H. Kwon, Thermal insulation foam having low adhesion for high explosives, US Patent 10 494 313, assigned to Agency for Defence Development, December 03, 2019.
50. F. Zou and T. Budtova, *Carbohydrate Polymers*, Vol. 266, p. 118130, 2021.
51. W. Sakuma, S. Yamasaki, S. Fujisawa, T. Kodama, J. Shiomi, K. Kanamori, and T. Saito, *ACS Nano*, Vol. 15, p. 1436, 2021.
52. A.C. Pierre and G.M. Pajonk, *Chemical Reviews*, Vol. 102, p. 4243, 2002.
53. K. Kanamori, *Journal of Materials Research*, Vol. 29, p. 2773, 2014.
54. G. Zu, T. Shimizu, K. Kanamori, Y. Zhu, A. Maeno, H. Kaji, J. Shen, and K. Nakanishi, *ACS Nano*, Vol. 12, p. 521, 2018.
55. Z. Zhao, Y. Cui, Y. Kong, J. Ren, X. Jiang, W. Yan, M. Li, J. Tang, X. Liu, and X. Shen, *ACS Applied Materials & Interfaces*, Vol. 13, p. 21286, 2021.
56. T. Zhuo, B. Xin, Z. Chen, Y. Xu, X. Zhou, and J. Yu, *European Polymer Journal*, Vol. 153, p. 110489, 2021.
57. X. Zhang, X. Ni, C. Li, B. You, and G. Sun, *J. Mater. Chem. A*, Vol. 8, p. 9701, 2020.
58. H. Yang, Z. Wang, Z. Liu, H. Cheng, and C. Li, *Polymers*, Vol. 11, p. 1899, Nov 2019.
59. J. Zhou and Y.-L. Hsieh, *Nano Energy*, Vol. 68, p. 104305, 2020.

6

Membranes

Methods for the preparation of polymeric membranes are collected in Table 6.1

Table 6.1 Polymeric membranes (1).

Polymer type	References
Poly(vinylidene fluoride)	(2)
Polyantimonic acid powder and organic binder	(3)
Hydrophilic poly(amide)	(4)
Maleic anhydride	(5)
Charge-modified poly(amide)	(6)
Interpenetrating polymer network	(7)
Poly(vinylidene fluoride)	(8)
Poly(vinylidene difluoride)	(9)
Poly(vinylidene fluoride)	(10)
Graft polymeric membranes	(11)
Polyolefin membrane	(12)
Polymer electrolyte membrane	(13)
Acid functional fluoropolymers	(14)
Extrudable fluoropolymer blends	(15)
Amphiphilic block copolymer	(16)
Poly(phenylene ether) copolymer	(17)

6.1 Cellulose Acetate

A continuous process for the production of cellulose ester membranes has been described. This process involves the following steps (18):

1. Coating a film of a *dope* of a cellulose ester on a web,
2. Permitting solvent to evaporate into the atmosphere, to cause incipient formation of a skin or *active* layer, and
3. Immersing the film in a hot aqueous bath, to gel the film in the form of an asymmetric membrane.

Furthermore, an asymmetric composite membrane that contains a polymeric matrix and carbon nanotubes within a single membrane layer has been developed (19).

Here, the carbon nanotubes are randomly oriented within the polymeric matrix and the composite membrane is formed by phase inversion.

Also, a method for producing such a composite membrane has been described. The method includes (19):

1. Coating a surface with a film of a polymer solution containing a polymeric matrix and carbon nanotubes dissolved in at least one solvent,
2. Immersing the coated surface in a non-solvent to affect solvent/non-solvent demixing resulting in phase inversion to form a carbon nanotube-containing membrane, and optionally,
3. Removing the carbon nanotube-containing membrane from the surface.

Polymers that are useful for the preparation of such membranes are collected in Table 6.2.

Among the materials that are employed to produce semi-permeable membranes, cellulose acetate (CA) is a relatively cheap, mechanically tough, chlorine resistant, hydrophilic abundant polymer that results from the acetylation of cellulose.

Cellulose acetate membranes were first developed in the 1960s to become the first high-flux asymmetric (anisotropic) membranes used in a large number of applications, including reverse osmosis, micro, ultra and nanofiltration, gases separation, water desalination, and wastewater treatment (19).

The water permeation rates and the salt rejection rates of CA membranes are affected by the degree of acetylation of the cellulose.

Table 6.2 Polymers for membranes (19).

Polymer type	Polymer type
Aromatic poly(amide)s	Aliphatic poly(amide)s
Cellulose acetates	Cellulose nitrate
Ethyl cellulose	Polyesters
Polycarbonates	Copolycarbonate esters
Polyethers	Polyetherketones
Polyetherimides	Polyethersulfones
Polyetheresters	Polysulfones
Polybenzimidazoles	Polybenzoxazoles
Polyacrylonitrile	Polyazoaroaromatics
Polyphenylene oxides	Polyureas
Polyurethanes	Polyhydrazides
Polyazomethines	Polyacetals
Polyquinoxaline	Poly(amide) imides
Poly(amide) esters	Polysiloxanes
Polyacetylenes	Polyphosphazenes
Polyolefines	Polyphenylenes
Polyimides	Polyacrylonitrile
Polyphthalamides	Polysulfonamides
Phenylene diamines	Poly(amide) hydrazide
Polyvinylidene fluoride	Poly(2,6-dimethylphenylene oxide)
Styrene-acrylonitrile copolymers	Brominated poly(xylene oxide)
Sulfonated poly(xylylene oxide)	

In general, a cellulose structure has three OH groups that can be partially or fully substituted. A completely acetylated cellulose, which is acetylated by three acetate groups (equivalent to 44.2% of acetyl groups) is known as cellulose triacetate. This material was found to have high salt rejection rates of 99.5% from seawater feed.

However, here the water flux is the lowest possible. On the other hand, lower acetylation percentage gave higher permeation rates but with lower salt rejection. The commercial cellulose acetate membranes used in RO desalination, for example, have 40% acetate, equivalent to acetyl content of 2.7 on the scale of 0 (the OH groups are not substituted) to 3 (the OH groups are fully substituted).

These membranes yield 98% to 99% salt rejection rates with reasonable water fluxes (19).

Regarding membrane fouling, CA membranes are relatively protected from biofoulants due to their hydrophilic smooth surface, which doesn't leave inter-membrane gaps or interfacial water-resistant areas for foulants to grow. Being chlorine resistant (up to 1 ppm of continuous exposure to free chlorine) helps in using them in high bacterial count water sources, pretreated with chlorine. CA membrane surfaces are also neutrally charged, thus reducing the charged macromolecules fouling.

The most common preparation method of CA membranes is known as phase inversion or the wet phase separation process. In this method, CA is first dissolved in a solvent. Then, on a substrate, usually but not always glass, the solution is homogeneously spread using, for example, a casting knife to maintain a uniform casting thickness. The casted film is then immersed into a coagulation bath filled with a CA non-solvent solution (usually water) that is highly miscible with the solvent used. At this point, liquid-liquid demixing takes places where the solvent's affinity towards the non-solvent is higher than towards CA.

Thus, the solvent is extracted into the medium. Meanwhile, water starts defusing to replace the solvent, while repelling CA into the casted solution borders until a point at which the concentration of CA becomes high enough for gelation to take place. The solidified CA is in the form of an asymmetric thin membrane with a polymer-rich dense layer and a polymer-lean porous sub-layer (19).

6.2 Poly(vinylidene fluoride)

A method for forming a microporous membrane from a polymer with a minimal content of 90% poly(vinylidene fluoride) (PVDF) has been described (2). The membrane has at least 60% of its volume taken up by intercommunicating capillary pores of an average size of of below about 14 *μm*.

Such synthesis can be performed as a a continuous process with the following steps (2):

1. Applying a layer of a solution consisting essentially of the polymer in acetone at a polymer concentration of from about 15% to about 25% by weight of the solution as a thin layer to a rigidly supported surface of a traveling backing belt,
2. Immersing the belt carrying the layer of polymer solution in a formation bath consisting essentially of a mixture of acetone and water, with the acetone forming from about 55% to about 85% by volume of the bath, the balance being water,
3. Maintaining the immersed layer and passing it through an elongate path in the formation bath between the point of entry of the belt and layer into the formation bath and the point of departure of the belt and layer from the bath, to permit replacement of the acetone in the layer with the formation bath liquid, until the layer has been converted to a microporous membrane having an average pore size in the range from about 25 *nm* to about 14 *μm*, wherein the pores intercommunicate, form at least 60% by volume of the membrane, and are of substantially uniform size, and the porous structure is essentially fully formed,
4. Continuously withdrawing formation bath liquid from the bath adjacent the entry point of the belt into the bath, continuously monitoring the composition of the withdrawn formation bath liquid, and adding the necessary amount of water to maintain the liquid at a composition within the specified range of composition for the bath, so that it is recirculable liquid,
5. Returning the recirculable liquid to the formation bath adjacent the departure point of the belt from the bath,

6. Separating the formed membrane from the belt,
7. Washing the membrane with water, apart from the formation bath, to extract any residual acetone or formation bath liquid from the membrane, and then
8. Drying the membrane to remove water.

Here, an important advantage is the use of acetone as solvent for the PVDF (2). Acetone is usually not thought of as a solvent for this particular polymer, because it is exceedingly difficult to dissolve any appreciable quantity of PVDF in acetone at room temperature. In order to dissolve a sufficient quantity to form an adequately viscous solution for use in practicing a continuous process, the acetone must be heated close to its boiling point, that is, at least to about 50°C.

However, acetone is completely compatible with water and a variety of like materials that are not solvents for the polymer but that are useful, just as water is, as an important constituent of the formation bath (2).

When acetone is used continuously not only as the solvent for the polymer, but also as an important, often major constituent of the formation bath, the engineering considerations relating to the process are very much simplified.

The recovery of the solvent for reuse is relatively easy. To maintain the composition of the formation bath reasonably constant, make up water is added as needed.

Since PVDF is hydrophobic, the removal of macromolecular contaminants in ultrapure water, such as bacterial lipopolysaccharides and endotoxins which contain hydrophobic groups, is possible by hydrophobic absorption. However, such a membrane may be converted to be hydrophilic by immersing the dried membrane in a solution of a water-insoluble surfactant, and then drying. Pluronic polyols grades L101, L121 and L122, were used as surfactants (2). These are difunctional block copolymers with hydroxyl end groups.

The properties of these materials are shown in Table 6.3.

6.2.1 *Grafted Phosphonium Poly(vinylidene fluoride)*

Poly(vinylidene fluoride) membranes have been developed that are positively charged with grafted polymer for the removal of undesirable contaminants from liquids during filtering (8).

Table 6.3 Properties of pluronic polyols.

Pluronic Grade	Mol weight of hydrophobic part/[Da]	Hydrophilic parts/[%]	Average mol weight/[Da]	Surface tension/ [$dyn\, cm^{-1}$]
L101	3250	10	3800	40.7
L121	4000	10	4400	35.0
L122	4000	20	5000	40.3

Here, positively charged organic phosphonium compounds are covalently bonded to the PVDF membrane. This is preferably triphenyl(vinyl)phosphonium bromide. The material is grafted to the membrane by γ-radiation to form covalent bonds. Triphenyl(vinyl)phosphonium bromide is shown in Figure 6.1.

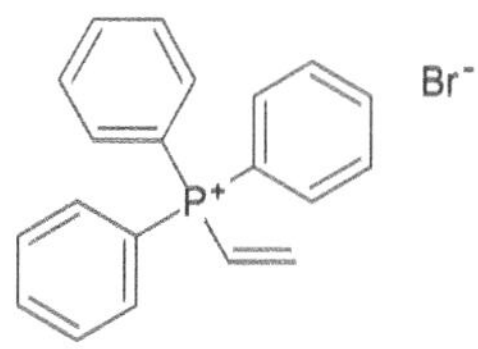

Figure 6.1 Triphenyl(vinyl)phosphonium bromide.

The synthesis of such a compound may be done as follows (8):

Preparation 6–1: An aqueous solution containing 2% of the triphenyl(vinyl)phosphonium bromide monomer and five strips of PVDF film (0.15 $mm \times 60\ mm \times 50\ mm$) having a mean pore size of 140 nm was degassed in a glass tube and then sealed. The tube was then irradiated in a ^{60}Co γ-ray facility. The total amount of irradiation was 0.61 $Mrad$ at a dose rate of 400 $Mrad\, min^{-1}$, as measured by a Fricke dosimeter. After irradiation, the grafted samples of medium were washed in a deionized water trough to remove unreacted monomer, and dried at 100°C for 10 min. The grafting of triphenyl(vinyl)phosphonium bromide onto PVDF was confirmed by infrared analysis.

The ζ-potential was calculated from the following equation:

$$\zeta = 4\pi\varepsilon\frac{\kappa}{\eta} \times \frac{\Delta E_s}{\Delta P}. \tag{6.1}$$

Here, ε, is the viscosity, ϵ is the dielectric constant, κ is conductivity of the solution, E_S is the streaming potential, and ΔP is the pressure drop.

The charge-modified membrane was characterized by a positive ζ potential in the pH ranges from 4 to 9.3.

The ζ-potential values of unmodified and charge-modified PVDF membranes are shown as a function of solution pH in Table 6.4.

Table 6.4 ζ-Potential vs. pH (8).

Sample	Radiation/ [*Mrad*]	pH 4	6	8	9
Unmodified PVDF	0.00	-4	-10	-15	-20
PVDF + 2% VPB	0.61	30	29	29	25
PVDF + 2% VPB	1.22	30	30	29	23

VBP: Triphenyl(vinyl)phosphonium bromide

6.2.2 *Hollow Fiber Poly(vinylidene fluoride)*

A process for producing a porous hollow fiber PVDF resin membrane has been developed (10).

The process for producing such a porous hollow fiber PVDF resin membrane, consists of blending a PVDF resin with an organic liquid and an inorganic particulate material, heat melting the resulting blend at a temperature of 60°C or more above the melting point of the resin by using a melt molded extruder equipped with an annular hollow fiber spinning die nozzle, extruding the molten blend in the form of a hollow fiber, passing the resultant hollow fiber-shaped extrudate through the air and taking up the so formed extrudate.

The here used organic liquid is required to be a liquid which is inactive in the melt molding process. Examples of the organic liquids having a solubility parameter of 8.4–10.5, include phthalic esters such as diethyl phthalate, dibutyl phthalate, dioctyl phthalate, and bis(2-ethylhexyl) phthalate, c.f. Figure 6.2. Bis(2-ethylhexyl) phthalate, dibutyl phthalate, c.f. Figure 4.19, and mixtures thereof are particularly preferred.

Among inorganic particulate materials, hydrophobic silica is preferred. The membrane has a unitary, three-dimensional, network

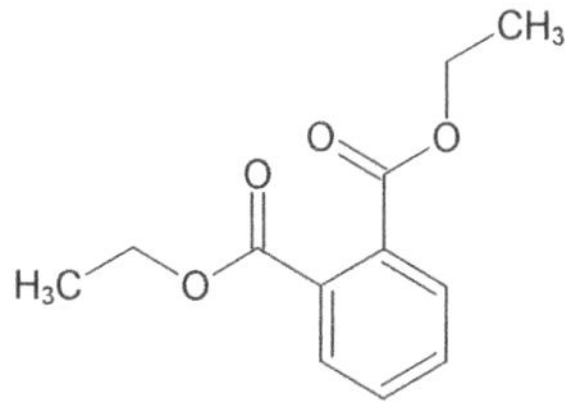

Diethyl phthalate

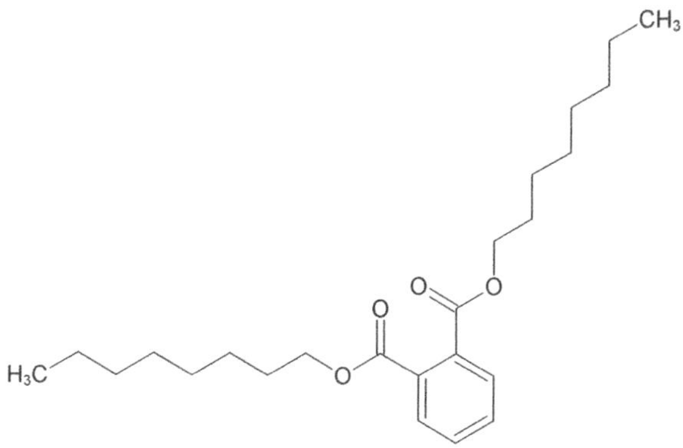

Dioctyl phthalate

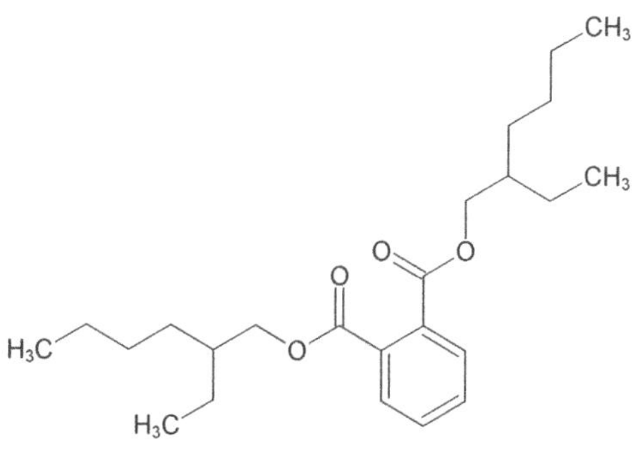

Bis(2-ethylhexyl) phthalate

Figure 6.2 Phthalic esters.

pore structure composed entirely of a PVDF resin and has an inside diameter of 1.5 PVDF to 5 *mm* and a wall thickness of 0.5 *mm* to 2 *mm*. The membrane is excellent in chemical resistance, mechanical properties and water permeability and is therefore suitable for the treatment of highly viscous liquids (10).

In membranes with a homogeneously porous structure at a given pore diameter, the wall thickness of the membrane and the water flux are inversely proportional. The relationship between membrane bubble-point and the water flux per unit membrane wall thickness is on or above the curve shown in Figure 6.3.

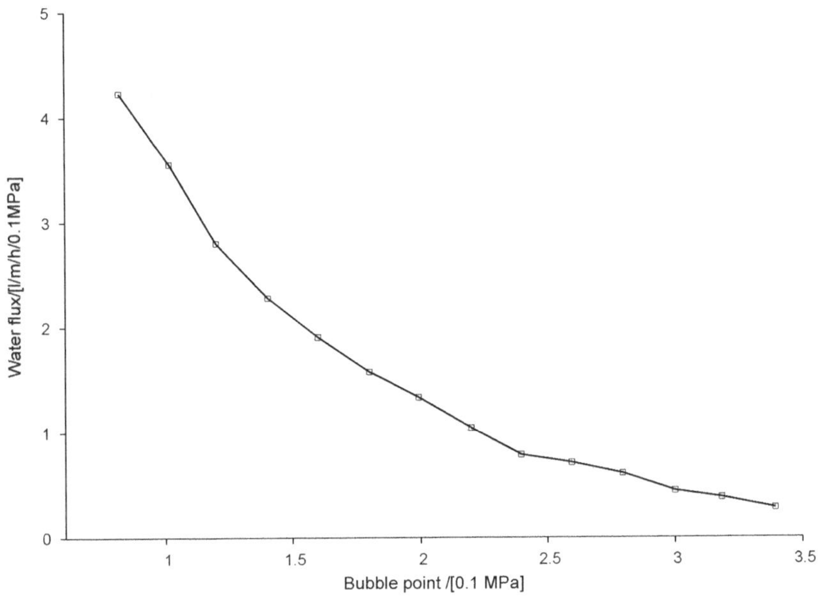

Figure 6.3 Membrane bubble-point vs. water flux (10).

6.2.3 *Casting Methods*

Synthetic polymeric membrane materials were formed from casting PVDF polymer solutions and/or dispersions (9). The so formed membranes were highly porous. Both internally isotropic membranes and highly asymmetric PVDF membranes could be fabricated.

In addition, the membranes may be hydrophobic or hydrophilic.

These membranes are useful in a variety of microfiltration and ultrathin applications (9).

6.3 Poly(amino acid)s

The functionalization of certain materials with poly(amino acid)s provides opportunities where electrostatic and conformational properties can be utilized for selective separations and controlled transport applications (20).

The influence of covalently attached poly(*l*-glutamic acid) on the performance characteristics of a microporous cellulosic support was investigated to determine its effect on the transport of water and charged solutes. The presence of these charged multifunctional groups promotes electrostatic interactions with ionic species far removed from the pore surface.

This allows for nonsteric ion exclusion involving membrane materials that offer less resistance to solvent (water) transport. In addition, the helix-coil transitions of this polypeptide and *core leakage* effects have been shown to affect both the permeability and solute retention in a reversible fashion upon variations in the solution pH.

After the attachment of poly(*l*-glutamic acid), the permeability could be adjusted from 50% to 90% of the value of the base support matrix ($3\text{-}6 \times 10^{-4}$ $cm^3 cm^{-2} s^{-1} bar^{-1}$). Experiments were performed on single and mixed electrolyte systems using model inorganic and organic solutes. Solute rejection values as high as 73% were reported for dilute Na_2SO_4 solutions using membranes with pore sizes ranging from 0.05 to 0.21 μm (20).

6.4 Hyper-crosslinked Polymers

Hyper-crosslinked polymers are an attractive class of porous materials that are prepared mainly by a Friedel-Crafts alkylation reaction (21–23).

The permanent porosity in hyper-crosslinked polymers is a result of extensive crosslinking reactions, which prevent the polymer chains from collapsing into a dense, nonporous state (24–26).

An *in-situ* crosslinking strategy can be used for the preparation of polymeric molecular sieve membranes. These membranes have a sandwich structure with hierarchical and tailorable porosity (27).

The structure consists of a macroporous core, a layer of dense micropores and a mesoporous surface composed of small polymer nanoparticles. The process uses nonporous poly(styrene) (PS) membranes as a precursor membrane template. Then, using a Friedel-Crafts reaction, PS membranes can be crosslinked *in-situ* to produce hyper-crosslinked porous polymeric membranes. In the crosslinking process, formaldehyde dimethylacetal is used as the crosslinking agent and $FeCl_3$ is used as catalyst.

This synthesis method offers both a tunable porosity and the formation of well-defined pore networks. These properties make these membranes amenable to quantitative modeling of their mass transport properties. Thus, the so obtained porous polymeric membranes exhibit properties that are promising for use in gas separations with high permeability and favorable selectivity. Also, they show great potential for applications in liquid separations and membrane catalysis (27).

6.5 Membrane for Specific Molecular Separation

The separation of shikimic acid is important for antiviral drug production. Current separation methods used for shikimic acid remain unsatisfied due to lack of specificity that can accurately and instantly anchor the target molecule (shikimic acid) from interfering impurities (28).

Membrane separation seems to be feasible with the advantages of low energy consumption and operating convenience.

Benefiting from the characteristic cis-diol group in shikimic acid, a smart and simple PVDF-based molecular imprinted membrane is synthesized by using 4-vinylbenzeneboronic acid as functional monomer to form prepolymers with shikimic acid. These compounds are shown in Figure 6.4.

The resulting shikimic acid-imprinted boronate affinity molecular imprinted membranes show an enhanced adsorption capacity (51.99 $mg\,g^{-1}$ compared with a common material). The formation of the boronate ester bond between functional monomer and shikimic

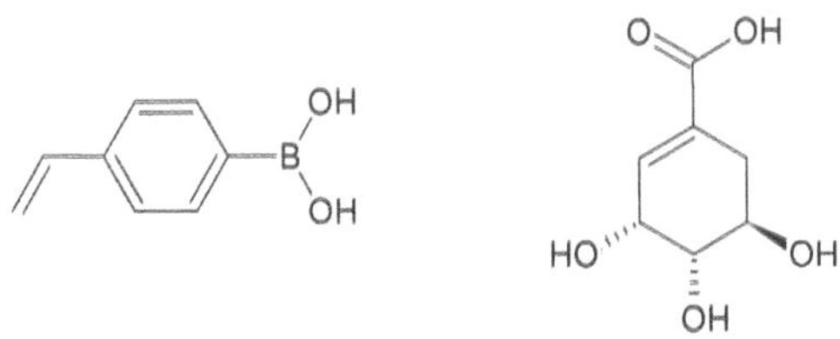

4-Vinylbenzeneboronic acid Shikimic acid

Figure 6.4 Componds for prepolymers .

acid can help to specifically capture the target, and the release of combined shikimic acid on the membrane becomes easy due to the reversible bind characteristic of boronate ester under different pH conditions.

Moreover, shikimic acid-imprinted boronate ester molecular imprinted membranes exhibit acceptable stability after ultrasonic treatment for 5 *h* and satisfactory regeneration performance after 8 cycles of adsorption-desorption experiments (93.02%). These convenient molecular imprinted membranes display great potential for rapid and specific separation of shikimic acid for the commercial synthesis of an antiviral (28).

6.6 Treatment of Water

6.6.1 *Ammonia Removal*

A modeling approach was developed for prediction of ammonia removal from water by means of porous membranes. The model was based on an adaptive neuro-fuzzy inference system (ANFIS) to simulate the stripping of ammonia from water by means of hollow-fiber membrane contactors (29). As controllable factors, the feed velocity, initial concentration of ammonia, pH, V_{acid} and salt concentration were used.

The predictions aimed to obtain optimum conditions for ammonia stripping using the Taguchi method. The initial concentrations of ammonia, pH of the ammonia solution, velocity of the feed, and the presence of excess ions in the ammonia feed solution were considered as the input properties.

On the other hand, the mass transfer coefficient was considered as output. The prediction results showed that the pH of the ammonia feed has a significant effect on the separation of ammonia from water. The results also showed that the prediction of the ANFIS model and experimental data match well. The maximum interaction was found between the velocity of ammonia feed and V_{acid}.

So, this model can be used for prediction of porous membranes. Furthermore, simulated annealing was also used to determine controllable conditions to find the highest mass transfer coefficient (29).

6.6.2 *Fine Pore Aeration*

The supply of oxygen for aeration is the single largest energy consumer at activated sludge wastewater treatment plants, representing 50–90% of total plant energy requirements (30, 31).

The replacement of less-efficient aeration systems with fine pore aeration devices can save up to 50% of aeration energy costs and has resulted in typical simple payback periods of 2–6 *y* (32).

As a result of these very impressive cost savings, a very large number somewhere between 1000–2000 municipal and industrial wastewater treatment facilities in the United States and Canada are using a fine pore aeration method (33).

6.6.3 *Water Contamination Treatment*

Water contamination is exacerbated by vital industries such as lead smelting, electroplating, petroleum and electronics, which discharge contaminants into the environment (34). To deal with this problem, various technologies have been developed to remove contaminants from waste streams, such as coagulation, ion exchange, chemical precipitation, membrane processes, and adsorption.

However, most of these methods suffer from significant drawbacks like high capital and operational costs, inappropriate efficiencies at usual discharge levels, and the production of residual toxic sludge and secondary wastes that are difficult and expensive to treat.

Several drinking water treatment technologies are available. However, they are not widely distributed and suitable for all types of water contamination because they have different treatment efficiencies and costs, and in some cases these technologies require trained personnel.

Biosorption has received considerable attention because biological materials are naturally available, cheap and harmless. However, the biological materials utilized must be processed in unique ways in order to maintain or maximize adsorption capacity while also promoting strength and durability.

So, the production and the application of porous nanocomposite polymers for the removal of chemical and biological contaminants from water has been described (34). These porous nanocomposite polymers contain nanomaterials, i.e., carbon or metal oxide nanoparticles, that have antimicrobial properties and adsorption capacities for heavy metals and other cations.

Metal oxide particles in these beads may have a photocatalytic capacity to enhance the photochemical degradation of organic and biological contaminants in the water.

Besides nanoparticles, these porous nanocomposites are made of natural biopolymers that are easily available, have low cost of fabrication and are biocompatible. The unique blend of polymers in these nanocomposites makes possible for the removal of both organic and inorganic chemicals, such as urea, humic acids, phosphates, nitrates, heavy metals and radioactive materials. The process of fabrication is facile and cost effective.

The nanocomposites can be used in fluidized bed reactors or can be packed in columns with different sizes or as coatings for membrane filtration, which will allow the treatment of different volumes of water.

Graphene oxide could be successfully incorporated into a chitosan-poly(acrylic acid) polymer matrix (34). These nanocomposite hydrogel beads have the ability to remove high levels of heavy metal from an aqueous solution. Such beads can be used in packed bed columns in different heights and flow rates to maximize the contaminant removal in diverse volumes of water.

In addition, graphene oxide could be successfully incorporated into a chitosan-poly(ethylene imine) polymer matrix or into a alginate-poly(ethylene imine) polymer matrix.

These nanocomposites were synthesized as porous hydrogel beads, colloidal suspensions, and solid hydrogels or as a coating material for membranes and filters. These nanocomposites can also be freeze-dried to form porous sponges.

The nanocomposites have the ability to remove high levels of heavy metals, total organic carbon, nitrates and phosphates. They can also be used in packed bed columns or in fluidized beds to maximize the contaminant removal in different volumes of water.

Chitosan is a biosorbent that can be used for removing heavy metals, since it contains amine and hydroxyl groups that can form complexes with various heavy metal ions (34). Chitosan is also an ideal material since it is derived from a naturally occurring and abundant biopolymer, chitin, which is freely available in large quantities from seafood processing waste. In order to use chitosan as an adsorbent, chitosan needs to be crosslinked to improve its acidic resistance and enhance mechanical strength. The crosslinking process, however, decreases the number of amine groups, which in return, reduces its adsorption capacity.

To overcome this issue, various polymers have been combined with chitosan via surface modification or interpenetrating network in order to impart additional functional groups and enhance chitosan sorption capacity. In particular, poly(acrylic acid) has been incorporated into chitosan because it contains many carboxyl groups and has an anionic polyelectrolyte form, which allows higher heavy metal removal (34).

Nanocomposites were prepared in different forms. It was possible to prepare the nanocomposite as a hydrogel block, hydrogel micro- and nanobeads, as a sponge or as a coating material for membrane filtration.

The hydrogel nanocomposites were made from 0.2% to 4% alginate or chitosan as supporting materials for the copolymers and nanomaterials to form solid structures, such as beads, hydrogels, and sponges. The concentrations of the nanomaterials, i.e., graphene oxide were in the range of 0 *ppm* to 5000 *ppm*. The copolymers had a concentration of 5% to 30% (w/v %). After preparing this mixture, glutaraldehyde was added to crosslink the components of the nanocomposite mixture.

For the filter preparation, the nanocomposite mixture with glutaraldehyde was used for the coating. For the filter, coagulation with NaOH or HCl was not necessary. To form the hydrogel blocks or beads, the coagulation of the nanocomposites was done with NaOH or HCl. For the preparation of the sponges the hydrogel blocks were freeze-dried between 6 *h* to 24 *h* depending on the size of the hydrogel blocks.

Therefore, this system, depending on the size of the column, will serve to treat water on a small to large scale. The uniqueness of this system is that the nanocomposite polymers produced are multifunctional and can be easily expanded and modified to remove different pollutants in water (34).

6.6.3.1 *Dye Removal from Wastewater*

The wastewater produced from various industries, such as textile, paper, rubber, leather, plastics, cosmetics, pharmaceuticals, contains many dyes (35). The most commonly found dyes in these effluents are Orange G and Methylene blue. These compounds are shown in Figure 6.5.

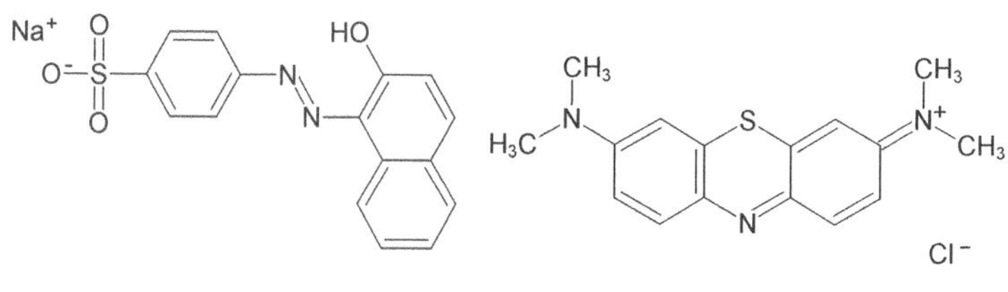

Orange G Methylene blue

Figure 6.5 Dyes in wastewater.

These dyes are toxic and are harmful to the environment as well as humans. Hence, they need to be removed from the wastewater. Poly(ethersulfone) (PES) membranes are most widely used in ultrafiltration separation processes. The enhancement in the permeability of the membrane and improving the rejection capabilities of the membrane is still a key challenge.

Here, the effect of barium chloride blending on the separation performance of PES membranes was investigated (35). Systematic experimental studies were used to fabricate PES-based membranes and to investigate their efficiency in removing Orange G and Methylene blue from synthetic wastewater.

Five different membranes, one of which is a pristine PES membrane and four others made of PES blended with different amounts of barium chloride, were synthesized by a phase inversion method. The characterization was carried out for these fabricated membranes for various properties like membrane morphology, surface hydrophilicity, porosity, and pure water flux.

From the experimental results it became evident that PES blended with $BaCl_2$ membranes showed a better dye separation efficiency in comparison to a pristine PES membrane. Also, out of the four blended membranes, the one with 2% $BaCl_2$ showed the highest pure water flux, permeate flux, dye rejection, lowest contact angle and the best porosity. The above experimental observations indicate that the UF PES membranes blended with 2% $BaCl_2$ can be a viable alternative membrane for the separation of dyes from industrial water (35).

6.6.3.2 *High-Salinity Wastewater*

High-performance thin film composite (TFC) forward osmosis (FO) membranes with a low degree of internal concentration polarization (ICP) are critical for concentrating high-salinity wastewaters (36).

The preparation of TFC FO membranes via a sacrificial approach was detailed. In order to improve the FO flux, hydrophilicity and morphology of the support membrane were mainly investigated. The hydrophilicity of the poly(sulfone) (PSf) substrate was tuned by blending with sulfonated poly(ether ether ketone) (SPEEK), and the resulting SPEEK blended PSf membrane was denoted as sulfonated poly(sulfone) (SPSf) substrate.

The pore structure of the SPSf membrane was tailored by the application of a co-casting technique, which yielded a TFC membrane with a structure parameter of 191 μm. In contrast, the TFC membranes based on the PSf and SPSf substrates through single layer casting showed S values of 527 μm and 361 μm, respectively. These results indicate that the combined hydrophilicity and open pore structure are responsible for the lowered S value (36).

Further application of the hydrophilic substrate based on TFC membranes in the treatment of high-salinity wastewaters (10%)

demonstrated the higher initial water flux (28.3 $l\,m^{-2}h^{-1}$ with a water recovery rate of 53.2%) in comparison to the TFC membrane based on the pristine PSf through the single layer casting. This new method paves a way to generate high-performing FO membranes (36).

6.6.3.3 *Bioreactor for Wastewater Treatment*

Usually, wastewater treatment plants include a biological process step in which a part of the wastewater containing solid matter, suspended and soluble organics and nutrients, is treated by an activated sludge that consists of mainly microorganisms (37). This process can take place either anaerobically or aerobically.

The most widely applied process for the aerobic treatment of wastewater involves air or oxygen that is introduced into a biological treatment reactor which contains a mixture of screened and sometimes primary treated sewage or industrial wastewater and purifying biomass. The mixed liquor suspended solids develop into a biomass-containing floc, which typically grows in suspended fluffy aggregates.

The subsequent settling tank, which is usually referred to as *final clarifier*, is used to allow the biological flocs to settle, thus separating the purifying sludge from the treated water. The settled sludge is recycled towards the biological process as return activated sludge (37).

Filtration Apparatus. Biological processes to treat contaminated water take many forms of biological oxygen demand (BOD). Generally, these processes involve the exposure of the waste stream to one or more forms of microorganisms that stabilize or digest the contaminants. The microorganisms are chosen to complement the waste stream both in terms of sewage contents and chemical environment, since any species of microorganism favors a particular environment with limited tolerance for variation. For example, the activated sludge process utilizes aerobic bacteria that remove the soluble BOD from wastewater. This involves conducting wastewater into an aeration basin containing a suspension of digestive microorganisms, thereby forming a *mixed liquor* that is aerated to furnish oxygen for respiration of biomass. The biomass sorbs, assimilates and metabolizes the BOD of the wastewater. After a suitable period of aeration, the mixed liquor is introduced into a clarifier in which the biomass settles, allowing the treated wastewater to overflow into an outlet effluent stream.

A process has been described for the treatment of wastewater that provides significant benefits in the digestion of sludge by reducing the costs and resources required for disposing of digested sludge. Here, a filtration system providing these benefits was provided that includes a membrane bioreactor combined with a biological sludge offline treatment vessel. The offline treatment vessel solubilizes the organic material by operating in a very low oxygen environment, and then the processed fluid is transferred back to the membrane bioreactor.

The apparatus employs a membrane bioreactor, two screens, a hydrocyclone, and an offline treatment vessel, The configuration of the filtration apparatus is shown in Figure 6.6.

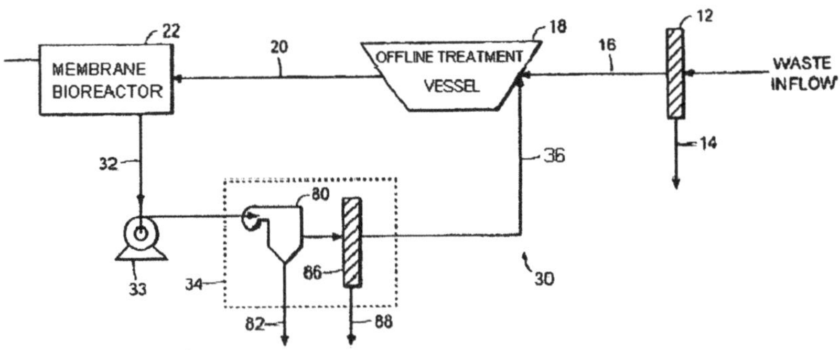

Figure 6.6 Filtration apparatus with a membrane bioreactor (1).

In this system, the waste inflow first encounters an optional screen 12 having an opening size of from about 6 *mm* or less to about 25 *mm* or more. The large items of trash accumulating on the receiving face of the screen 12 are periodically removed, as indicated by the arrow 14. The screened liquid is conducted along a conduit 16 to the offline treatment vessel 18, where waste-digestive organisms are allowed to digest its biodegradable components. Preferably, the waste liquid contains at least two different forms of waste-digestive microorganism, each requiring a different chemical environment for survival or at least optimum performance. The different forms of microorganism can also be complementary in the sense that each degrades a different type of waste. If the waste liquid lacks a desirable form of digestive organism, this can be introduced directly into offline treatment vessel 18.

After a start-up period that depends on the concentration of digestible waste in the influent stream, the mixed liquor is continuously conducted from the offline treatment vessel 18 over a conduit 20 to a membrane bioreactor 22, where filtration takes place, as new influent reaches offline treatment vessel 18 via conduit 16. Biological solids (so-called *activated sludge*) are continuously withdrawn from an outlet point near the bottom of the membrane bioreactor 22 and recycled to the offline treatment vessel 18 via an internal recycle loop 30, while permeate is continuously or periodically withdrawn, and mixed liquor is periodically withdrawn or allowed to overflow from an outlet point near the top of the membrane bioreactor 22, thereby maintaining a suitable concentration of biosolids within the system.

The recycle loop 30 consists of a first conduit 32 leading from membrane bioreactor 22 to a pump 33, which conveys activated sludge to a removal subsystem 34 and a second conduit 36 leading back to the head of offline treatment vessel 18. The removal subsystem 34, consisting of a screen, optional surge tank, and hydrocyclone, is designed to remove inert and nondegradable materials from the biosolids return stream, thereby improving the efficiency of waste treatment. This general system configuration is suitable for use with an offline treatment vessel 18 configured for multi--zone waste treatment, as described immediately below; however, because of the general utility of removal subsystem 34, it can also be employed with entirely conventional (i.e., single-zone) treatment processes. Similarly, the utility of multi-zone treatment is not limited to recycle configurations that include a removal system in accordance herewith (1).

A general layout for a nitrification/denitrification membrane bioreactor is shown in Figure 6.7.

Heterotrophs and Nitrifiers. An alternative to the treatment of wastewaters in activated sludge systems is the usage of a high-capacity biological system, such as hybrid bioreactors.

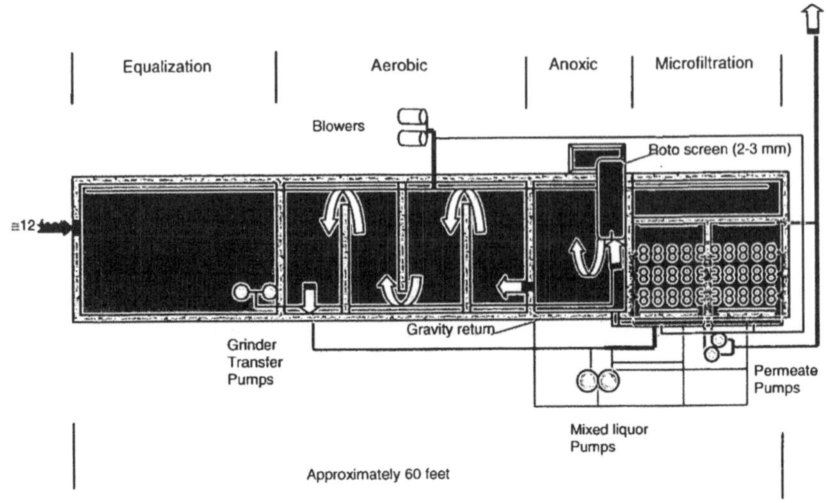

Figure 6.7 Nitrification/denitrification membrane bioreactor (1).

A membrane-assisted hybrid bioreactor has been developed for the treatment of wastewater.

A schematic view of the bioreactor is shown in Figure 6.8.

This system was composed by a hybrid circulating bed reactor coupled in series to an ultrafiltration membrane module (38).

The ultrafiltration membrane module (Zenon, ZW-1) was in a vessel of 1.2 l, which separated the permeate, i.e., the effluent from retentate (sludge), which was recycled to the reactor. The membrane module has an average pore size of 0.045 μm, a nominal surface area of 0.093 m^2 and typical operating transmembrane pressure of 10–50 kPa.

Biomass was maintained growing both in suspension and in biofilms onto small rough plastic particles. The growth of nitrifiers in the hybrid system was promoted into biofilms while heterotrophs were in suspension. This made it feasible to operate the unit at higher solid retention time for nitrifiers than for heterotrophs. This resulted in the presence of a higher fraction of nitrifiers in the reactor.

For testing, the hybrid bioreactor was used to treat two industrial wastewaters from a tannery and a fish canning factory with a high nitrogen and organic matter content (38).

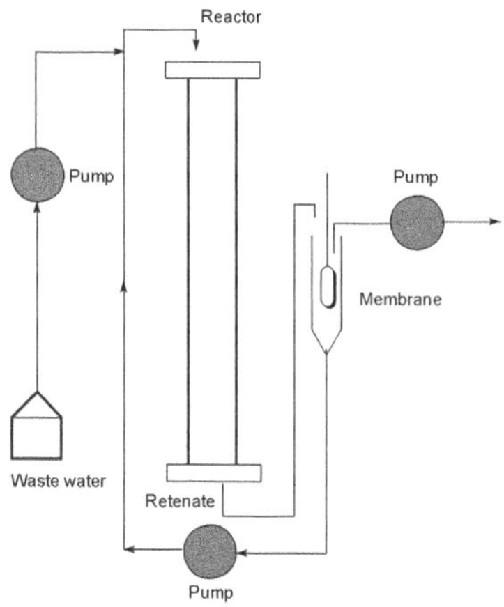

Figure 6.8 Hybrid bioreactor (38).

The removal of organic substances of around 99% was obtained at an organic loading rate of 6.5 $kg\,m^{-3}d^{-1}$ of organic substances and a nitrogen loading rate (NLR) of 1.8 $kg\,m^{-3}d^{-1}$ during the treatment of the fish canning wastewater.

The efficiency for the removal of organic substances was 95%; while up to 97% of ammonia removal was obtained, the concentration of ammonia in the effluent was low, 10 $mg\,l^{-1}$.

For both industrial wastewaters, the biofilm nitrifying activity was not affected by the variations in the organic loading rates, concentration of organic substances in the influent or characteristics of the influent (38).

In another study, the hybrid biofilm-suspended biomass membrane bioreactor was used to treat two wastewater streams that were generated in a fish canning factory (39).

During a first stage, the wastewater generated during tuna cooking with brine was treated (39). The chemical oxygen demand (COD) was between 7.8–11.8 $g\,l^{-1}$ and the nitrogen content was 1.2–1.8 $g\,l^{-1}$. The salt concentration was up to 84 $g\,l^{-1}$. The efficiency of COD removal was affected by the salinity, but after 73 operating days an adaptation of the sludge to the hypersaline conditions took place. A COD efficiency of 92% was obtained after adaptation of the sludge to the salinity.

During a second stage, wastewaters generated during tuna cooking by steam injection were treated. HRT was fixed at 5 d. Organic loading rate was up to 4 $kg\,m^{-3}d^{-1}$. COD concentration in the permeate was lower than 100 $mg\,l^{-1}$ throughout this period. Salt concentration below 15 $g\,l^{-1}$ did not affect the nitrification.

The NLR was gradually increased up to 0.7 $kg\,m^{-3}d^{-1}$. Total nitrogen concentration in the permeate was lower than 100 $mg\,l^{-1}$, and nitrate was below 65 $mg\,l^{-1}$ during the whole experimental period. Moreover, very low nitrate concentrations of less than 1 $mg\,l^{-1}$ were observed during the period in which a NLR up to 0.55 $kg\,m^{-3}d^{-1}$ was applied (39).

Endocrine-disrupting compounds. In recent years, the occurrence of toxic organic compounds, called endocrine-disrupting compounds, in wastewater and their fate during wastewater treatment in a plant has become an important environmental and health issue. Endocrine-disrupting compounds are exogenous substances or mixtures that alter the normal function of the endocrine system and, consequently, cause adverse effects on populations of living organisms or n their progeny.

A process and a bioreactor has been described for reducing the bisphenol compound content of wastewater using a membrane bioreactor (40, 41). The bioreactor contains a biomass in a volatile suspended solid form, where the biomass in the membrane bioreactor is operating continuously to reduce the content of bisphenol when the bisphenol compound volumetric load is below 108 $g\,m^{-3}d^{-1}$.

A schematic diagram of such a submerged membrane bioreactor system is shown in Figure 6.9. The pilot unit is comprised of ZeeWeed® (ZW-1) hollow-fiber membranes of 0.04 μm-rated pore diameter and 0.047 m^2 total filtering surface area, assembled vertically. The membrane unit is provided by Zenon Environmental Inc. (Oakville, ON, Canada).

The ZW-1 membrane, whose module contains 80 fibers of 20 cm length, may be either operated in closed loop or in continuous mode of operation, through a stirred 6 l tank containing 19.1 cm water. Filtration may be achieved by drawing water through the membrane under a slight vacuum generated by a peristaltic pump (11) and measured by a manometer (10) located just after the membrane on the aspiration line. Air may be supplied through an extended

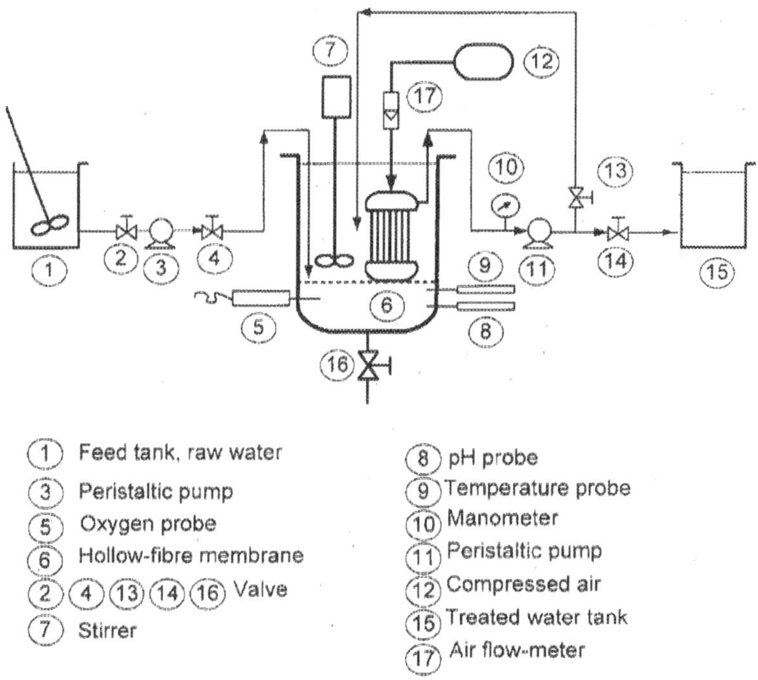

Figure 6.9 Membrane bioreactor system (41).

aeration tube located inside the membrane module and also used to attach it vertically.

The filtration of the effluent may be carried out intermittently (5 *min* of filtration followed by a pause of 30 *s*) but air should be continuously injected into the submerged membrane bioreactor system. An oxygen probe (5) may be connected to an oxymeter and installed in the bioreactor tank to measure dissolved oxygen concentration. The air flow rate may be measured using a flow meter (17). The pH and temperature can also be measured during the experiments by using pH (8) and temperature (9) probes installed in the bioreactor tank.

Bisphenol compounds are collected in Table 6.5. Bisphenol A and Bisphenol S are shown in Figure 7.21.

Table 6.5 Bisphenol compounds (41).

Compound	Shortcut
2,2-Bis(3-methyl-4-hydroxyphenyl)propane	
Bis(4-hydroxyphenyl)-2,2-dichlorethylene	
2,2-Bis(4-hydroxyphenyl)propane	Bisphenol A
1,1-Bis(4-hydroxyphenyl)-1-phenyl-ethane	Bisphenol AP
2,2-Bis(4-hydroxyphenyl)hexafluoropropane	Bisphenol AF
2,2-Bis(4-hydroxyphenyl)butane	Bisphenol B
Bis-(4-hydroxyphenyl)diphenylmethane	Bisphenol BP
1,1-Bis(4-hydroxyphenyl)ethane	Bisphenol E
Bis(4-hydroxydiphenyl)methane	Bisphenol F
2,2-Bis(4-hydroxy-3-isopropyl-phenyl)propane	Bisphenol G
1,3-Bis(2-(4-hydroxyphenyl)-2-propyl)benzene	Bisphenol M
1,4-Bis(2-(4-hydroxyphenyl)-2-propyl)benzene	Bisphenol P
Bis(4-hydroxyphenyl)sulfone	Bisphenol S
1,1-Bis(4-hydroyphenyl)-3,3,5-trimethyl-cyclohexane	Bisphenol TMC
1,1-Bis(4-hydroxyphenyl)-cyclohexane	Bisphenol Z)

6.6.3.4 *Removal of Lead Ions*

The maximum adsorption capacities by various sorbents reported in the literature are shown in Table 6.6.

It can be observed that the binding energy of the sorption system increases with the addition of graphene oxide and that the binding

Table 6.6 Maximum adsorption capacities (34).

Material	Maximum adsorption capacity/[$mg\,g^{-1}$]	References
Amino modified MWCNTs	58.26	(42)
Dispersed MWCNTs	92.3	(43)
Undispersed MWCNTs	74.5	(43)
CNTs/MnO_2	78.55	(44)
CNTs/HNO_3	35.6	(45)
Coated bentonite	95.88	(46)
Activated carbon	51.81	(47)
Chitosan beads	34.98	(48)
Activated carbon	43.85	(49)
Magnetic nanoparticles	40.10	(50)
PVA-PEI nanofibers	90.03	(51)
CS-PAA beads	109.89	(48)
GO1 beads (1% graphene oxide)	116.28	
GO5 beads (5% graphene oxide)	138.89	

energy increased with the increase of the graphene oxide content (34).

A chitosan-aminopropylsilane graphene oxide nanocomposite hydrogel was synthesized and tested as a hydrophilic additive in different dosages of 0.5%, 1%, 2% and 5% by weight for the fabrication of porous PES membranes via the induced phase inversion process by immersion precipitation method for heavy metal ion and dye removal.

The modified membranes were characterized using ATR-FTIR, AFM, SEM, water contact angle, overall porosity and mean pore radius evaluations and zeta potential measurement.

The addition of chitosan-aminopropylsilane graphene oxide nanocomposite hydrogel to PES doping solutions enhanced the hydrophilicity of the membranes and so the permeability for pure water flux. The filtration performance of the chitosan propylsilane graphene oxide embedded membranes showed promising antifouling properties during a bovine serum albumin filtration test (FRR> 90%) and 1% membranes showed the highest pure water flux of 123.8 $l\,m^{-2}h^{-1}$ with bovine serum albumin rejection of more than 98% and removal capability of more than 82% for lead (II) ion, 90.5% and 98.5% for C.I. Reactive Blue 50 and C.I. Reactive Green

19, respectively. These compounds are shown in Figures 6.10 and 6.11.

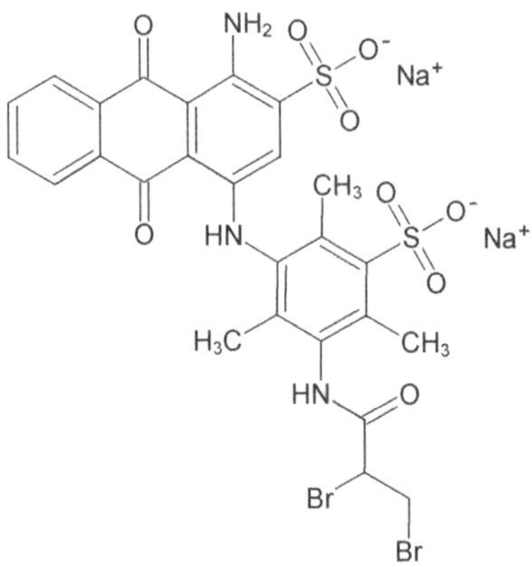

Figure 6.10 Reactive Blue 50.

Therefore, a chitosan-aminopropylsilane graphene oxide nanocomposite hydrogel blend used to modify PES-based membranes, has a noticeable potential in improving filtration performance of blended membranes (52).

6.6.3.5 *Removal of Uranium*

A solid porous adsorbent with amino and carboxyl has been developed via a one-pot Friedel-Crafts alkylation reaction (53). Here, tryptophan is used. Tryptophan is shown in Figure 6.12.

The resulting materials are highly chemically stable and the surface area can be changed by different ratios of initial aromatic monomers. The critical factors affecting adsorption performances, such as temperature, contact time, ionic strength and pH, were studied by using batch adsorption experiments.

The adsorption process is fast and can be well simulated with the pseudo-second-order model, yielding a rate constant of 1.31×10^{-3} min^{-1}.

The sorption isotherms were correlated with the Langmuir isotherms, according to which the highest adsorption amount (46.74 $mg\,s^{-1}$) was measured at a pH of 8 under 308 K. According to

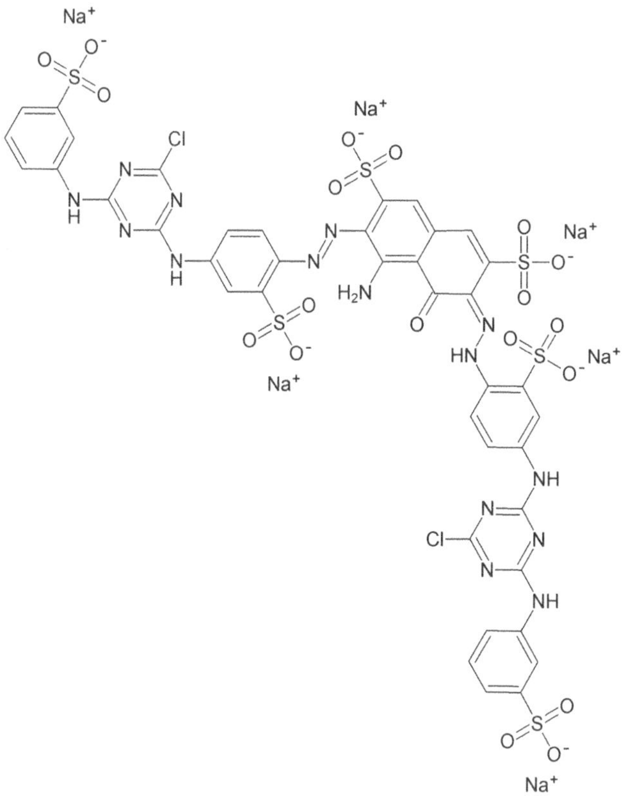

Figure 6.11 Reactive Green 19.

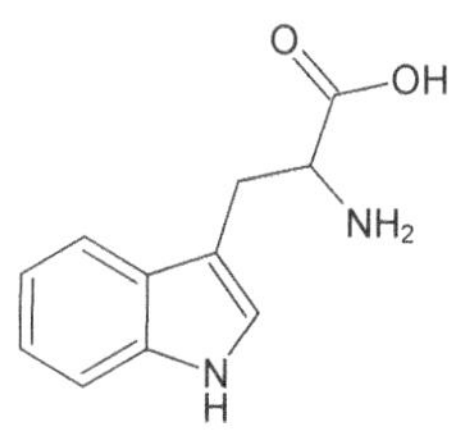

Figure 6.12 Tryptophan.

Fourier transform infrared spectroscopy (FTIR) and X-ray photoelectron spectroscopy analysis, uranium is chelated by the amino and the carboxyl groups. This indicates that the addition of tryptophan significantly increases the adsorption capacity. These results provide a new design strategy for the development of novel adsorbents for the recovery of uranium (VI) (53).

Traditional adsorption resins that are prepared by a low efficiency method have a low porosity and a low surface area. These properties limit the adsorption performance of such resins. Therefore, a novel porous resin was developed by a structural design of the resins through post-modification with HPA and a post-crosslinked method (54). The novel porous post-modification resin was successfully prepared by the multi-step method.

Meanwhile, the high surface area and abundant micropores in hyper-crosslinked resins can improve the adsorption rate by accelerating the diffusion of uranium in the external diffusion stage. The removal capacity of hyper-crosslinked resins increased 366%. The adsorption amount of the material dropped only slightly even after 8 cycles. The short equilibrium time and the outstanding removal efficiency makes it a novel candidate for further extraction of U(VI) from seawater (54).

6.7 Enzyme Reactors

6.7.1 Thermoresponsive Enzyme Reactor

6.7.1.1 Acrylic Block Polymers

An enzyme reactor based on a thermoresponsive magnetic porous block copolymer membrane was designed and constructed. Reversible addition-fragmentation chain transfer polymerization was used to synthesize the block copolymer, poly(maleic anhydride-styrene-*N*-isopropylacrylamide), with poly(*N*-isopropylacrylamide) as the thermoresponsive moiety.

The self-assembly property of the block copolymer was used for preparation of magnetic porous thin film matrices with iron oxide nanoparticles. By covalent bonding of glutaminase onto the surface of the membrane matrices and changing the temperature to tune the nanopore size, an enhanced enzymolysis efficiency was observed due to the confinement effect.

The apparent Michaelis-Menten constant and the maximum rate of the enzyme reactor were determined (K_m = 32.3 *mM*, V_{max} = 33.3 *mM min*$^{-1}$) by a chiral ligand exchange capillary electrochromatography protocol with l-glutamine as the substrate.

Compared with free glutaminase in solution, the proposed enzyme reactor exhibits higher enzymolysis efficiency, greater stability, and greater reusability. Furthermore, the enzyme reactor was applied for a glutaminase kinetics study. The tailored pore sizes and the thermoresponsive property of the block copolymer result in the designed porous membrane-based enzyme reactor having great potential for high enzymolysis performance (55).

6.7.1.2 *Styrenic Block Copolymers*

The use of a thermoresponsive porous polymer membrane for the construction of switchable enzyme reactors has been described (56).

The reactor can change its membrane surface properties in diverse temperature environments. The enzyme reactor consists of block copolymers, poly(styrene-maleic anhydride-oligo methacrylate-methoxyethoxy ethyl methacrylate), magnetic nanoparticles, and covalently bonded *D*-amino acid oxidase.

By varying the temperature between 25°C and 39°C, the self--assembly properties of the enzyme reactor could be controlled. Using DL-Met as the substrate, the enzymolysis efficiency of the thermoresponsive porous polymer membrane enzyme reactor was evaluated by the chiral capillary electrophoresis technique.

At higher temperatures, *D*-amino acid oxidase was embedded in the curled poly(methoxyethoxy ethyl methacrylate) chains, resulting in improved enzymolysis efficiency owing to the confinement effect. At lower temperatures, inefficient enzymolysis occurred because of the stretching of the poly(methoxyethoxy ethyl methacrylate) chains, which weakened the confinement effect.

Remarkably, the enzymolysis rate of the thermoresponsive porous polymer membrane enzyme reactor at 37°C (3.1 *mM min*$^{-1}$) was 45 times higher than that at 25°C (0.066 *mM min*$^{-1}$) (56).

6.7.1.3 *Covalently Immobilized Enzyme Reactor*

A *D*-amino acid oxidase reactor for enhanced enzymolysis efficiency was developed. Magnetic polymer matrices, composed of iron oxide nanoparticles and a porous polymer membrane, i.e., poly(styrene-*co*-maleic anhydride), were prepared (57). The covalent bonding of *D*-amino acid oxidase on the surface of the matrices and characterization of scanning electron microscope and vibrating sample magnetometer, demonstrated that the membrane enzyme reactor was successfully constructed.

The enzymolysis efficiency of the enzyme reactor was evaluated and the apparent Michaelis-Menten constants of *D*-amino acid oxidase were determined (K_m was 1.10 mM, V_{max} was 23.8 $mM\,min^{-1}$) by a chiral ligand exchange capillary electrophoresis protocol with methionine as the substrate. Methionine is shown in Figure 6.15.

The results indicated that this enzyme reactor could exhibit a good stability and an excellent reusability. Because the enzyme and the substrate could be confined to the pores of the matrices, the enzyme reactor displayed an improved enzymolysis efficiency due to the confinement effect. Furthermore, the prepared enzyme reactor was applied for *D*-amino acid oxidase inhibitors screening (57).

6.7.2 *Reversible pH-Control*

The challenge for polymeric enzyme reactors is to selectively control the enzymolysis rate under complex conditions (58). In addition, the fabrication methodology is hindered by complex processes, especially for achieving diverse stimuli responsiveness and functions.

A pH-sensitive polymer, poly(styrene-*co*-maleic anhydride-acrylic acid)-based hybrid enzyme reactor was developed (58). The system consisted of magnetic nanoparticles and a pH-sensitive poly(styrene-*co*-maleic anhydride-acrylic acid) porous polymer membrane made by the breath-figure method. The enzyme *l*-asparaginase could covalently bond on the surface of the pH-sensitive porous polymer membrane.

The resultant enzyme reactor was characterized by FTIR spectroscopy and a vibrating sample magnetometer. The apparent Michaelis-Menten constants (K_m and V_{max}) of the *l*-asparaginase enzyme reactor at different pH values were determined by a chiral ligand exchange capillary electrophoresis method with *l*-asparagine as the substrate.

Michaelis-Menten kinetics is a well-known model of enzyme kinetics (59).

The model uses an equation that describes the rate of enzymatic reactions, by relating the rate v of the formation of a product $[P]$ to the concentration of a substrate $[S]$.

$$v = \frac{d[P]}{dt} = V_{max} \frac{[S]}{K_M + [S]} \tag{6.2}$$

V_{max} is the maximum rate that can be achieved by the system. This happens at a saturated substrate concentration for a given enzyme concentration. The value of the Michaelis constant K_M is numerically equal to the substrate concentration at which the reaction rate is half of V_{max} (59).

The V_{max} value of the *l*-asparaginase enzyme reactor (0.67 mM/min) was almost 3-fold that of the free *l*-asparaginase (0.23 $mM\,min^{-1}$) at a pH of 8.2 (58). Asparagine is shown in Figure 6.13.

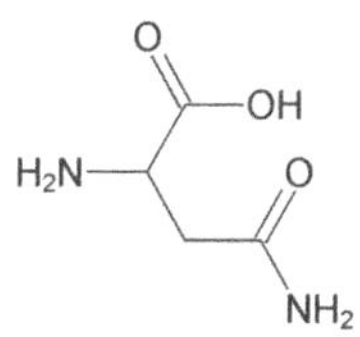

Figure 6.13 Asparagine.

Its ability to precisely control the enzymolysis rate in complex conditions is triggered primarily by the pH of the buffer solution, allowing controlled enzymatic reactions and displaying excellent stability and reusability of the proposed pH-sensitive porous polymer membrane.

This strategy for porous polymer membrane enzyme reactor fabrication has established a platform for enzyme efficiency adjusting. These valve-like distinguished features highlight the outstanding potential of stimuli-responsive enzyme reactor applied for enzyme immobilization and enzyme-related disease treatment (58).

6.7.3 *UV-Responsive Enzyme Reactor*

Polymer materials that enable controllable enzymolysis based on ultraviolet-visible responsive properties, have hitherto remained unexplored. An enzyme reactor was fabricated by immobilization of *l*-asparaginase on a UV-responsive porous polymer membrane, which consisted of poly(styrene-maleicanhydride-4-[(4-methacryloyloxy)phenylazo] benzoic acid) (60). The controllable enzymolysis of this reactor was tested.

By controlling the *on/off* switch of 365 *nm* UV irradiation, the configuration of polymer membrane surface changed to improve and tune the enzymolysis. Using *l*-asparagine as substrate, the enzymatic efficiency of the UV-responsive porous polymer membrane was evaluated by a chiral capillary electrophoresis technique.

Upon UV irradiation, the poly((4-methacryloyloxy)phenylazo benzoic acid) moiety in the membrane changed from a *trans*-configuration to a *cis*-configuration and encapsulated the enzyme and substrate into a narrow cavity, further improving the enzymatic efficiency due to the confinement effect.

The *trans*-configuration and the a *cis*-configuration is shown in Figure 6.14.

It was found that the enzymatic reaction rate with the UV-responsive porous polymer membrane under UV irradiation (13.3 $mM\,min^{-1}$ was 4.5 times higher than that of UV irradiation was off (2.94 configuration). In addition, the low cytotoxicity and excellent UV-responsivity of UV-responsive porous polymer membrane were verified in the living cells and serum samples (60).

6.7.4 *Kidney Mimicking*

A porous polymer membrane-based *D*-amino acid oxidase reactor was developed that mimicked the enzymatic activity in a renal ischemia model (61). Using glycidyl methacrylate as a biocompatible reactive monomer, poly(styrene-glycidyl methacrylate) was synthesized via a reversible addition fragment chain transfer polymerization technique.

The so prepared porous polymer membrane was used as a support to effectively immobilize *D*-amino acid oxidase. Compared to a *D*-amino acid oxidase modified on a nonporous polymer membrane

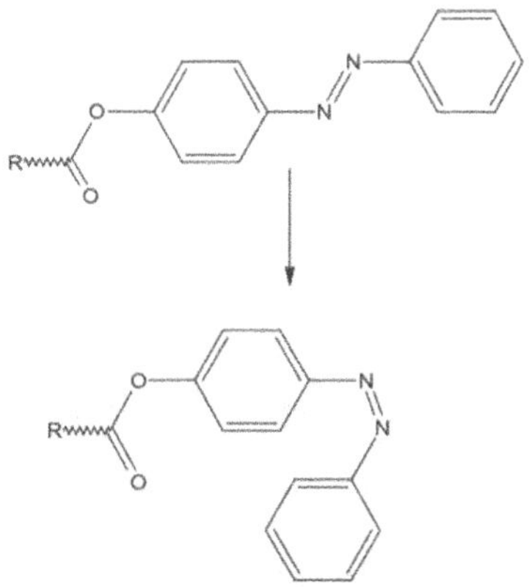

Figure 6.14 Trans-configuration and cis-configuration of the 4-methacryloyloxy)phenylazo] benzoic acid moiety (60).

and free *D*-amino acid oxidase in solution, the constructed porous polymer membrane-based *D*-amino acid oxidase enzyme reactor displayed an increase in enzymolysis efficiency.

In addition, a chiral ligand exchange capillary electrophoresis system for *D*-amino acid oxidase was used to study the enzymatic kinetics of *D*-amino acid oxidase with *d,l*-ine as the substrate. Methionine is shown in Figure 6.15.

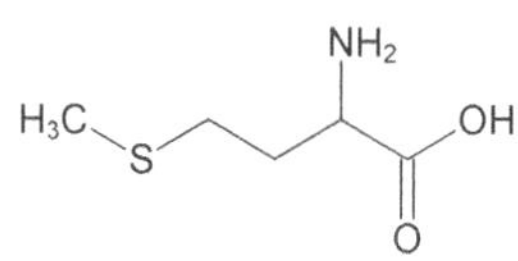

Figure 6.15 Methionine.

The porous polymer membrane-based enzyme reactor showed an excellent performance both on reproducibility and stability. Moreover, the enzyme reactor was successfully applied to mimic *D*-amino acid oxidase activity in a renal ischemia model. These results demonstrated that the enzyme could be efficiently immobilized onto a porous polymer membrane as an enzyme reactor and has great potential in mimicking the enzymatic activity in kidney (61).

6.8 Electrolyte Membranes

6.8.1 *Membranes for Fuel Cells*

6.8.1.1 *Polymer Electrolyte Fuel Cells*

A polymer electrolyte fuel cell is a potential power source for new-generation automobiles like fuel cell vehicles that run on hydrogen energy (62). The most important characteristics of polymer electrolyte fuel cells are their low operation temperature and quick start and shutdown.

To develop high performance fuel cell vehicles and fuel cell systems, a lot of knowledge is required, such as the development of innovative catalysts and electrolyte materials, optimization of catalyst layer, measurement of actual porous structure, simulation of mass transport and internal reaction, and understanding the mechanism of electrochemical reaction and mechanical durability.

In order to obtain a high electrochemical activity, a Pt particle catalyst is used as both anode and cathode of polymer electrolyte fuel cells. Thus, the most important obstacle to be overcome is the high cost of polymer electrolyte fuel cells.

The mechanism of carbon black agglomeration in the catalyst layer was examined based on the experimental results of the particle diameter distribution of carbon black in catalyst layer ink with or without ionomers and the transmission electron microscopy images of carbon black dispersion. The formation of large pores was considered by calculating the net repulsive potential energy of the carbon black particles in a slurry (62).

6.8.1.2 *Poly(tetrafluoroethylene) Fibrils*

A porous body with a high-porosity can be obtained by stretching a molded article produced by paste extrusion molding of a poly(tetrafluoroethylene) (PTFE) fine powder. This PTFE porous body consists of nodes (knots) and fibrils (fibers) and allows gas such as water vapor to pass therethrough, but does not allow water drops to pass therethrough owing to strong water repellency of PTFE. This stretched porous body can be used as a sealing material without being sintered, or can be used in clothes or separation membranes after being sintered and formed into a tough, continuously stretched sheet or tube.

The PTFE is preferably a PTFE obtained by copolymerizing tetrafluoroethylene and a trace comonomer. The trace comonomer may be any monomer copolymerizable with tetrafluoroethylene.

Examples are perfluoroolefins such as hexafluoropropylene, chlorofluoroolefins such as chlorotrifluoroethylene, hydrogen-containing fluoroolefins such as trifluoroethylene and vinylidene fluoride, perfluorovinyl ethers, and (perfluoroalkyl)ethylenes and ethylene. One trace comonomer may be used, or multiple trace comonomers may be used. The trace comonomer may be perfluoro(methyl vinyl ether).

Examples of the trace comonomer other than perfluoro(methyl vinyl ether) include fluoroolefins such as hexafluoropropylene and chlorotrifluoroethylene, fluoro(alkyl vinyl ethers) having a alkyl group with one or three carbon atoms, fluorinated cyclic monomers such as fluorodioxole, perfluoroalkyl ethylenes and ω-hydroperfluoroolefins. Some of these compounds are shown in Figure 6.16.

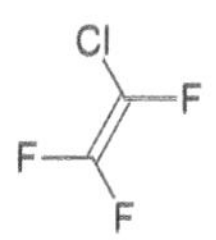

Chlorotrifluoroethylene

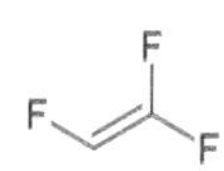

Trifluoroethylene

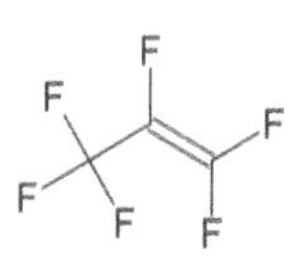

Hexafluoropropylene

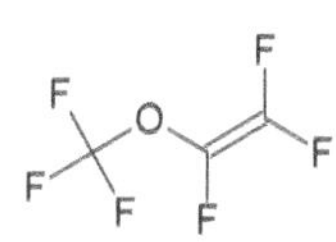

Perfluoro(methyl vinyl ether)

Figure 6.16 Trace comonomers.

The PTFE may be unsintered PTFE or may be semi-sintered PTFE. For simple processing or easy control of the thickness and the pore size, unsintered PTFE is preferred. For high strength or a small pore size of biaxially stretched membranes, semi-sintered PTFE is preferred. The PTFE may have a core-shell structure. The core-shell structured PTFE may be, for example, a modified PTFE whose particles each include a core of a high molecular weight PTFE and a shell of a lower molecular weight PTFE or of a modified PTFE.

A method of preparation and determining the extrusion pressure is as follows (63):

Preparation 6–2: First, 100 *g* of PTFE fine powder is blended with a lubricant (trade name: IP1620, Idemitsu Petrochemical Co., Ltd.) in an amount of 18.0% based on the sum of the amounts of the PTFE fine powder and the lubricant, and the components are mixed in a glass bottle for 3 *min* at room temperature. Next, the glass bottle is left to stand for at least 1 *h* at room temperature (25°C) before extrusion, and thereby a lubricated resin is obtained.

The lubricated resin is paste-extruded through an orifice (diameter: 2.5 *mm*, land length: 5 *mm*, introduction angle: 30°) at room temperature and a reduction ratio of 100:1, and thereby uniform beading (extrudate) is obtained. The extrusion rate, i.e., ram speed, is 51 $cm\,min^{-1}$. The extrusion pressure is a value determined by measuring a load when the extrusion load reaches equilibrium during the paste extrusion, and then dividing the measured load by the cross-sectional area of a cylinder used in the paste extrusion. In order to provide a porous body having higher strength, a smaller pore size, and better homogeneity, the PTFE preferably has a breaking strength of 20 *N* or higher. The breaking strength is more preferably 28 *N* or higher, still more preferably 32 *N* or higher, particularly preferably 36 *N* or higher.

In the case of stretching at a high stretch ratio, the breaking strength is preferably within the above range. The upper limit of the breaking strength may be any value, and may be 70 *N*.

The method of polymerization and fabrication has been given in much detail (63).

The microstructure preferably has five or more fused points within an area of 5 $\mu m \times 5\ \mu m$, as can be observed in an electron microscopic image. As shown in the scanning electron microscopic image of a conventional porous body in Figure 6.17, it has a microstructure including nodes 200 and fibrils 300, but the fibrils only three-dimensionally cross each other.

In contrast, as shown in the scanning electron microscopic image of Figure 6.18, the here produced porous body has a characteristic microstructure including 100 fused points where fibrils fused with each other.

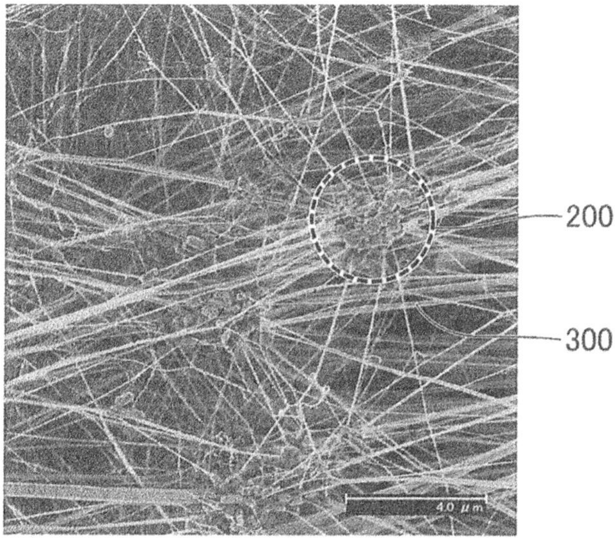

Figure 6.17 Conventional porous body (63).

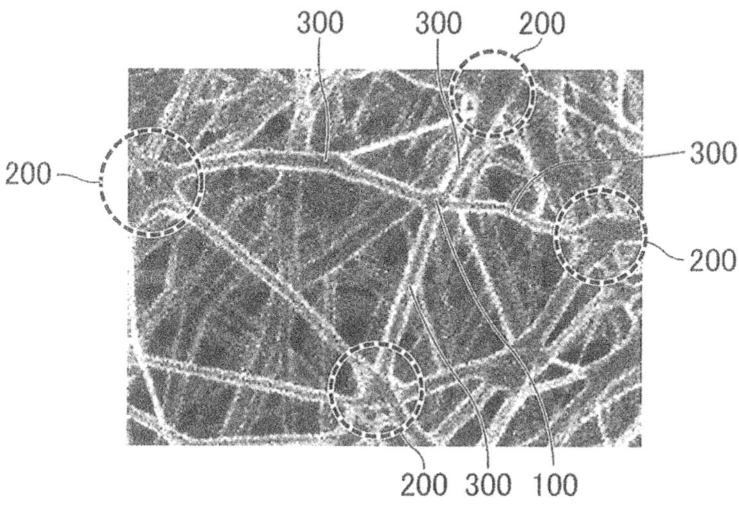

Figure 6.18 Produced porous body (63).

6.8.1.3 *Electrode Substrate with Carbon Fibers*

Polymer electrolyte fuel cells are required to have high conductivity and excellent current collection capability while exhibiting mechanical strength to withstand various operations (64). Also, the diffusion of substances that contribute to electrode reactions needs to be excellent in such fuel cells. To respond to such requirements, carbonized sheets are generally used as electrode substrates.

Fuel cell applications that have attracted attention in recent years are those for automotives where high-power density is required. In such applications, fuel cells are operated in regions of high current density, and amount of water generated per unit reaction area increases. Accordingly, efficient discharge of water produced by reactions is the issue, and high dewatering capability is thereby required for carbonized sheets used as gas diffusion materials in fuel cells. Therefore, controlling pore distributions in the electrode substrate has been attempted for the enhancement of dewatering capability.

A porous electrode substrate has been described in which carbon fibers dispersed in the structure thereof have a fiber diameter of 3-15 microns and a fiber length of 2-30 *mm*, and are bound to one another by carbonized resin such that, when a pore distribution in the porous electrode is determined with a mercury intrusion porosimeter, such that when a pore distribution curve is plotted on a graph having a common logarithmic scale on the horizontal axis, and a 1-100 micron pore diameter range of the pore distribution curve includes 80 or more measurement points at equal intervals along the common logarithmic scale, the pore distribution has a skewness S of $-2.0<S<-0.8$ and a kurtosis K of $3.5<K<10$ in the 1-100 micron pore diameter range (64).

A porous electrode substrate can be produced by the following steps (64):

1. **Sheet material production process**: A step for producing a carbon-fiber sheet material by dispersing carbon fibers in water;
2. **Resin impregnation process**: A step for forming a resin-impregnated sheet by impregnating a thermosetting resin in the carbon-fiber sheet material;

3. **Hot-pressing process**: A step for hot pressing the resin-impregnated sheet after step 2 at a temperature range of 100°C to 400°C. to obtain a resin-cured sheet; and
4. **Carbonization process**: A step for carbonizing the resin-cured sheet after step 3 at a temperature of 1000°C. or higher to obtain a porous electrode substrate.

In detail, a preparation can be done as follows (64):

Preparation 6–3: Poly(acrylonitrile) (PAN)-based carbon fibers (product name: Pyrofil TR50S, average fiber diameter: 7 *μm*, made by Mitsubishi Rayon Co., Ltd.) cut to have an average fiber length of 3 *mm*, poly(vinyl alcohol) (PVA) fibers (product name: VPB105-1, fiber length: 3 *mm*, made by Kuraray Co., Ltd), and poly(ethylene) pulp (product name: SWP, made by Mitsui Chemicals Inc.). PAN-based carbon fibers, poly(ethylene) pulp and PVA fibers at a ratio of 10:3:2 were charged in the slurry tank of a continuous wet papermaking machine with a short wire mesh; water was further added to the tank and the fibers were evenly dispersed and opened. When the fibers were fully dispersed, the web was discharged, passed onto the short wire mesh, and dried with a blower. Accordingly, a rolled carbon-fiber sheet material was obtained to have a width of 1000 *mm* and a basis weight of 20 $g\,m^{-1}$.

Next, the carbon-fiber sheet material was immersed in a methanol solution with a phenolic resin (product name: Phenolite J-325, made by DIC Corporation) so as to adhere 60 parts by mass of the phenolic resin to 100 parts by mass of the carbon-fiber sheet material. The sheet material was then slit to a width of 850 *mm* and a resin-impregnated sheet with adhered phenolic resin was obtained. Two of the resin-impregnated sheets were press-molded using a double-belt pressing machine.

Preheating was conducted under conditions set to have a hot air temperature of 150°C and preheating roll temperature of 235°C, while the vertical clearance of preheating rolls was adjusted to be 30% of the total thickness of two resin-impregnated sheets. Furthermore, hot pressing was conducted at a roll temperature of 260°C and a pressure of 6.7 *MPa*. Here, the preheating roll temperature indicates the hot pressing temperature defined in the present application. As a result, an 850 *mm* wide and 100 *m* long resin-cured sheet was obtained.

6.8.1.4 High Temperature Polymer Electrolyte Membranes

The introduction of a porous structure into high temperature polymer electrolyte membranes is one of the effective pathways to increase their proton conductivity at elevated temperature. The proton transport behavior of a series of porous PSf membranes under elevated temperature was investigated (65).

The functional triazole ring in the framework of porous triazole-PSf acts a proton acceptor to form acid-base pair with phosphoric acid (PA). In addition, the proton diffusion coefficient and proton conductivity of PA-doped porous triazole-PSf is an order of magnitude higher than that of the PA-doped dense triazole-PSf membrane. Percolation theory calculation convinces that the high proton conductivity of PA-doped porous triazole-PSf is due to the formation of continuous long-range proton diffusion channels under high pore connectivity and porosity. On the contrary, excessive pore connectivity also results in high gas permeability, leading to a decrease of the open circuit voltage and cell performance of the single cell. The optimum porosity for the PA-doped porous triazole-PSf membrane is 75% for fuel cell operating with the maximum peak power density of 550 $mW\,cm^{-2}$ and a great durability for 120 h under a temperature of 140°C (65).

Finite Element Simulation. Porous polymer electrolyte membranes for fuel cell applications have multiple benefits, including enhancements in thermal removal, electron conduction, and two-phase flow control. The intrusion of the gas diffusion layer into the gas flow channel and potential delamination at the gas diffusion layer and microporous layer interface upon land compression and their depressions by adoption of the porous media flow field have been investigated.

A finite element simulation was used to predict the gas diffusion layer deformation and stresses for both porous media and traditional hollow gas flow channels (66). The model predictions are compared with other literature data of gas diffusion layer intrusion/deformation for traditional hollow gas flow channels.

The results showed that gas diffusion layer intrusion may significantly reduce the gas flow channel space and gas diffusion layer deformation, causeing a large spatial variation of porosity and permeability of traditional gas flow channels. Adoption of the porous media gas flow channels significantly depresses the gas diffusion layer intrusion and deformation and hence the porosity/permeability spatial variation. In addition, the shear and tensile stresses at the gas diffusion layer-microporous layer interface are evaluated, which are major causes for delamination.

It is found that both compressive and tensile stresses may be induced by assembly pressure for traditional gas flow channels. The shear stress can be significantly depressed and tensile stress can be avoided by porous media gas flow channels, thereby mitigating the risk of gas diffusion layer/microporous layer delamination (66).

6.8.1.5 *Stretchable Fuel Cells*

Ceramic membranes can serve as viable alternatives to less mechanically stable polymeric membranes utilized in microbial fuel cells (67).

A series of polymer-derived ceramic proton exchange composite membranes with a large ion exchange capacity (IEC), high cation transport numbers, and low oxygen diffusion coefficients were synthesized at various pyrolysis temperatures using a pressing technique.

These materials were composed of a polysiloxane matrix mixed with proton-conducting fillers such as montmorillonite and $H_3PMo_{12}O_{40}/SiO_2$ at different ratios.

By tuning the average pore sizes of the membranes between 0.1 μm and 1 μm and their hydrophilic and hydrophobic characteristics, the maximum IEC of 0.6072 $mequiv\ g^{-1}$ and cation transport number of 0.6988 were obtained, which is 67% and 72% of polymeric Nafion performance, respectively. In addition, the minimal oxygen mass transfer coefficient achieved by this approach was equal to $5.62 \times 10^{-4}\ cm\ s^{-1}$, which is very close to the commercial Nafion membrane value.

The fabricated polymer-derived ceramic composite membranes meet all the essential criteria required for their use in microbial fuel cell applications and represent a high potential to overcome limitations of polymeric membrane (67).

6.8.1.6 *Stretchable Fuel Cells*

Skin-like energy devices can be conformally attached to the human body, which are highly desirable to power soft wearable electronics. Here, a skin-like stretchable fuel cell based on ultrathin gold nanowires and polymerized high internal phase emulsion scaffolds was demonstrated (68).

The polymerized high internal phase emulsions can offer a high porosity of 80% yet with an overall thickness comparable to human skin. Upon impregnation with electronic inks containing ultrathin (2 *nm* in diameter) and ultrahigh aspect-ratio (>10 000) gold nanowires, skin-like strain-insensitive stretchable electrodes could be successfully fabricated.

With such designed strain-insensitive electrodes, a stretchable fuel cell is fabricated. The resulting epidermal fuel cell can be patterned and transferred onto skin as tattoos, yet can offer a high power density of 280 $\mu W\,cm^{-2}$ and a high durability performance retention under stretching, compression, and twisting.

The here shown results demonstrate that this skin-thin, porous, yet stretchable electrode is essentially multifunctional, simultaneously serving as a current collector, an electrocatalyst, and a fuel host, indicating potential applications to power future soft wearable 2.0 electronics for remote healthcare and soft robotics (68).

6.8.1.7 Composite Polymer Electrolyte Membrane

A block copolymer composite membrane of sulfonated poly(arylene ether sulfone ketone) decorated with porous hygroscopic titanium oxide nanotubes was designed and fabricated for low relative humidity operating polymer electrolyte fuel cells (69).

The sulfonated poly(arylene ether sulfone ketone)-titanium oxide nanotubes composite membrane in a polymer electrolyte fuel cell operated at 100% relative humidity, and 80°C resulted in about 1.3 and 1.1-folds higher power density in comparison with Nafion (NRE-212) and pristine sulfonated poly(arylene ether sulfone ketone) membranes at 0.6 *V*.

Furthermore, operating under 30% relative humidity and 80°C, the sulfonated poly(arylene ether sulfone ketone)-titanium oxide nanotubes composite membrane generates 3.8 and 2.8-folds improved power density in comparison with the pristine sulfonated poly(arylene ether sulfone ketone) and commercial NRE-212 membranes, respectively, at 0.6 *V* (69).

An outstanding improvement of polymer electrolyte fuel cell performance using sulfonated poly(arylene ether sulfone ketone)-titanium oxide nanotubes composite membrane also maintains relative to a sulfonated poly(arylene ether sulfone ketone)-TiO_2 nanoparticles composite membrane, operating under 100 and 30% relative humidity. Moreover, the sulfonated poly(arylene ether sulfone ketone)-titanium oxide nanotubes composite membrane shows a stable operating potential of more than 200 *h* at 30% relative humidity and 80°C, so confirming the durable polymer electrolyte fuel cell operation.

The enhanced polymer electrolyte fuel cell performance under dry conditions is mainly the result of improving water management by the titanium oxide nanotube filler in the membrane and cathode catalyst utilization, significantly suppressing both ohmic and mass transport overpotentials (69).

6.9 Membranes for Batteries

6.9.1 *Membranes for Lithium-Ion Batteries*

6.9.1.1 *3D Porous Hosts*

Li metal anodes have regained major attention due to their high theoretical specific capacity (70). However, Li anodes tend to exhibit a dendritic growth and constant electrolyte consumption upon cycling, which lead to safety concerns, low coulombic efficiency, and short battery lifetime.

Both conductive and non-conductive 3D porous hosts were coupled with a viscous (melt) polymer electrolyte. The cross section of the hosts showed a good contact between porous hosts and the melt polymer electrolyte before and after extensive cycling. This indicates that the viscous electrolyte successfully refilled the space upon Li stripping. Upon deep Li deposition/stripping cycling (5 $mA\,h\,cm^{-2}$), the non-conductive host with the viscous electrolyte successfully cycled, while the conductive host allowed rapid short circuiting. Post-mortem cross-sectional imaging showed that the Li deposition was confined to the top layers of the host.

The COMSOL simulation results of conductive hosts and non-conductive hosts at equilibrium, showed that the concentration gradient and the current density is constant. The COMSOL simulations indicated that current density was higher and more restricted at the top of the conductive host with the polymer electrolyte than the liquid electrolyte. This resulted in quicker short circuiting of the polymer electrolyte cell during deep cycling. Thus, the non-conductive 3D host is preferred for coupling with the melt polymer electrolyte (70).

Zirconium oxide. A composite microporous polymer electrolyte membrane has been described (71). The membrane is an optimized composition of poly(vinylidene fluoride-*co*-hexafluoropropylene) with ZrO_2. The material was prepared by a preferential polymer dissolution process.

The effect of the porosity on the electrochemical properties of polymer electrolyte membranes, such as ionic conductivity and lithium transference number, was studied by electrochemical AC-impedance spectroscopy and steady state current method, respectively. The results show that the formation of the microporous membrane by the removal of PVA in the membrane matrix can enhance the ionic conductivity and lithium transference number when polymer membrane is soaked in 1 *M* $LiClO_4$ electrolyte solution containing a 1:1 (v/v) ratio of ethylene carbonate and diethyl carbonate (71).

It was also found that the ionic conductivity follows the Arrhenius equation. Charge-discharge studies were carried out by fabricating a cell with carbon as the anode and $LiMg_{0.10}Mn_{1.90}O_4$ as the cathode and composite microporous polymer electrolyte membrane as the separator.

The excellent performance, such as high ionic conductivity, high lithium transference number, good compatibility with electrode materials and higher discharge capacity of the poly(vinylidene fluoride-*co*-hexafluoropropylene) ZrO_2-based polymer electrolyte membrane suggest that the polymer electrolyte can be used as a candidate for high-performance lithium-ion batteries (71).

Titanium oxide. Nano-TiO_2 was selected as inorganic filler to fabricate a reduced crystalline composite polymer membrane. The porous polymer matrix employed in this study was a poly(vinylidene fluoride-*co*-hexafluoropropylene) (P(VDF-HFP)) membrane, synthesized via non-solvent evaporate method. The electrolyte solution uptake was carried out in the glove box to avoid a moisture contamination. The effect of nano-TiO_2 on the crystallinity of porous polymer membrane was studied (72).

It was found that blending with nano-TiO_2 helps to reduce the crystallinity of the polymer membrane. Electrochemical impedance spectroscopy showed that the room temperature conductivity of P(VDF-HFP)/TiO_2 composite-porous polymer electrolyte increased up to 1.5×10^{-3} $S\,cm^{-1}$. Scanning electron microscopy micrographs showed that the micropores and nano-TiO_2 particles were observed in the polymer membrane.

The composite-porous polymer electrolyte was stable up to 4.5 V (vs. Li/Li^{+}) and the $LiFePO_4$/composite-porous polymer electrolyte/Li coin cell showed an excellent rate capability, also a fine discharge capacity. The cell had about 6% capacity loss when it discharged at 1C for 50 cycles (72).

6.9.1.2 *Poly(ether imide)*

Until now, the safety and cycle stability problems of lithium-oxygen batteries have not been completely resolved. A non-solvent induced phase separation method was shown (73). The phase separation method is combined with the *in-situ* hydrolysis modification strategy of tetrabutyl titanate, a type of a poly(ether imide) (PEI) separator with typical finger-shaped pores and honeycomb-shaped supporting layer pore structure.

After being infiltrated by an ionic liquid electrolyte, 1-Ethyl-3-methylimidazolium-tetrafluoroborate, c.f. Figure 6.19, the ionic conductivity of the separator can reach 0.75 $mS\,cm^{-1}$.

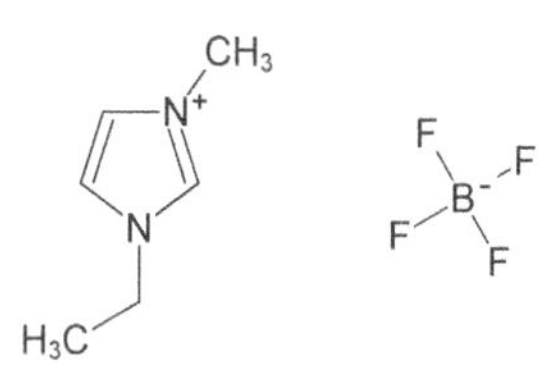

Figure 6.19 1-Ethyl-3-methylimidazolium-tetrafluoroborate.

Due to the low crystallinity, high thermal stability, and stable ion transport pore structure of the separators, the assembled lithium-oxygen batteries can stably cycle for more than 75 cycles under the condition of a limit of 1000 $mA\,h\,cm^{-1}$. In addition, under the strategy of high-performance cathode catalysts and electrolyte additives, the application of the porous PEI separators in lithium-oxygen batteries can be further expanded (73).

6.9.1.3 *Carbonized Bacterial Cellulose Film*

A freestanding lithium-sulfur electrode was designed and prepared by loading sulfur into a hybrid material that was constructed by growing Mo_3Ni_3N and Mo_2C nanoparticles on a carbonized bacterial cellulose film (74).

A high sulfur content up to 70% and a high areal sulfur loading of 15.5 $mg\,cm^{-2}$ in the membrane could be achieved. Reversible specific capacities of 1218 $mA\,h\,g^{-1}$ and 823 $mA\,h\,g^{-1}$ were shown for current rates of 0.1 puC and 1.0 C, respectively. A high capacity of more than 500 $mA\,h\,g^{-1}$ was retained after 500 cycles at 1.0 C.

The superior electrochemical performance of this composite could be attributed to the unique structure feature and the synergistic effect of the Mo_3Ni_3N and Mo_2C nanoparticles. The nanoparticles could suppress the dissolution and shuttling of the poly(sulfide)s and Mo_2C could improve the electronic conductivity of the composite (74).

6.9.1.4 *Benzimidazole-Containing Polymers*

Modified Poly(benzimidazole) Membranes. Poly(benzimidazole) (PBI) is a polymeric material with very high levels of heat resistance, strength, and chemical stability and may be used in many applications such as heat-resistant fibers, coating varnishes, etc.

Modified PBI membranes have been described that are fabricated from these polymers (75). These membranes have a high ionic conductivity and are suitable for solid polymer electrolytes in electrochemical applications, especially for high temperature polymer electrolyte membrane fuel cells.

Poly(benzimidazole) and poly(2,2′-(*p*-oxydiphenylene)-5,5′-bibenzimidazole) are shown in Figure 6.20.

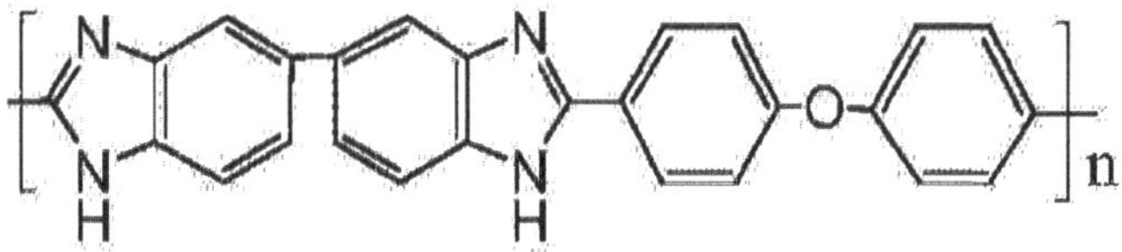

Poly(2,2′-(*p*-oxydiphenylene)-5,5′-bibenzimidazole)

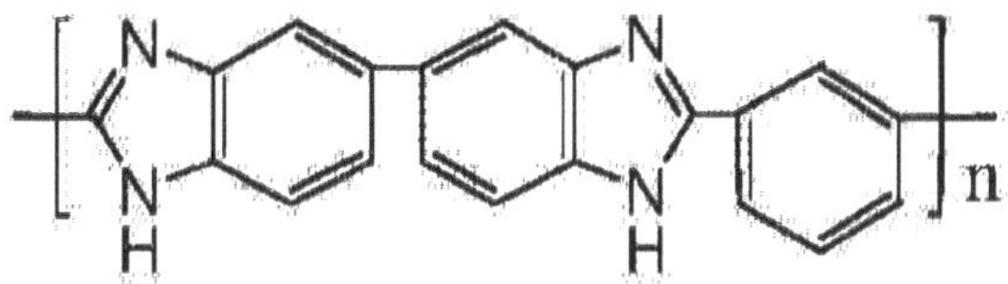

Poly(benzimidazole)

Figure 6.20 Benzimidazole-containing polymers.

Poly(2,2′-(*p*-oxydiphenylene)-5,5′-bibenzimidazole). A simple and controllable approach has been developed for the preparation of a series of poly(2,2′-(*p*-oxydiphenylene)-5,5′-bibenzimidazole) (OPBI) porous membranes with a controlled porosity and pore size (76).

The micropores were formed by simply extracting poly(ethylene glycol) (PEG) from the dry OPBI/PEG blend membranes with water. The effects of PEG molecular weight (M_w = 350–10,000 *Da*) and PEG weight fraction (OPBI/PEG = 1:1–1:5) on membrane morphology and properties were investigated (76).

It was found that a higher M_w and weight fraction of PEG in polymer blends tended to give a higher porosity but a lower mechanical strength. The porous membrane, M-4, prepared from PEG10000 (M_w = 10,000) at a weight ratio of OPBI/PEG10000 = 1:5 exhibited a high porosity of 71%, a high ionic conductivity of 1.3 $mS\,cm^{-1}$ at room temperature, and a reasonably high tensile strength of 10 *MPa*. Furthermore, the M-4 showed only a ~5% thermal shrinkage after heating at 200°C for 1 *h* and good fire-retardant properties. The charge-discharge cycle test and C-rate performance test results revealed that M-4 was a promising candidate as a separator for lithium-ion batteries (76).

An OPBI nonwoven fabric was prepared by electrospinning as a highly safe separator of lithium-ion batteries (77). The characterization and analysis of safety mechanism in terms of thermostability, flame resistance, electrochemical stability, as well as dendrite growth, were investigated. The OPBI membrane shows excellent stability, such as no shrinkage at 200°C, remaining weight above 54% at 800°C, even no combustion on fire, and electrochemical stability window exceeding 5.75 V vs. Li^+/Li.

The $Li/LiFePO_4$ cells using OPBI membrane deliver an outstanding cycling stability, a specific capacity retention rate of 94% over 300 cycles and an excellent rate capacity (77).

6.9.1.5 *Lithium bis(trifluoromethane) sulfonylimide*

Solid polymer electrolytes composed of poly(ε-caprolactone) (PCL) and a lithium bis(trifluoromethane) sulfonylimide salt, c.f. Figure 6.21, were prepared for investigating their electrochemical properties (78). The optimized solid polymer electrolyte showed ionic conductivities of 2.5×10^{-5} and $1.6 \times 10^{-4} S\,cm^{-1}$ at room temperature and 55°C, respectively. Dimensional stability was improved by infiltrating PCL-based solid polymer electrolyte into the porous polymer membranes prepared from PEI, PAN and nonwoven poly(propylene) (PP) membranes.

Due to the highly porous nature and good compatibility of the PEI membrane toward the PCL-based solid polymer electrolyte, a solid-state $Li/LiNi_{0.6}Co_{0.2}Mn_{0.2}O_2$ cell assembled with a solid polymer electrolyte employing a PEI membrane exhibited the best cycling performance. Based on these results, this highly porous PEI membrane is proposed as a promising supporting membrane for enhancing the mechanical stability of amorphous solid polymer electrolytes for applications in all-solid-state lithium batteries.

6.9.1.6 *Hexaazatrinaphthylene*

The commercialization of lithium-sulfur batteries is hampered by bottlenecks like low sulfur loading, high cathode porosity, uncontrollable Li_2S_x deposition and sluggish kinetics of Li_2S activation (79).

A densely stacked redox-active hexaazatrinaphthylene polymer was developed with a surface area of 302 $m^2\,g^{-1}$ and a very high bulk

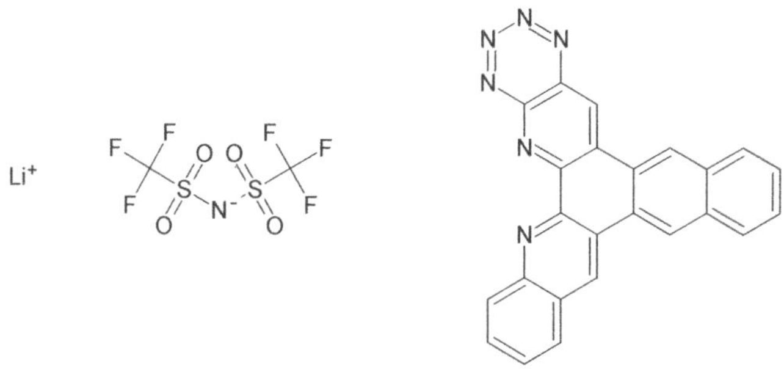

Lithium bis(trifluoromethane) sulfonylimide

Hexaazatrinaphthylene sulfonylimide

Figure 6.21 Nitrogen-containing electrolytes (78, 79).

density of ca. 1.60 $g\,m - 3$ (79). Hexaazatrinaphthylene is shown in Figure 6.21.

Uniquely, the hexaazatrinaphthylene polymer has a similar redox potential window to S, which facilitates the binding of Li_2S_x and its transformation chemistry within the bulky polymer host, leading to fast Li_2S/S kinetics.

The compact polymer sulfur electrode showed a high sulfur loading of ca. 15 $mg\,m^{-2}$ (200 μm thickness) with a low cathode porosity of 41%. It delivers a high areal capacity of ca. 14 $mAh\,cm^{-2}$ and has a good cycling stability (200 cycles) at an electrolyte-to-sulfur (E/S) ratio of 5 $\mu l\,mg^{-1}$. The assembled pouch cell delivers a cell-level high energy density of 303 $Wh\,kg^{-1}$ and 392 $Wh\,l^{-1}$ (79).

6.9.1.7 Poly(vinylidene fluoride-co-hexafluoropropylene)/Poly(1,3-dioxolane) Polymer Electrolyte

The application of solid polymer electrolytes has been impeded by the low ionic conductivity, narrow oxidation potential and poor mechanical strength (80).

An *in-situ* P(VDF-HFP)/Poly(1,3-dioxolane) polymer electrolyte was designed. The compound contains dual-salts and functional plasticizers of succinonitrile/fluoroethylene carbonate. One of the dual-salts, i.e., lithium difluoro(oxalato)borate, could initiate the polymerization of 1,3-dioxolane *in-situ* in the porous P(VDF-HFP) host to improve the interfacial compatibility. Succinonitrile is beneficial to the improvement of the ionic conductivity and the stability of high voltage cathodes, while the fluoroethylene carbonate can generate the LiF-rich section on the Li metal anode. Some of the above-mentioned compounds are shown in Figure 6.22 and Figure 6.23.

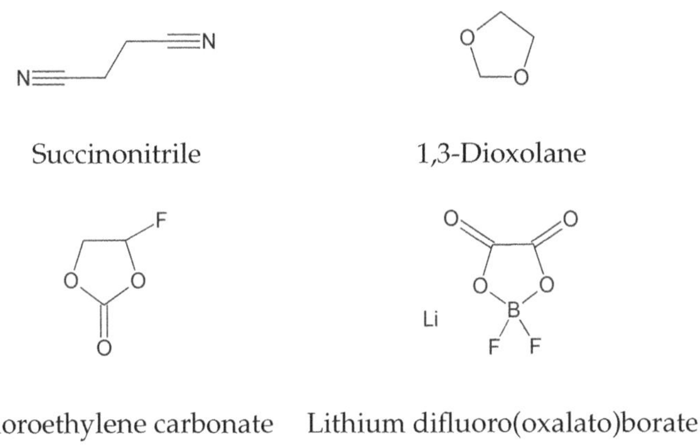

Figure 6.22 Compounds for polymer electrolytes.

The P(VDF-HFP) host can further restrict the motion of anions and increase the strength of the solid polymer electrolyte. Therefore, the as-obtained solid polymer electrolyte exhibits a thin thickness of ca. 32 μm, a high ionic conductivity of $1.06 \times 10^{-4}\ s\,cm^{-1}$ at room temperature, an enlarged electrochemical window of 5.86 V versus Li/Li^+ and a high lithium-ion transference number of 0.72 (80).

Due to this synergetic effect, the $Li/LiFePO^4$ cell delivers a high initial capacity of 141.7 $mA\,h\,g^{-1}$ at 2C with a capacity retention of 94.99% after 1000 cycles.

These three-dimensional porous film-derived *in-situ* polymer electrolytes provide a simple strategy for designing polymer electrolytes, exhibiting attractive potential in flexible and high-energy density solid-state batteries (80).

6.9.2 *Membranes for Sodium-Ion Batteries*

A porous composite membrane with P(VDF-HFP) as host and non-woven materials for sodium-ion batteries as reinforcement could be prepared (81). The monomers are shown in Figure 6.23.

Figure 6.23 Monomers for poly(vinylidene fluoride-*co*-hexafluoropropylene).

The porous P(VDF-HFP)) membrane was prepared by phase inversion according to a previous report (82).

Preparation 6–4: P(VDF-HFP) (500 *mg*) was dissolved in a mixture of distilled water and acetone (12.7 *ml*) (w(water)/w(acetone) = 9/100) at 40°C and vigorously stirred for 2 *h* to yield a homogeneous solution. Cooled to room temperature, the clear solution (10 *ml*) was poured into a glass plate (d = 15 *cm*). Afterwards, a PP nonwoven membrane (d = 14.6 *cm*) was covered on the surface of the acetone solution. A resulting composite membrane (thickness: 55 μm) incorporating nonwoven with porous P(VDF-HFP) was obtained after the solvents were completely evaporated on an electric heating plate at 40°C. There was no cover on the glass plate when casting the membrane.

The tensile strength of the composite was found to be very close to that of a Celgard 2730 separator in both wet and dry state. Celgard® products were initially developed in the mid-1960s at the Celanese Research Labs in Summit, New Jersey, USA (83). Now, Celgard is a wholly-owned subsidiary of the company Asahi Kasei. Celgard is highly active in the fabrication of, for example, improved battery separators, base films or membranes (84).

The Na^+ ionic conductivity of the given gel-type composite membrane (0.82 $mS\ cm^{-1}$) absorbed $NaClO_4$ organic liquid electrolyte at room temperature at a rate which was found to be four times higher than that of the commercial separator saturated with the same electrolyte (0.16 $mS\,cm^{-1}$) (81). It also showed enhanced transference number of Na^+ ions and good electrolyte retention even at 110°C. The electrochemical reversibility for $Na_4Mn_9O_8$ cathode has been confirmed by cyclic voltammetry by using the aforementioned gel-polymer electrolyte. The porous composite exhibits great application prospects for sodium-ion batteries with a high energy density (81).

Sodium ion batteries are considered to be a highly powerful next-generation energy storage device (85). However, the conventionally used glass fiber separators for sodium-ion batteries do not meet the necessary standards due to their randomly distributed pore structure, which causes severe safety problems and capacity decay.

A strategy was introduced to control the inner porous nanostructure of a glass fiber via non-solvent induced phase separation engineering for the gel-polymer electrolyte in sodium-ion batteries (85).

Here, it was reported how different types of porous polymer gel electrolyte inside the glass fiber matrix can affect performance of sodium-ion batteries in terms of both pore nanostructure and vertical position of polymer layer. As a result of the non-solvent induced phase separation, the optimized gel-polymer electrolyte in glass fiber facilitates increases in ionic conductivity via a modified ion transport and displays superior cell characteristics with excellent stability (85).

Room temperature sodium-sulfur (Na-S) batteries are promising energy storage devices for meeting increasing energy demands (86). An electrolyte with reasonably high ionic conductivity ($\geq 10^{-4}$ $S\,cm^{-1}$) is one of the key requirements for room temperature Na-S batteries.

A sodium ion conducting flexible porous gel-polymer electrolyte membrane of TiO_2 dispersed poly(vinylidene fluoride-hexafluoropropylene) soaked in a liquid electrolyte of $NaPF_6$ (sodium hexafluorophosphate) in ethylene carbonate and propylene carbonate has been prepared (86). The electrolyte is obtained through phase inversion method and shows an ionic conductivity of 1.3×10^{-3} $S\,cm^{-1}$.

A prototype o the Na-S battery has been fabricated and characterized using the prepared electrolyte membrane. The battery shows an open circuit voltage of ~2.1 V and a discharge capacity of 231

$mA\,h\,g^{-1}$. Unfortunately, the battery shows capacity fading during charge-discharge cycles (86).

6.9.3 *Vanadium Redox Flow Batteries*

Vanadium redox flow batteries are a technical solution for storing fluctuating renewable energies on a large scale (87). A vanadium redox flow battery device was patented in 1988 (88). Here, a sulfonated poly(ethylene) membrane or a poly(styrene sulfonic acid) membrane was selected as ionically conducting separator.

An overview about flow battery targeted membranes in the years from 1995 to 2020 has been collected. Here, more than 200 membrane samples are sorted into fluoro-carbons, hydro-carbons or *N*-heterocycles according to the basic polymer used (87).

Several publications that deal with vanadium redox flow battery membranes are shown in Table 6.7.

Table 6.7 Vanadium redox flow battery membranes (87).

Main focus	References
Historical overview of materials research	(89)
Porous membranes	(90)
Ion exchange membranes	(91)
Selectivity of ion exchange membranes	(92)
Basic scientific issues	(93)
Membranes that have received attention	(94)
Nanocomposite membranes	(95)
Technological and scientific limitations	(96)
Developments in the synthesis	(97)
Current development trends	(98)

An asymmetric porous sulfonated poly(ether sulfone) membrane with ultrahigh ion selectivity was prepared via an ionic-liquid-induced phase separation (99). The characteristics of the prepared porous membrane, including microstructure, proton conductivity, VO^{2+} permeability, mechanical strength and single cell performance, were studied.

The influence of the ionic liquid on the porous structure of the membrane was investigated. The results showed that the vanadium ion permeability of the porous sulfonated poly(ether sulfone) membrane membranes decreases by an order of magnitude as compared with Nafion212, and its proton conductivity exceeds 20 $mS\,cm^{-1}$.

With a similar energy efficiency to Nafion212, the cell with the porous sulfonated poly(ether sulfone) membrane possesses higher coulombic efficiency. At 100 $mA\,cm^{-2}$, the coulombic efficiency and energy efficiency of the single cell equipped with the porous sulfonated poly(ether sulfone) membrane are 98.8% and 84.1%, respectively, and the membranes show excellent cycling stability (99).

6.9.3.1 *Sulfonated Poly(phthalazinone ether ketone)s*

Sulfonated poly(phthalazinone ether ketone)s containing pendant phenyl moieties were synthesized and characterized (100). These membranes are thermally, chemically and mechanically stable.

The membranes are impermeable to the bulky hydrated VO^{2+} ion and exhibit a low VO^{3+} ion permeability (100).

According to a previous study (101), 4-(3,5-Dimethyl-4-hydroxyphenyl)(2H)-phthalazin-1-one, 4,4′-difluorobenzophenone, and 4-(4-hydroxyphenyl)(2H)-phthalazin-1-one were employed to synthesize a poly(ether ketone) containing 3,5-dimethyl phthalazinone moieties via a condensation polymerization process. Similarly, a poly(phthalazinone ether ketone) was prepared from 4-(4-hydroxyphenyl)(2H)-phthalazin-1-one and 4,4′-difluorobenzophenone in an equimolar ratio (102). Figure shows 6.24 4-(4-hydroxyphenyl)(2H)-phthalazin-1-one and 4,4′-difluorobenzophenone.

6.10 pH-Sensitive Gating in Membranes

Benzyl glutamate was graft-polymerized onto a porous poly(tetrafluoroethylene) membrane in order to study the effects of the pH and the ionic strength on the rate of permeation (103).

The membrane was first glow-discharged in the presence of ammonia in order to produce amino groups on the surface. After the graft polymerization the graft chains were hydrolyzed to yield poly(glutamic acid). Glutamic acid is shown in Figure 6.25.

The rate of water permeation through this poly(glutamic acid)-grafted polymer membrane was pH-dependent and found

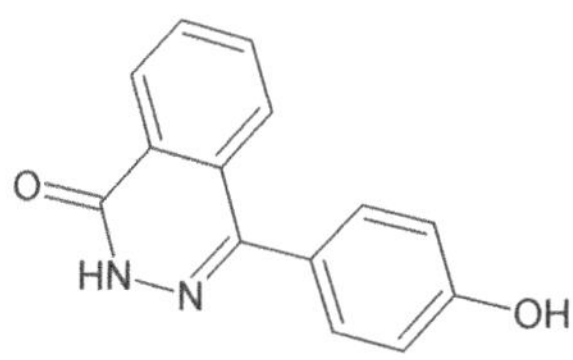

4-(4-Hydroxyphenyl)(2H)-phthalazin-1-one

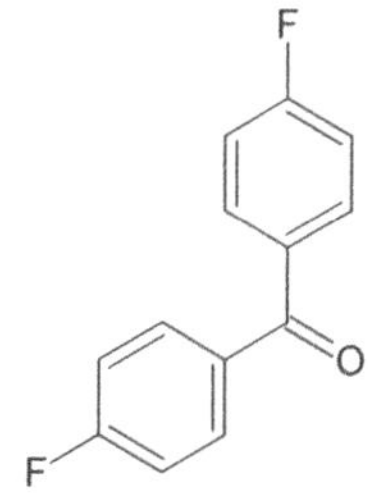

4,4′-Difluorobenzophenone

Figure 6.24 Monomers for a poly(ether ketone) membrane.

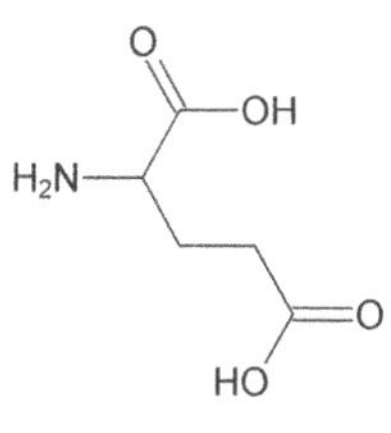

Glutamic acid

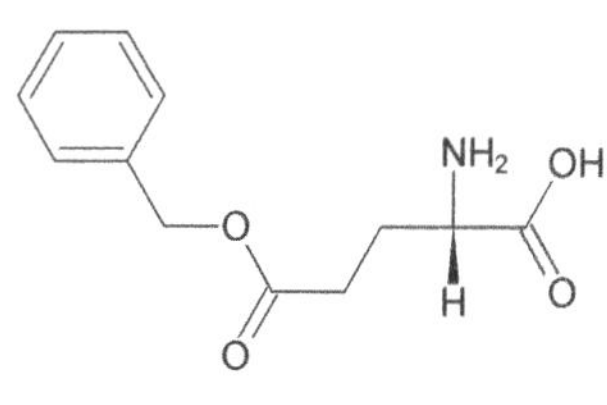

Benzyl glutamate

Figure 6.25 Glutamic compounds.

to be slow under high pH conditions and fast under low pH conditions. Under high pH conditions, randomly coiled graft chains extend to close the pores.

The chains form a helix structure and open the pores under low pH conditions. The magnitude of the permeation rate was dependent upon the length and density of graft chains. Also, the ionic strength affected the rate of permeation (103).

References

1. E.J. Jordan, Filtration apparatus comprising a membrane bioreactor and a treatment vessel for digesting organic materials, US Patent 7387723, assigned to Siemens Water Technologies Corp., June 17, 2008.
2. I.J.D. Grandine, Processes of making a porous membrane material from polyvinylidene fluoride, and products, US Patent 4203848, assigned to EMD Millipore Corp., May 20, 1980.
3. R.F.R. Leysen, P. Vermeiren, L.H.J.M. Baetsle, G.J.F. Spaepen, and J.-B.H. Vandenborre, Method of preparing a membrane consisting of polyantimonic acid powder and an organic binder, US Patent 4253936, assigned to Studiecentrum voor Kernenergie SCK, March 03, 1981.
4. D.B. Pall, Process for preparing hydrophilic polyamide membrane filter media and product, US Patent 4340479, assigned to Pall Corp., July 20, 1982.
5. T. Yamaguchi, Polymeric membrane having maleic anhydride residues, US Patent 4407975, assigned to Agency of Industrial Science and Technology, October 04, 1983.
6. P.J. Degen, I.B. Joffee, and T.C. Gsell, Charge modified polyamide membrane, US Patent 4702840, assigned to Pall Corp., October 27, 1987.
7. J. Anthony E. Allegrezza and E.C. Bellantoni, Porous membrane formed from interpenetrating polymer network having hydrophilic surface, US Patent 5158721, assigned to EMD Millipore Corp., October 27, 1992.
8. S. Raghavan, D. Jan, and R. Chilkunda, Modification of polyvinylidene fluoride membrane and method of filtering, US Patent 5531900, assigned to University of Arizona, July 02, 1996.
9. I.-F. Wang, J.F. Ditter, and R. Zepf, Highly porous polyvinylidene difluoride membranes, US Patent 6146747, assigned to USF Filtration and Separations Group Inc., November 14, 2000.

10. M. Takamura and H. Yoshida, Porous polyvinylidene fluoride resin film and process for producing the same, US Patent 6 299 773, assigned to Asahi Kasei Corp, October 09, 2001.
11. C. Stone, A.E. Steck, and B. Choudhury, Graft polymeric membranes and ion-exchange membranes formed therefrom, US Patent 6 723 758, assigned to Ballard Power Systems Inc., April 20, 2004.
12. P.D. Radovanovic and S.D. Thomas, Puncture-resistant polyolefin membranes, US Patent 6 096 213, assigned to 3M Innovative Properties Co., August 01, 2000.
13. S.J. Hamrock and D.A. Ylitalo, Polymer electrolyte membranes from mixed dispersions, US Patent 6 277 512, assigned to 3M Innovative Properties Co., August 21, 2001.
14. S.J. Hamrock, N. Jing, S.S. Mao, and L.C. Hardy, Acid functional fluoropolymer membranes and method of manufacture, US Patent 6 423 784, assigned to 3M Innovative Properties Co., July 23, 2002.
15. T.J. Blong, Extrudable fluoropolymer blends, US Patent 6 790 912, assigned to 3M Innovative Properties Co., September 14, 2004.
16. A.P. Berzinis, P. Bajaj, R.E. Halbfinger, and M. Bikel, Amphiphilic block copolymer; composition, membrane, and separation module thereof; and methods of making same, US Patent 10 358 517, assigned to SABIC Global Technologies BV, July 23, 2019.
17. A.P. Berzinis, J. de Grooth, J. de Brouwer, M. Zhou, P. Bajaj, R.E. Halbfinger, and K.J. Narang, Porous membranes and associated separation modules and methods, US Patent 10 307 717, assigned to SABIC Global Technologies BV, June 04, 2019.
18. B. Brown and E. Ray, Process for manufacturing cellulosic reverse osmosis membranes using a very high temperature initial aqueous quench, US Patent 3 792 135, assigned to Eastman Kodak Co., February 12, 1974.
19. N.A.A.H. El Badawi, A.M.K. Esawi, and A.R. Ramadan, Polymer-carbon nanotube nanocomposite porous membranes, US Patent 10 898 865, assigned to American University in Cairo, January 26, 2021.
20. A.M. Hollman and D. Bhattacharyya, *Langmuir*, Vol. 18, p. 5946, 2002.
21. J. Germain, J. Hradil, J.M. Fréchet, and F. Svec, *Chemistry of Materials*, Vol. 18, p. 4430, 2006.
22. M.P. Tsyurupa and V.A. Davankov, *Reactive and Functional Polymers*, Vol. 66, p. 768, 2006.
23. B. Li, R. Gong, W. Wang, X. Huang, W. Zhang, H. Li, C. Hu, and B. Tan, *Macromolecules*, Vol. 44, p. 2410, 2011.
24. C.D. Wood, B. Tan, A. Trewin, F. Su, M.J. Rosseinsky, D. Bradshaw, Y. Sun, L. Zhou, and A.I. Cooper, *Advanced Materials*, Vol. 20, p. 1916, 2008.

25. Y. Luo, B. Li, W. Wang, K. Wu, and B. Tan, *Advanced Materials*, Vol. 24, p. 5703, 2012.
26. S. Xu, Y. Luo, and B. Tan, *Macromolecular Rapid Communications*, Vol. 34, p. 471, 2013.
27. Z.-A. Qiao, S.-H. Chai, K. Nelson, Z. Bi, J. Chen, S.M. Mahurin, X. Zhu, and S. Dai, *Nature Communications*, Vol. 5, p. 3705, 2014.
28. Y. Zhu, Z. Pan, J. Rong, K. Mao, D. Yang, T. Zhang, J. Xu, F. Qiu, J. Pan, and H. Yu, *Applied Surface Science*, Vol. 581, p. 152349, 2022.
29. M. Rezakazemi, A. Ghafarinazari, S. Shirazian, and A. Khoshsima, *Polymer Engineering & Science*, Vol. 53, p. 1272, 2013.
30. G.M. Wesner and United States. Environmental Protection Agency, *Energy Conservation in Municipal Wastewater Treatment*, number 32 in Energy Conservation in Municipal Wastewater Treatment, Office of Water Program Operations, 1978.
31. K. Carns, *Proceedings of the Water Environment Federation*, Vol. 2005, p. 7650, 2005.
32. Risk Reduction Engineering Laboratory (U.S.), *Design Manual: Fine Pore Aeration Systems*, EPA 625/1, U.S. Environmental Protection Agency, Office of Research and Development, Center for Environmental Research Information, 1989.
33. N.K. Shammas, Fine pore aeration of water and wastewater in L.K. Wang, Y.-T. Hung, and N.K. Shammas, eds., *Advanced Physicochemical Treatment Technologies*, pp. 391–448. Humana Press, Totowa, NJ, 2007.
34. D.F. Rodrigues, Porous nanocomposite polymers for water treatment, US Patent Application 20 150 353 381, assigned to University of Houston System, December 10, 2015.
35. S. Ashok Kumar, G. Srinivasan, and S. Govindaradjane, *Environmental Nanotechnology, Monitoring & Management*, Vol. 12, p. 100238, 2019.
36. G. Chen, R. Liu, H.K. Shon, Y. Wang, J. Song, X.-M. Li, and T. He, *Desalination*, Vol. 405, p. 76, 2017.
37. T.W.T. Peeters, Hybrid wastewater treatment, US Patent 9 758 405, assigned to HaskoningDHV Nederland BV, September 12, 2017.
38. P. Artiga, V. Oyanedel, J.M. Garrido, and R. Méndez, *Desalination*, Vol. 179, p. 171, 2005. Membranes in Drinking and Industrial Water Production.
39. P. Artiga, G. García-Toriello, R. Méndez, and J.M. Garrido, *Desalination*, Vol. 221, p. 518, 2008. European Desalination Society and Center for Research and Technology Hellas (CERTH), Sani Resort 22–25 April 2007, Halkidiki, Greece.
40. P. Drogui, B. Seyhi, G. Buelna, and J.-F. Blais, Submerged membrane bioreactor system and biological methods for removing bisphenol compounds from municipal wastewater, US Patent Application

20110315629, assigned to Centre de Recherche Industrielle du Québec CRIQ, Institut National de La Recherche Scientifique INRS, December 29, 2011.

41. P. Drogui, B. Seyhi, G. Buelna, and J.-F. Blais, Submerged membrane bioreactor system and biological methods for removing bisphenol compounds from municipal wastewater, US Patent 9199200, assigned to Centre de Recherche Industrielle du Québec CRIQ, Institut National de La Recherche Scientifique INRS, December 01, 2015.
42. G.D. Vuković, A.D. Marinković, S.D. Škapin, M. Ristić, R. Aleksić, A.A. Perić-Grujić, and P.S. Uskoković, *Chemical Engineering Journal*, Vol. 173, p. 855, 2011.
43. Y. Tian, B. Gao, V.L. Morales, L. Wu, Y. Wang, R. Muñoz-Carpena, C. Cao, Q. Huang, and L. Yang, *Chemical Engineering Journal*, Vol. 210, p. 557, 2012.
44. S.-G. Wang, W.-X. Gong, X.-W. Liu, Y.-W. Yao, B.-Y. Gao, and Q.-Y. Yue, *Separation and Purification Technology*, Vol. 58, p. 17, 2007. Application of Nanotechnologies in Separation and Purification.
45. Y.-H. Li, Z. Di, J. Ding, D. Wu, Z. Luan, and Y. Zhu, *Water Research*, Vol. 39, p. 605, 2005.
46. M. Randelović, M. Purenović, A. Zarubica, J. Purenović, B. Matović, and M. Momčilović, *Journal of Hazardous Materials*, Vol. 199-200, p. 367, 2012.
47. S.Z. Mohammadi, M.A. Karimi, D. Afzali, and F. Mansouri, *Desalination*, Vol. 262, p. 86, 2010.
48. R.P. Medina, E.T. Nadres, F.C. Ballesteros, and D.F. Rodrigues, *Environ. Sci.: Nano*, Vol. 3, p. 638, 2016.
49. J. Acharya, J.N. Sahu, C.R. Mohanty, and B.C. Meikap, *Chemical Engineering Journal*, Vol. 149, p. 249, 2009.
50. Y. Tan, M. Chen, and Y. Hao, *Chemical Engineering Journal*, Vol. 191, p. 104, 2012.
51. X. Zheng, C. Ni, W. Xiao, Y. Liang, and Y. Li, *Chemical Engineering Journal*, Vol. 428, p. 132111, 2022.
52. S. Amiri, A. Asghari, V. Vatanpour, and M. Rajabi, *Journal of Environmental Management*, Vol. 294, p. 112918, 2021.
53. J. Bai, X. Ma, H. Yan, J. Zhu, K. Wang, and J. Wang, *Microporous and Mesoporous Materials*, Vol. 306, p. 110441, 2020.
54. J. Bai, X. Ma, C. Gong, Y. Chen, H. Yan, K. Wang, and J. Wang, *Journal of Molecular Liquids*, Vol. 320, p. 114443, 2020.
55. L. Zhao, J. Qiao, M.H. Moon, and L. Qi, *Analytical and Bioanalytical Chemistry*, Vol. 410, p. 5211, 2018.
56. J. Shen, J. Qiao, and L. Qi, *ACS Applied Bio Materials*, Vol. 4, p. 966, 2021.
57. J.F. Jiang, J. Qiao, X.Y. Mu, M.H. Moon, and L. Qi, *Talanta*, Vol. 165, p. 251, 2017.

58. J. Qiao, J. Jiang, L. Liu, J. Shen, and L. Qi, *ACS Applied Materials & Interfaces*, Vol. 11, p. 15133, 2019.
59. Wikipedia contributors, Michaelis-Menten kinetics — Wikipedia, the free encyclopedia, https://en.wikipedia.org/w/index.php?title=Michaelis%E2%80%93Menten_kinetics&oldid=1043460338, 2021. [Online; accessed 18-September-2021].
60. J. Shen, J. Qiao, D.-P. Kim, and L. Qi, *Talanta*, Vol. 234, p. 122676, 2021.
61. L. Liu, J. Qiao, H. Zhang, and L. Qi, *Talanta*, Vol. 216, p. 120963, 2020.
62. G. Inoue and M. Kawase, *International Journal of Hydrogen Energy*, Vol. 41, p. 21352, 2016.
63. S. Chaen, T. Yamanaka, K. Yasuda, M. Yamada, N. Uraoka, T. Konishi, K. Inui, Y. Shibuya, and M. Ono, Porous body, polymer electrolyte membrane, filter material for filter, and filter unit, US Patent 10 688 448, assigned to Daikin Industries Ltd., June 23, 2020.
64. H. Tatsuno and M. Nakamura, Porous electrode substrate, membrane-electrode assembly using same, and polymer electrolyte fuel cell using same, US Patent 10 727 497, assigned to Mitsubishi Chemical Corp., July 28, 2020.
65. C. Dong, X. Xu, J. Zhang, H. Wang, Y. Xiang, H. Zhu, M. Forsyth, and S. Lu, *International Journal of Hydrogen Energy*, p. in press, 2022.
66. B. Zheng, Z. Wang, and Y. Wang, *International Journal of Energy Research*, Vol. n/a, p. 1, 2022.
67. V. Ahilan, M. Wilhelm, and K. Rezwan, *Ceramics International*, Vol. 44, p. 19191, 2018.
68. S. Gong, S. Du, J. Kong, Q. Zhai, F. Lin, S. Liu, N.R. Cameron, and W. Cheng, *Small*, Vol. 16, p. 2003269, 2020.
69. S. Shanmugam, K. Ketpang, M.A. Aziz, K. Oh, K. Lee, B. Son, and N. Chanunpanich, *Electrochimica Acta*, Vol. 384, p. 138407, 2021.
70. B. Park, C. Oh, S. Yu, B. Yang, N.V. Myung, P.W. Bohn, and J.L. Schaefer, *Journal of The Electrochemical Society*, Vol. 169, p. 010511, January 2022.
71. A. Subramania, N.T. Kalyana Sundaram, A.R. Sathiya Priya, and G. Vijaya Kumar, *Journal of Membrane Science*, Vol. 294, p. 8, 2007.
72. P. Yan, Z. Huang, Y. Lin, X. Wu, Y. Yang, D. Wang, F. Chen, C. Zhang, and D. He, *Ionics*, Vol. 21, p. 593, 2015.
73. J. Liu, C. Wang, Q. Li, J. Li, Y. Chen, F. Song, Z. Hong, M. Liu, L. Bai, and F. Zeng, *International Journal of Energy Research*, Vol. 45, p. 16603, 2021.
74. Y.-S. Liu, Y.-L. Bai, X. Liu, C. Ma, X.-Y. Wu, X. Wei, Z. Wang, K.-X. Wang, and J.-S. Chen, *Chemical Engineering Journal*, Vol. 378, p. 122208, 2019.
75. I. Cabasso, Y. Yuan, and F.E. Johnson, Modified polybenzimidazole (PBI) membranes for enhanced polymer electrochemical cells,

US Patent 6 987 163, assigned to Research Foundation for the State University of New York, January 17, 2006.
76. N. Liang, J. Fang, and X. Guo, *J. Mater. Chem. A*, Vol. 5, p. 15087, 2017.
77. X. Liu, B. Zhang, Y. Wu, J. Chen, M. Fang, L. Wang, and L. Wang, *Journal of Power Sources*, Vol. 475, p. 228624, 2020.
78. Y. Seo, Y.-C. Jung, M.-S. Park, and D.-W. Kim, *Journal of Membrane Science*, Vol. 603, p. 117995, 2020.
79. X. Wang, Y. Yang, C. Lai, R. Li, H. Xu, D.H.S. Tan, K. Zhang, W. Yu, O. Fjeldberg, M. Lin, W. Tang, Y.S. Meng, and K.P. Loh, *Angewandte Chemie International Edition*, Vol. 60, p. 11359, 2021.
80. Y. Liu and Y. Xu, *Chemical Engineering Journal*, Vol. 433, p. 134471, 2022.
81. Y. Zhu, Y. Yang, L. Fu, and Y. Wu, *Electrochimica Acta*, Vol. 224, p. 405, 2017.
82. W. Pu, X. He, L. Wang, C. Jiang, and C. Wan, *Journal of Membrane Science*, Vol. 272, p. 11, 2006.
83. Celgard, Battery fabrication, Electronic: https://www.celgard.com/, 2021.
84. X. Zhang, G.P. Rumierz, K.F. Humiston, C.E. Haire, T.S. Fields, M.A. Braswell, R.A. Proctor, and R.E. Smith, Separators, batteries, systems, vehicles, and related methods, US Patent 10 944 087, assigned to Celgard LLC, March 09, 2021.
85. J.I. Kim, K.Y. Chung, and J.H. Park, *Journal of Membrane Science*, Vol. 566, p. 122, 2018.
86. H. Verma, K. Mishra, and D. Rai, *Materials Today: Proceedings*, Vol. 28, p. 346, 2020. International Conference on Advanced Materials and Nanotechnology.
87. D. Düerkop, H. Widdecke, C. Schilde, U. Kunz, and A. Schmiemann, *Membranes*, Vol. 11, 2021.
88. M. Skyllas-Kazacos, M. Rychick, and R. Robins, All-vanadium redox battery, US Patent 4 786 567, assigned to Unisearch Ltd., November 22, 1988.
89. A. Parasuraman, T.M. Lim, C. Menictas, and M. Skyllas-Kazacos, *Electrochimica Acta*, Vol. 101, p. 27, 2013.
90. Z. Yuan, H. Zhang, and X. Li, *Chem. Commun.*, Vol. 54, p. 7570, 2018.
91. X. Li, H. Zhang, Z. Mai, H. Zhang, and I. Vankelecom, *Energy Environ. Sci.*, Vol. 4, p. 1147, 2011.
92. T. Luo, S. Abdu, and M. Wessling, *Journal of Membrane Science*, Vol. 555, p. 429, 2018.
93. B. Schwenzer, J. Zhang, S. Kim, L. Li, J. Liu, and Z. Yang, *ChemSusChem*, Vol. 4, p. 1388, 2011.
94. H. Prifti, A. Parasuraman, S. Winardi, T.M. Lim, and M. Skyllas-Kazacos, *Membranes*, Vol. 2, p. 275, 2012.
95. S.-H. Cha, *Journal of Nanomaterials*, Vol. 2015, p. 207525, 2015.

96. J.R. Varcoe, P. Atanassov, D.R. Dekel, A.M. Herring, M.A. Hickner, P.A. Kohl, A.R. Kucernak, W.E. Mustain, K. Nijmeijer, K. Scott, T. Xu, and L. Zhuang, *Energy Environ. Sci.*, Vol. 7, p. 3135, 2014.
97. S. Maurya, S.-H. Shin, Y. Kim, and S.-H. Moon, *RSC Adv.*, Vol. 5, p. 37206, 2015.
98. L. Gubler, *Current Opinion in Electrochemistry*, Vol. 18, p. 31, 2019.
99. X. Zhou, R. Xue, Y. Zhong, Y. Zhang, and F. Jiang, *Journal of Membrane Science*, Vol. 595, p. 117614, 2020.
100. L. Chen, S. Zhang, Y. Chen, and X. Jian, *Journal of Power Sources*, Vol. 355, p. 23, 2017.
101. S. Zhang, B. Zhang, D. Xing, and X. Jian, *Journal of Materials Chemistry A*, Vol. 1, p. 12246, 2013.
102. Y. Chen, S. Zhang, Q. Liu, and X. Jian, *RSC Advances*, Vol. 9, p. 26097, 2019.
103. Y. Ito, Y. Ochiai, Y.S. Park, and Y. Imanishi, *Journal of the American Chemical Society*, Vol. 119, p. 1619, 1997.

7

Separation Methods

7.1 Chromatography

7.1.1 Solid Phase Extraction

Porous polymer sorbents for solid phase extraction (SPE) emerged in the 1980s to overcome the limitations of silica and carbon-based sorbents. They have continued to be one of the materials most used for SPE because they have high retention features and a broad availability of interactions (1).

Solid phase extraction is a chromatographic technique which is widely used, e.g., for pre-concentration and cleanup of analytical samples, for purification of various chemicals, and for removal of toxic or valuable substances from aqueous solutions. SPE is usually performed using a column or cartridge containing an appropriate resin (2).

The SPE procedures have been developed using sorbents which can interact with analytes by hydrophobic, ion exchange, chelation, sorption, and other mechanisms, to bind and remove the analytes from fluids.

Since different SPE applications can require different sorbents, there is a need for sorbents that have unique selectivities (2).

A porous resin can be formed by copolymerizing at least one hydrophobic monomer and at least one hydrophilic monomer so as to form a copolymer, and subjecting the copolymer to a sulfonation reaction so as to form a sulfonated copolymer comprising at least one ion exchange functional group, at least one hydrophilic component and at least one hydrophobic component (2). Preferably, the porous resin is a sulfonated poly(divinylbenzene-*co*-*N*-vinylpyrrolidone).

The preparation of such materials is similar to those described in Section 7.1.2.1.

Mass balance experiments were performed on SPE fractions using several model compounds such as acetaminophen, caffeine, toluamide, procainamide, amphetamine, methamphetamine, and toluidine. As sorbents, three different compounds were used. A 1 *ml* sample was loaded which consisted of either spiked phosphate buffered saline, or spiked porcine plasma. The samples were spiked to 10 $\mu g\, mL^{-1}$ with acetaminophen, toluamide, caffeine and procainamide, and to 20 $\mu g\, mL^{-1}$ with amphetamine, methamphetamine, and toluidine. Some of these compounds are shown in Figure 7.1.

The experiments demonstrated that sulfonated poly(divinylbenzene-*co*-*N*-vinylpyrrolidone) can be used with a generic procedure for SPE of basic compounds. In addition, it can be used as a tool to perform class fractionation of neutral and basic compounds (2).

7.1.2 *Liquid Chromatography*

Liquid chromatography columns have been extensively developed and are used routinely in both analytical and preparative chromatography. The separation in a chromatography column of a sample, also termed an analyte or solute, that contains a mixture of components is achieved by dissolving the sample in a liquid mobile phase and passing the mobile phase through a stationary phase typically packed within a tubular column.

This causes the sample to separate into its components due to different partitioning between the mobile and stationary phases of the different components, i.e., the components have different partition coefficients (3). The stationary phase is typically in the form of a bed of particles packed within the column.

7.1.2.1 *Porous Copolymers*

A porous material has been described that consists of a random copolymer of a least one hydrophobic monomer and at least one

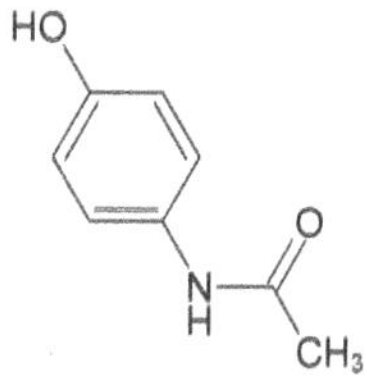

Acetaminophen

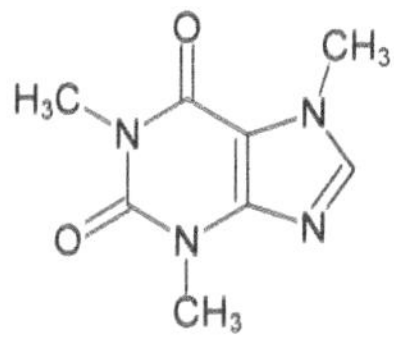

Caffeine

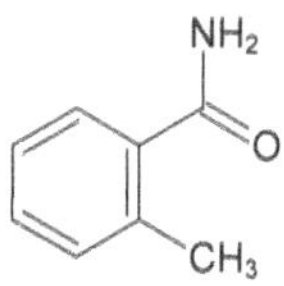

Toluamide

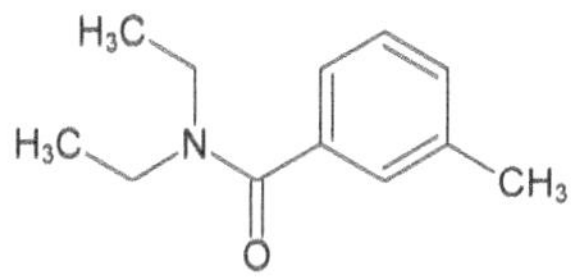

Diethyltoluamid

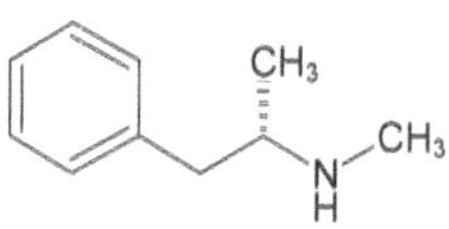

Methamphetamine

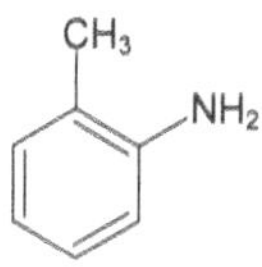

o-Toluidine

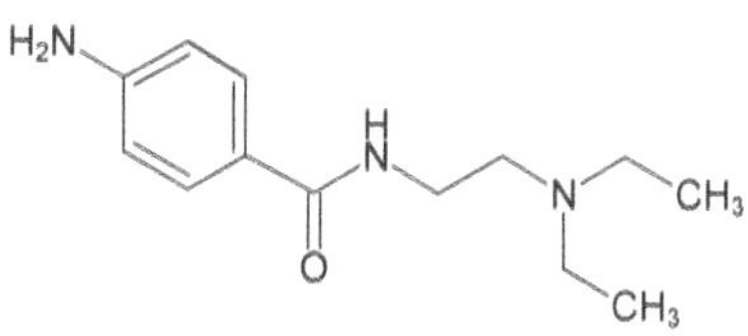

Procainamide

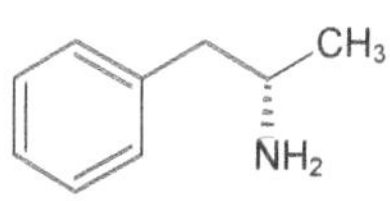

Amphetamine

Figure 7.1 Compounds for spiking.

hydrophilic monomer (4). For example, the copolymer can be poly(divinylbenzene-*co*-*N*-vinylcaprolactam).

A ion exchange functional moiety is added to improve the properties. The ion exchange functional moiety can be formed from a substituted acyclic amine or a substituted cyclic amine. The substitution can be at any of the ring atoms, including heteroatoms.

Compounds for such an acyclic amine substituted with an electron-withdrawing group are shown in Table 7.1. Some of these compounds are shown in Figure 7.2.

Cyclic secondary amines substituted with an electron-withdrawing group are shown in Table 7.2. Some cyclic secondary amines are shown in Figure 7.3.

Table 7.1 Acyclic amines (4).

Compound	Compound
Benzyl amine	*N*-Methylbenzyl amine
N-Ethylbenzyl amine	*N*-Propylbenzyl amine
N-Butylbenzyl amine	*N*-Pentylbenzyl amine
N-Hexylbenzyl amine	*N*-Heptylbenzyl amine
N-Octylbenzyl amine	*N*-Nonylbenzyl amine
N-Decylbenzyl amine	*N*-Undecylbenzyl amine
N-Dodecylbenzyl amine	*N*-Tridecylbenzyl amine
N-Tetradecylbenzyl amine	*N*-Pentadecylbenzyl amine
N-Hexadecylbenzyl amine	*N*-Heptadecylbenzyl amine
N-Octadecylbenzyl amine	Dibenzyl amine
N-Methylaniline	*N*-Ethylaniline
N-Propylaniline	*N*-Butylaniline
N-Pentylaniline	*N*-Hexylaniline
N-Heptylaniline	*N*-Octylaniline
N-Nonylaniline	*N*-Decylaniline
N-Undecylaniline	*N*-Dodecylaniline
N-Tridecylaniline	*N*-Tetradecylaniline
N-Pentadecylaniline	*N*-Hexadecylaniline
N-Heptadecylaniline	*N*-Octadecylaniline
Bis(2,2,2-trifluoromethyl)amine	Phenethylamine
N-Methylphenethylamine	4-Methylphenethylamine
3-Phenylpropylamine	1-Methyl-3-phenylpropylamine
N-Isopropylbenzyl amine	4-Phenylbutylamine

The porous materials can be prepared via a number of processes and mechanisms, including chain addition and step condensation

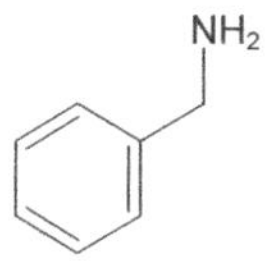

Benzyl amine

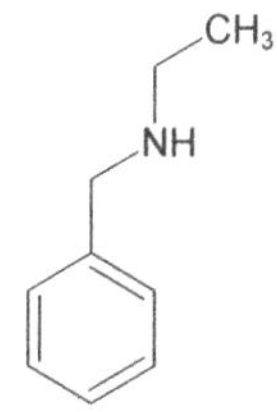

N-Ethylbenzyl amine

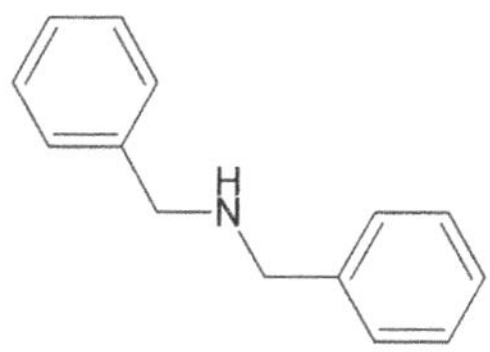

Dibenzyl amine

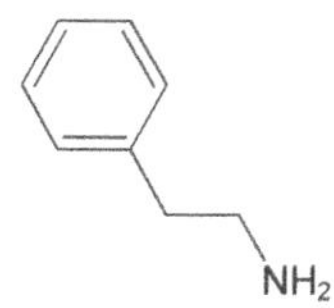

Phenethylamine

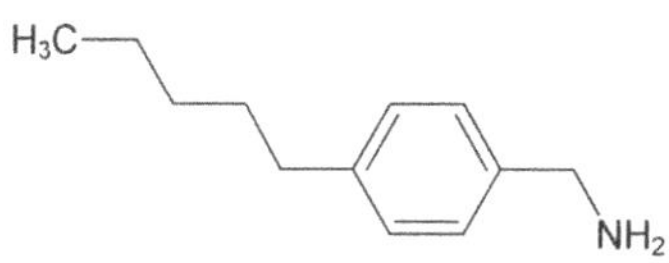

1-(4-Pentylphenyl)methanamine

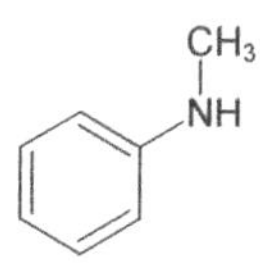

N-Methylaniline

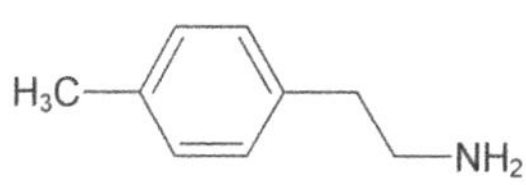

4-Methylphenethylamine

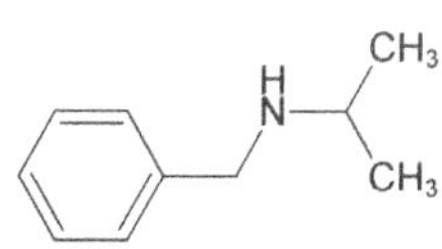

4-Phenylbutylamine

Figure 7.2 Acyclic amines.

Table 7.2 Cyclic secondary amines (4).

Compound	Compound	Compound	Compound
Oxazetane	Oxazolane	Oxazinane	Oxazepane
Oxazocane	Oxazonane	Oxazecane	Thiazetane
Thiazepane	Thiazocane	Thiazonane	Thiazecane

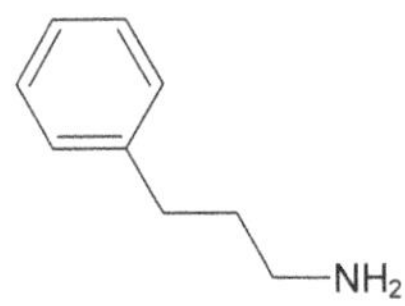

3-Phenylpropylamine

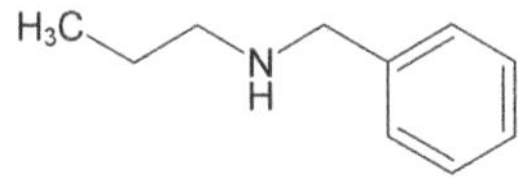

N-Propylbenzyl amine

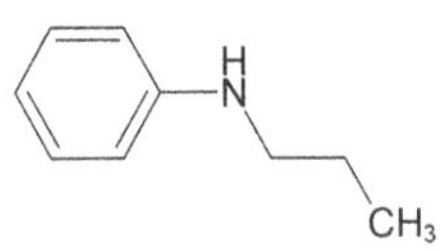

N-Propylaniline

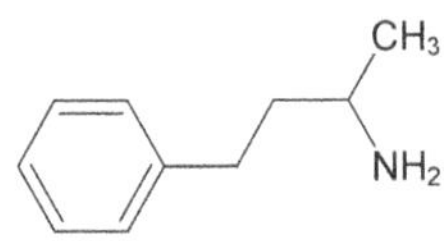

1-Methyl-3-phenylpropylamine

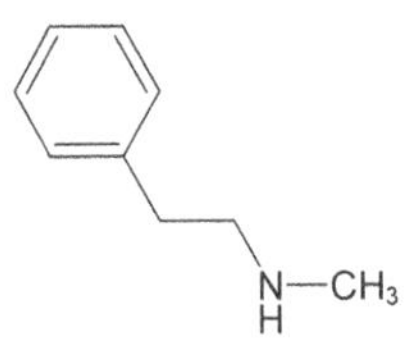

N-Methylphenethylamine

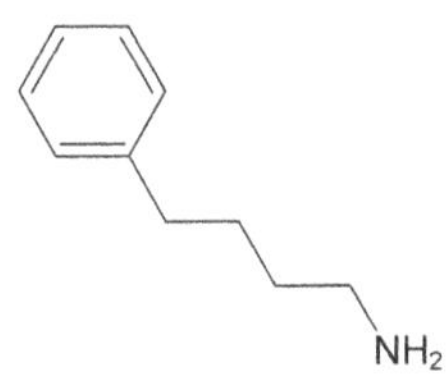

N-Isopropylbenzyl amine

Figure 7.2 (cont) Acyclic amines.

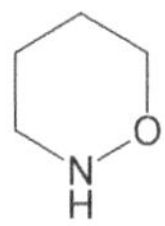

1,2-Oxazinane

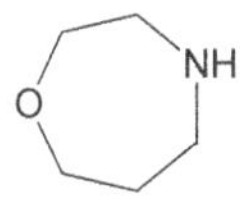

1,4-Oxazepane

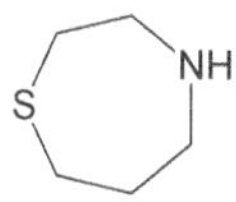

1,4-Thiazepane

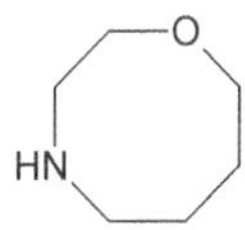

1,4-Oxazocane

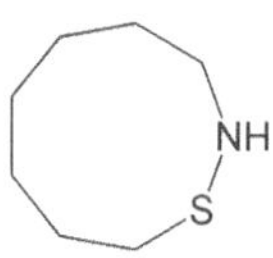

Thiazonane

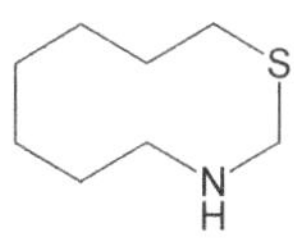

1,3-Thiazecane

Figure 7.3 Cyclic secondary amines (4).

processes, radical, anionic, cationic, ring-opening, group transfer, metathesis, and photochemical mechanisms (4). The copolymer can be prepared via standard synthetic methods.

In addition, a ion exchange functional moiety can be formed by the reaction of the materials with hydrogen peroxide.

Several examples of the synthesis of porous particles have been detailed (4).

A material using *N*-vinylcaprolactam with a high nitrogen content and a high standard average pore diameter was prepared as follows (4):

Preparation 7–1: First, 5 *g* of Methocel E-15 (a low molecular weight hydroxypropyl methylcellulose thickener from DuPont) was dissolved in 1 *l* of water at 90°C, then cooled to room temperature. In a separate flask, 174.5 *g* of divinylbenzene, 139.2 *g* of *N*-vinylcaprolactam, and 1.9 *g* of 2,2′-azobisisobutyronitrile were dissolved and mixed in 243 *g* of toluene. The organic and aqueous phases were combined in a 3 *l* glass kettle and stirred at 400 *rpm* with an overhead stirrer, while running through a static mixing loop for 30 *min*. The emulsion droplet size was checked after 30 *min*.

Once the desired droplet size range had been reached (by adjusting the rate of flow through the static mixer), the static mixing loop was turned off. Then, the mixture was heated at 70°C for 16 *h* and then cooled to room temperature. The mixture was filtered using a 20 *μm* polyester filter cloth, washed 3 times with 600 *ml* of methanol and dried for 16 *h* in vacuum at 80°C. This material has nitrogen content of 3.34%, a surface area of 562 $m^2 g^{-1}$, and a Brunauer-Emmett-Teller (BET) average pore size of 88.5 Å.

Methocel is shown in Figure 7.4. The here described porous materials have certain chemical affinities or attractions between the materials and certain molecules, particularly biological or biochemical molecules such as proteins, peptides, hormones, oligonucleotides, polynucleotides, vitamins, cofactors, metabolites, lipids and carbohydrates. As such, these materials may be used to selectively adsorb and isolate certain biomolecules for analysis and or quantification (4).

For testing, a series of polythymidine oligonucleotides were analyzed for recovery. Thymidine is shown in Figure 7.5. Also, some proteins were tested with good results (4).

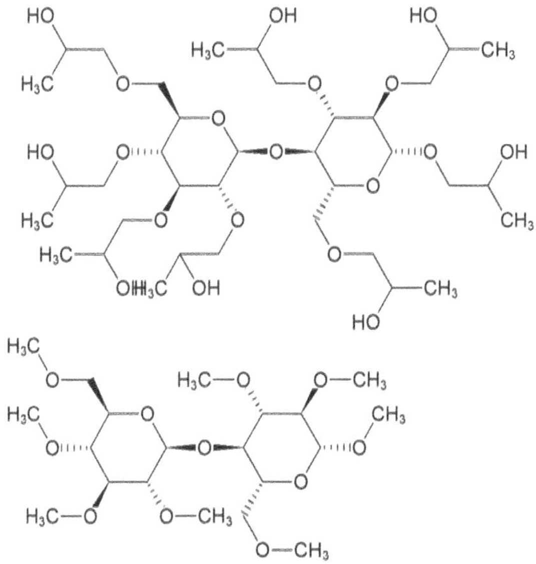

Figure 7.4 Methocel.

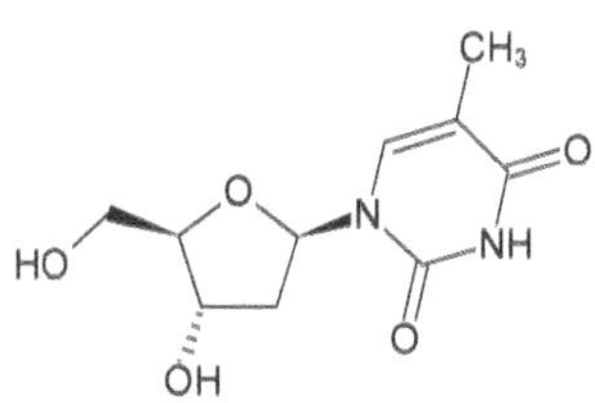

Figure 7.5 Thymidine.

7.1.2.2 *Superficially Porous Silica*

Porous silica particles are commonly used, which contain a network of pores to increase the surface area of the stationary phase and thus improve the capacity of the separation. The pores may be micropores of less than 2 *nm* size, mesopores of 2 *nm* to 50 *nm*, or macropores of greater than 50 *nm*.

A method for the preparation of superficially porous silica is a coacervation method (5). The excellent kinetic properties of these column packings allow stable, high-resolution gradient chromatography of polypeptides, proteins, nucleic acids, or DNA fragments. Here, solid, i.e., nonporous, silica microspheres were coated by a coacervate of a polymer and silica sol, with the polymer being subsequently removed by heating at a high temperature.

Similarly, superficially porous particles were formed by a two-step synthesis. This process runs as (6):

1. Providing a substantially nonporous hybrid core material, and
2. Applying to the core material one or more layers of porous shell material to form a superficially porous material.

Furthermore, superficially porous silica particles were synthesized by a one-pot process (3). This process consists of hydrolyzing and condensing a silica precursor with a functional group to form superficially porous particles. The particles are useful as a stationary phase in liquid chromatography. They allow a fast mass transfer and a fast separation of the used samples.

The following chemicals and reagents were used: 3-Mercaptopropyl trimethoxysilane (MPTMS), c.f. Figure 7.30, (95%), Cetyl trimethylammonium bromide (CTAB) (=98%), c.f. Figure 7.30, Ammonium hydroxide solution (reagent grade, 28-30% NH_3 basis), Hydrophilic polymers: Poly(vinyl alcohol) (PVA) (M_w 10K), poly(*N*-vinyl-2-pyrrolidone) (M_w 8K), hydroxypropyl methyl cellulose (M_w 10K), and methanol.

The preparation of spheres-on-sphere silica has been detailed as follows (3):

Preparation 7–2: The hydrophilic polymer (0.25 *g*) and CTAB (0.1 *g*) were dissolved in 5 *g* water. To this solution, 8 *ml* methanol was added while stirring. The ammonium hydroxide solution (0.5 *g*) was diluted by mixing with 2 *ml* water and then 2 *ml* of the diluted ammonia solution was added into the reaction mixture. After stirring for 15 *min*, 0.5 *ml* MPTMS was added dropwise over a 30 *s* period. The concentrations of polymer and surfactant in these examples were referred to as the volume of water. The reaction was stirred for 24 *h* at room temperature.

The resulted silica particles were collected by centrifuging the suspensions. These collected silica particles were calcined in a furnace (Carbolite, CWF1500) to remove the organic template and other organic components such as the functional groups. The calcining condition: heat at 1°C/*min* in air to 550°C–1250°C, hold for 300 *min*, and then cool down to room temperature at 5°C/*min*.

7.1.2.3 Chiral Stationary Phases

Several examples of polymeric types of chiral selectors that can be used for chromatography have been discussed, such as polysaccharides (7, 8).

Isopropyl-substituted cyclofructan chiral stationary phases were synthesized according to procedures also described for high-performance liquid chromatography (HPLC) (9, 10).

Cyclofructans can be used as chiral selectors in their native form or in a derivatized format (11, 12). The derivatization of the cyclofructan molecules can take place before or after they are immobilized on the superficially porous particle silica. The derivatizing groups are either alkanes or aromatic groups.

Ligand exchange chiral stationary phases (CSP) were synthesized by derivatizing penicillamine with dodecylbenzoyl chloride. These compounds are shown in Figure 7.6

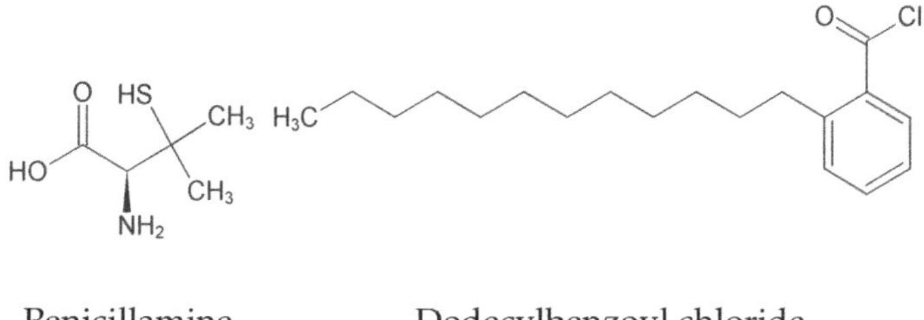

Penicillamine Dodecylbenzoyl chloride

Figure 7.6 Compounds for ligand exchange chiral stationary phases.

This purified product was then irreversibly adsorbed onto C_{18} superficially porous particles and C_{18} fully porous particles for comparison. Ion exchange CSPs were made by immobilizing quinine or quinidine via the method of Lammerhofer and Lindner (13).

Examples of brush-type chiral selectors are shown in Table 7.3.

Table 7.3 Brush-type chiral selectors (12).

Selector Class	Examples
Oligosaccharides, cyclic oligosaccharides, and their derivatives	Cyclodextrins Derivatized cyclodextrin Cyclofructans Derivatized cyclofructans
Peptides, glycopeptides, macrocyclic glycopeptides, and their derivatives	Teicoplanin Vancomycin Teicoplanin aglycone Ristocetin A Dalbavancin Boromycin
pi-Complexes	DNB-phenylglycine DNB-diphenylethylene diamine DNB-N-(1,2,3,4-tetrahydro-phenanthren-4-yl)benzamide DNB-1,2-diaminocyclohexane
Chiral crown ethers	3,3′-diphenyl-binaphthyl functionalized 18-crown-6
Ligand exchangers	Proline, Penicillamine, Hydroxyproline, Quinine
Ion exchangers	Derivatized quinine Quinidine Derivatized quinidine

Examples of macrocyclic glycopeptide (antibiotic) chiral selectors are shown in Figure 7.7.

β-Cyclodextrin was functionalized by a derivation of its reactive hydroxyl group on the ring due to its special chiral environment and structural characteristics (14). This property can be used to identify or separate a variety of chiral substances.

A series of excellent CSP for high-performance liquid chromatography were developed for enantioseparation by using anhydride modified β-cyclodextrin bearing chiral (R/S)-α-phenethylamine or (S)-(+)-2-amino-1-propanol .

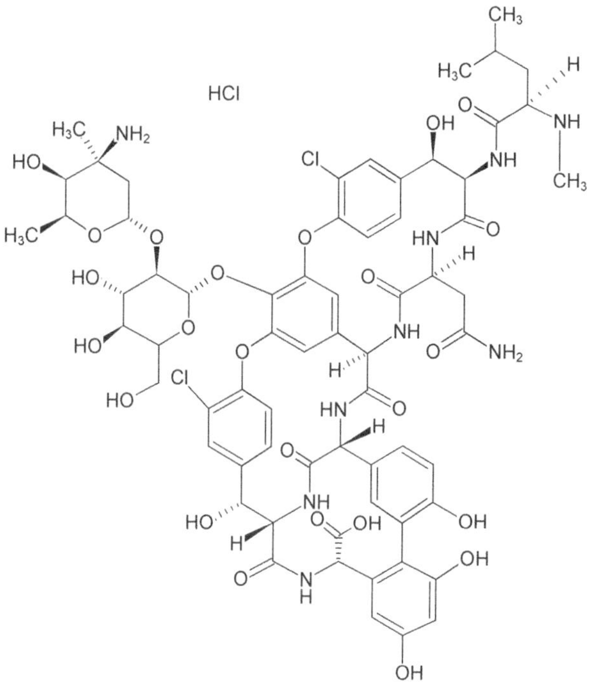

Vancomycin

Figure 7.7 Macrocyclic chiral selectors (12).

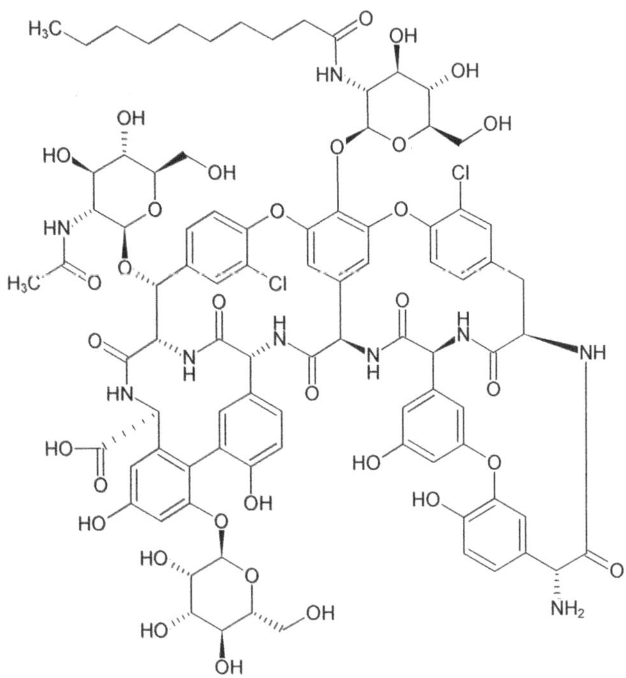

Teicoplanin

Figure 7.7 (cont) Macrocyclic chiral selectors (12).

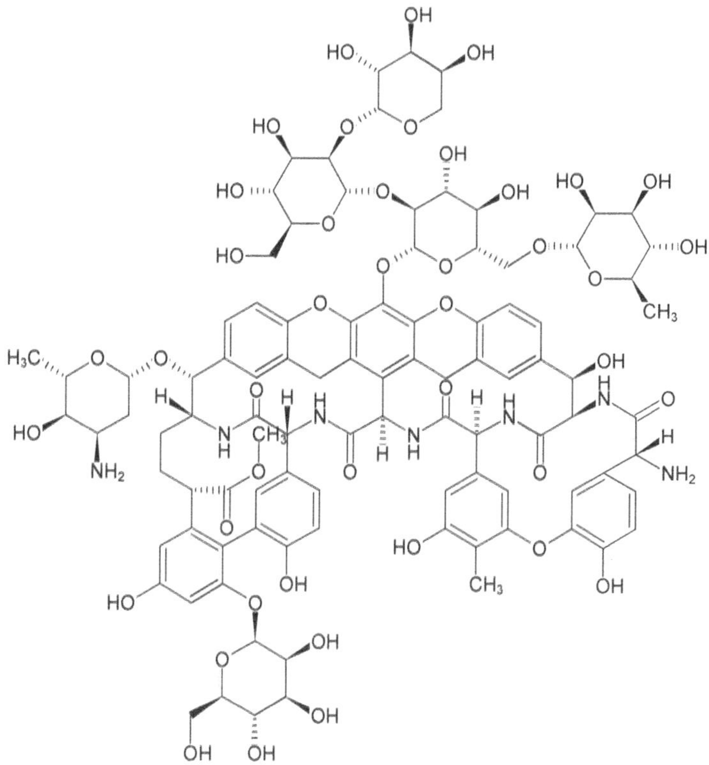

Ristocetin

Figure 7.7 (cont) Macrocyclic chiral selectors (12).

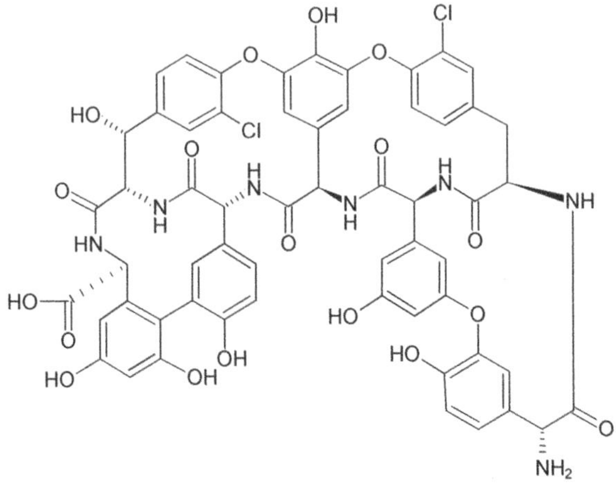

Teicoplanin aglycon

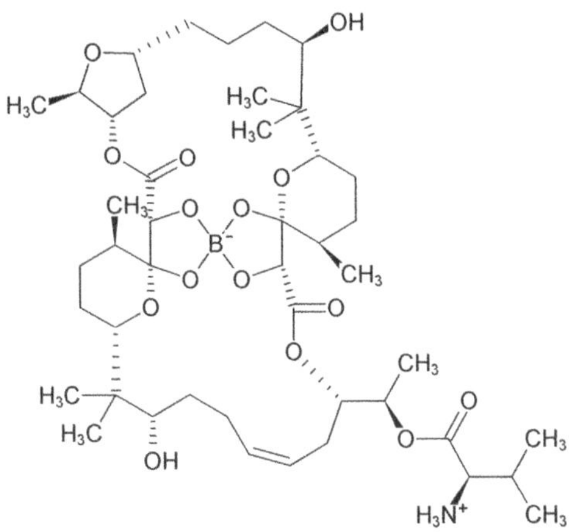

Boromycin

Figure 7.7 (cont) Macrocyclic chiral selectors (12).

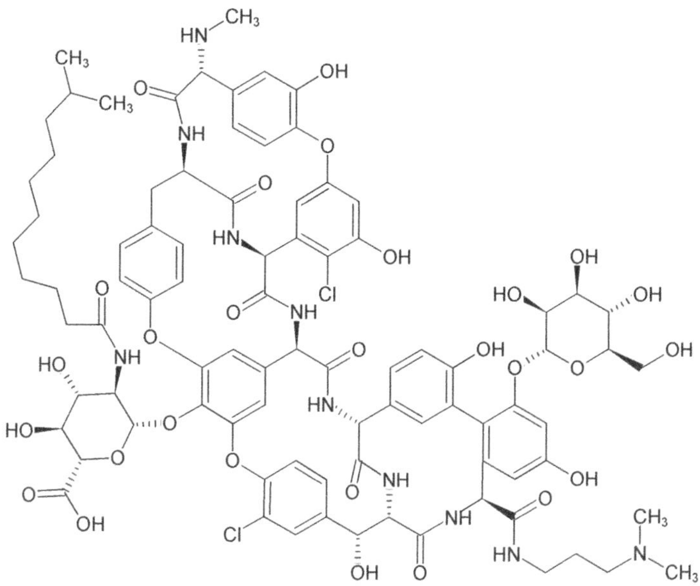

Dalbavancin

Figure 7.7 (cont) Macrocyclic chiral selectors (12).

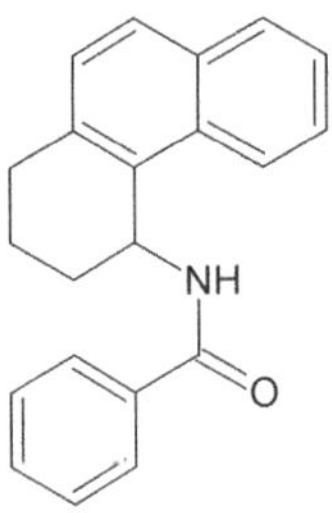

N-(1,2,3,4-Tetrahydro-
phenanthren-4-yl)benzamide

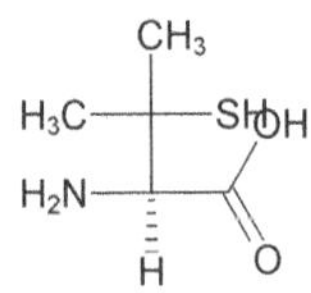

Penicillamine

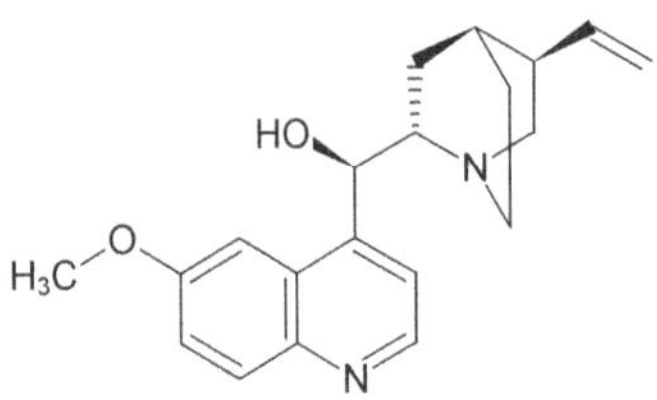

Quinine

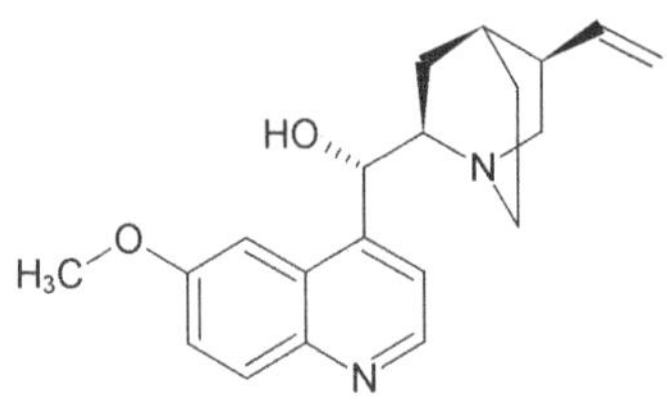

Quinidine

Figure 7.7 (cont) Macrocyclic chiral selectors (12).

These chiral stationary phases showed a good resolution and repeatability. Around 17 kinds of enantiomers were effectively separated. Most of the enantiomers were separated better than those with the same normal and reversed-phase modes (14).

7.1.3 *Thin-Layer Chromatography*

Monolithic 50 μm thin layers were made from copolymers of 4-methylstyrene, chloromethylstyrene, and divinylbenzene (15). These monomers are shown in Figure 7.8. For polymer fabrication,

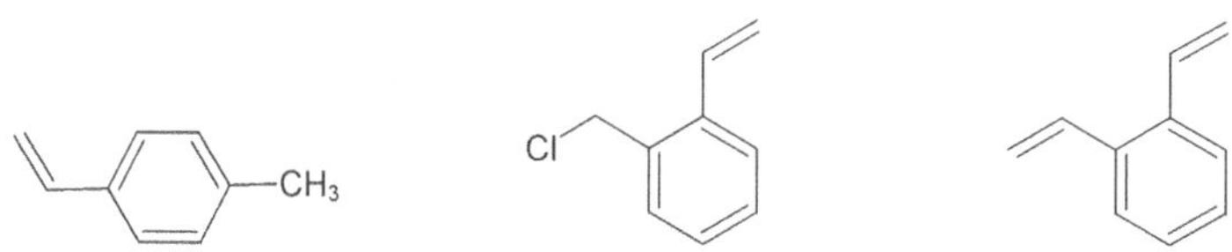

4-Methylstyrene   Chloromethylstyrene   1,2-Divinylbenzene

Figure 7.8 Monomers for monolithic thin layers.

a thermally initiated polymerization process was used.

The layers were attached to 6.0 $cm \times 3.3$ cm glass plates. These layers had a well-defined porous structure with a globular morphology demonstrated with scanning electron microscopy images and exhibited superhydrophobic properties characterized with a water contact angle of 157°.

These materials were used for thin-layer chromatography of peptides and proteins fluorescently labeled with fluorescamine. Fluorescamine is shown in Figure 7.9.

The spots of the individual separated compounds were visualized with UV light. Their identities were confirmed with a matrix-assisted laser desorption/ionization time-of-flight mass spectrometry.

The presence of the chloromethylstyrene units in the polymer enabled a hyper-crosslinking reaction via a Friedel-Crafts alkylation reaction. This resulted in monoliths with much larger surface areas, which were suitable for the separation of small dye molecules (15).

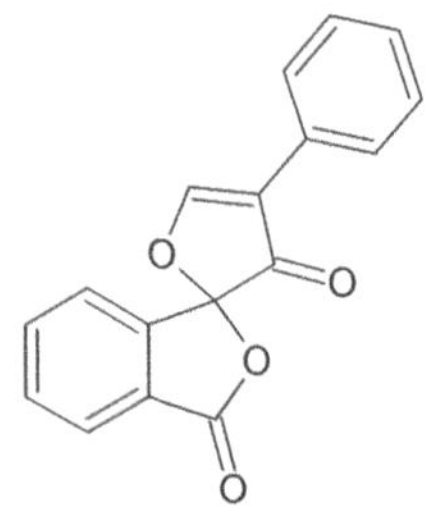

Figure 7.9 Fluorescamine.

7.1.3.1 *Ultrathin Stationary Phases*

Ultrathin and nanostructured stationary phases allow faster separations over shorter distances with a higher sensitivity than what is possible with thin-layer chromatography or high-performance thin-layer chromatography (16). Innovative stationary phases have been reported in a chapter of a monograph that are prepared from monolithic silica gels, monolithic polymers, electrospun nanofibrous polymer mats, nanostructured columnar thin films, and structure templates onto carbon nanotubes. Furthermore, general fabrication approaches were summarized for each of the above-mentioned methods (16).

7.1.4 Gas Chromatography

Modern gas chromatography (GC) was invented by Martin and James in 1952 (17). It has become one of the most important and widely applied analytical techniques in modern chemistry (18). The first stationary phases for GC were a varied collection of hydrocarbon and silicone oils and greases, esters and polymers.

Porous polymer beads and their usage in GC have been described (19). Such materials are available in a range of polarities, according to the method of preparation, which allow certain differences in the order of elution. With operating temperatures from subambient to ca. 250°C, the range of samples for analysis to which they can be applied is large.

A review with 63 references from 1982 to 1992 has been given on the development of porous polymeric beads as gas chromatography stationary phase (20).

7.1.4.1 *Silicon-Containing Norbornene Polymers*

The recent approaches to the synthesis of highly permeable polymers by metathesis and addition polymerization of silicon-containing norbornenes for GC applications have been reviewed (21). $(CH_4)_3Si$–substituted exo-tricyclononenes can be used as highly reactive norbornene-type monomers for polymerization. The basic relationships between polymer structures and their gas permeability characteristics were considered. Also, the transport properties of the polymers by modification of their structures was shown. Calixarenes and cyclodextrins, with various substituents, were chosen as filling agents.

Norbornene polymers with an intrinsic porosity were investigated as candidates for polymeric stationary phases in capillary GC (22). The kinetic and the thermodynamic properties of the stationary phases were evaluated.

Poly(trimethylsilyl norbornene) shows very large solubility coefficients S similar to those of poly(trimethylsilyl propyne) (23)

In another study, it was shown that only a polymer with a vicinal distribution of trimethylsilyl groups provided properties, making it a potential candidate for application in capillary GC. Columns with stationary phase based on this polymer demonstrated unique selectivity, moderate thermostability and a better efficiency than previously studied columns with a stationary phase based on poly[(1-trimethylsilyl)-1-propyne] (22, 24).

7.1.4.2 *Enantiomeric Separation*

The resolution of chiral compounds into optically pure enantiomers is very important in various fields such as pharmaceutical, chemical, agricultural, and food industries (25). Chiral GC is one of the efficient methods for enantioseparations of volatile compounds. In recent years, porous materials as stationary phases for chromatographic separations have achieved increasing attention.

Porous organic cages (POCs) are an emerging class of porous molecular materials which are self-assembled by discrete, shape-persistent, cage-like molecules (26). They have received intensive interest in various fields.

A hydroxyl-functionalized homochiral POC was synthesized by the condensation of 2-hydroxy-1,3,5-triformylbenzene with (1R,2R)-diaminocyclohexane and coated on a capillary column for GC separation (26). The monomers are shown in Figure 7.10.

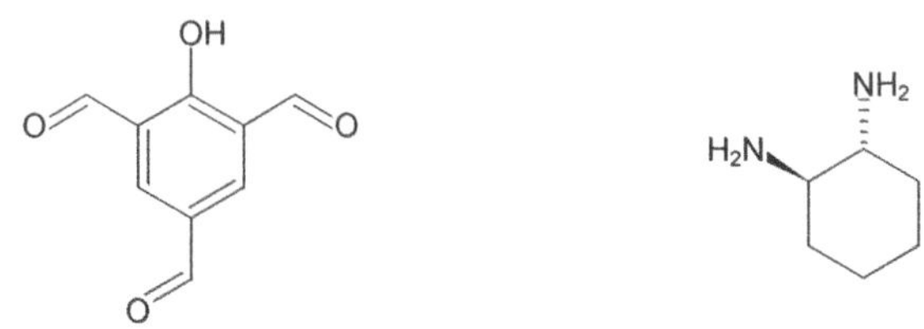

2-Hydroxy-1,3,5-triformylbenzene (1R,2R)-Diaminocyclohexane

Figure 7.10 Monomers for porous organic cage materials.

Several pairs of enantiomers belonging to various classes could be resolved on the column, including alcohols, diols, esters, lactones, halohydrocarbons, ethers, epoxides, ketones and sulfoxides.

Compared with a β-cyclodextrin (c.f. Figure 7.25) derivative-based commercial β-DEX 120 column, previously reported chiral POCs and (CC_9 and CC_{10})-based columns, there are 12, 27 and 19 pairs of studied enantiomers that cannot be resolved on a β-DEX 120 column, CC_9 column and CC_{10} column, respectively.

Both separation factors and the resolution values of some racemates are higher on this column than those on β-DEX 120 column, CC_9 column and CC_{10} column.

The results demonstrate that the column exhibits good chiral recognition complementarity or superior chiral resolution capability to those columns which were used for comparison, and the introduction of the hydroxyl group enhances the hydrogen-bonding interaction to some racemates, leading to a better enantioselectivity.

The column also shows excellent separation performance toward positional isomers of dichlorobenzene, dibromobenzene, nitrotoluene, chlorotoluene, nitrobromobenzene and nitrochlorobenzene. The retention time and selectivity of analytes have no significant changes after more than 500 injections and 260°C conditioned for 6 *h*, showing the good repeatability and thermal stability of the column. This work indicates the promising prospect of POCs for GC separation, especially for the separation of enantiomers and also demonstrates the significance of design and synthesis of more functionalized chiral POCs for GC enantioseparation that broaden the enantiomer separation scope and applicability of POCs-based columns through their chiral recognition complementarities (26).

7.1.4.3 *Microporous Organic Polymers*

Microporous organic polymers are a class of functional porous materials with unique characteristics and potential uses in diverse areas. However, the field of microporous organic polymers for GC separations has not been well explored (27).

In a study, a microporous organic polymer, i.e., KAPs-1, was dynamic coated onto a capillary column. The fabricated column exhibited a nonpolar nature and the column efficiency for *n*-dodecane was up to 7769 plates m^{-1}. The KAPs-1 coated column showed high GC separation performance for a series of volatile organic compounds, including the challenging ethylbenzene and xylene isomers, which could not be resolved at baseline on the commercial 5% phenyl polysiloxane stationary phase (27).

Moreover, the relative standard deviations for five replicate determinations of the studied analytes were 0.0–0.6%, 0.9–3.2%, 1.1–5.9%, and 0.8–3.7% for retention time, peak area, peak height and peak width, respectively. To investigate the interaction between some analytes and the stationary phase, thermodynamic and kinetic parameters were also evaluated.

The results of the study revealed that it is very promising to utilize microporous organic polymers as stationary phases for capillary GC (27).

7.1.5 *Gel Permeation Chromatography*

7.1.5.1 *Thermo-controllable Chromatography*

Poly(*N*-isopropylacrylamide) (PIPA) with a carboxyl group at its end was coupled with amino group-carrying porous glass beads by the active-ester method (28). Gel permeation chromatography

(GPC) of dextran was carried out by using a column packed with the PIPA-carrying glass beads.

The chromatograms of dextrans with various molecular weights are strongly affected by temperature: The elution time of dextran is largely changed between 25°C and 32°C due to a change in the effective pore size by the transition of the PIPA chains from coils to globules on the surface of the pores of the glass beads.

These results suggest the usability of PIPA-carrying glass beads as devices for thermo-controllable chromatography (28).

7.1.5.2 Ethylene glycol dimethacrylate/Glycidylmethaerylate microspheres

Polar polymeric porous microspheres of ethylene glycol dimethacrylate/glycidylmethaerylate (EGPM) have been synthesized by suspension copolymerization in the presence of diluent with different nature and composition (29). The correlation of phase separation with crosslinking agent and diluent, which are the key points of pore formation during polymerization, has been discussed.

The physicochemical properties and the porous structure of the microspheres were characterized by apparent density, specific surface area, skeleton density, pore radius distribution, solvent regain, pore volume, bed volume, degree of swelling and the scanning electron microscopy. Also, the molecular weight vs. elution volume calibration curve and the chromatogram of dextran separation of the microspheres were presented.

The evaluation of EGPM showed that it possesses the characteristics of hydrophilic stationary phase for modern gel permeation chromatography. Furthermore, the molecular weight polydispersity and the solute mass transfer diffusion of the water-soluble polymers were investigated (29).

7.1.5.3 Microparticulates

High-performance GPC was performed by using packed columns of μ-Styragel, a porous polymer of styrene and divinylbenzene, and μ-Bondagel E, a silica-based chromatographic packing. The latter is compatible with aqueous and nonaqueous mobile phases (30).

By using particle sizes of about 10 μm, packed columns yield 4,000–10,000 plates per 30 *cm* for total permeating samples depending upon mobile phase velocity and viscosity.

The times of analysis are typically 5–20 *min*; μ-Bondagel E Linear columns with linear calibration curves from 2 million to several thousand of molecular weight were prepared and were shown to be valuable general purpose columns for the separation of aqueous and nonaqueous soluble polymers.

So, the value of high-performance GPC for the rapid characterization of aqueous and nonaqueous soluble polymers, as a simple preparative tool and as a useful sample cleanup method for subsequent analysis, could be demonstrated (30).

7.1.5.4 *Study of the Porous Structure*

Gel permeation chromatography (GPC) was used to characterize the porous structure of macroporous copolymers (31). By using macroporous glycidyl methacrylate copolymers as an example, the relationships between the exclusion limit in GPC and the pore size and porosity of the sorbent, the distribution coefficient and the pore geometry and size, and the peak widening and morphology of the porous structure were established.

Characteristic quantities are compared with results provided by other methods, especially electron microscopy, mercury porosimetry and nitrogen sorption. It was demonstrated that the porous structures of strongly crosslinked macroporous copolymers in the swollen state and in the dry state differ (31).

7.1.6 *High-Performance Liquid Chromatography*

Long alkyl chains for a soybean trypsin inhibitor have been prepared for use in high-performance liquid chromatography (HPLC) separations in the reversed-phase and affinity modes (32).

A copper(I)-catalyzed (3 + 2) azide-alkyne cycloaddition was used, which is an element of *click chemistry*. The ligands were attached via a triazole ring to size monodisperse porous beads containing either alkyne or azide pendant functionalities.

Alkyne-containing beads prepared by direct copolymerization of propargyl acrylate with ethylene glycol dimethacrylate were allowed to react with azidooctadecane to give a reversed-phase sorbent.

Azide-functionalized beads were prepared by chemical modification of glycidyl methacrylate particles.

The subsequent reaction with a terminal aliphatic alkyne produced a reversed-phase sorbent similar to that obtained from the alkyne beads. The soybean trypsin inhibitor was functionalized with *N*-(4-pentynoyloxy)succinimide to carry alkyne groups and then allowed to react with the azide-containing beads to produce an affinity sorbent for trypsin.

The above-mentioned compounds are shown in Figure 7.11. Ethylene glycol dimethacrylate is shown in Figure 4.1.

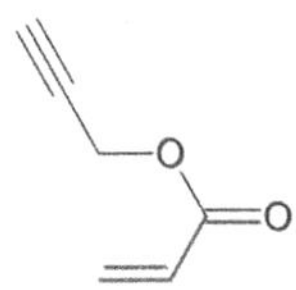

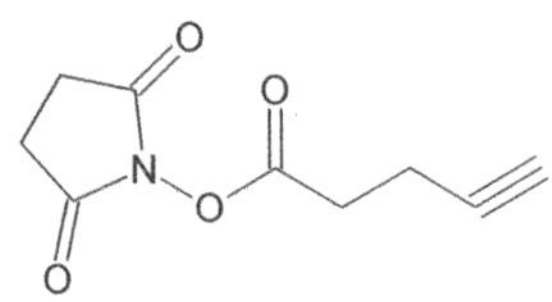

Propargyl acrylate   *N*-(4-Pentynoyloxy)succinimide

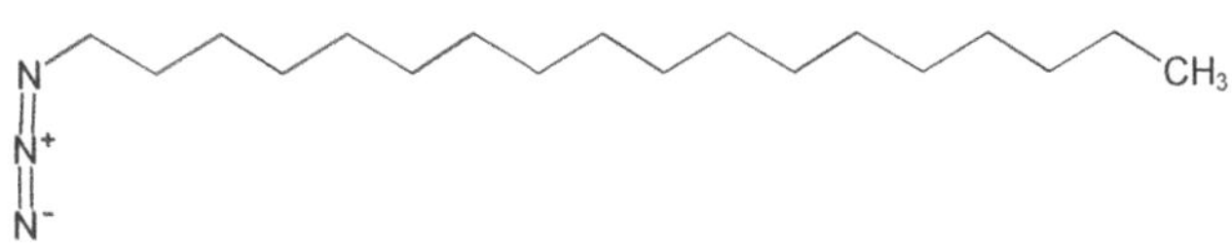

Azidooctadecane

Figure 7.11 Substances of a sorbent.

The performance of these stationary phases was demonstrated with the HPLC separations of a variety of peptides and proteins (32).

Azidooctadecane was also used in a similar study (33). Here, the chromatographic performances of the C_8/C_{18} monoliths were evaluated through the separation of a mixture of five proteins such as ribonuclease A, soybean trypsin inhibitor, cytochrome C, c.f. Figure 7.12, bovine-hemoglobin and bovine serum albumin.

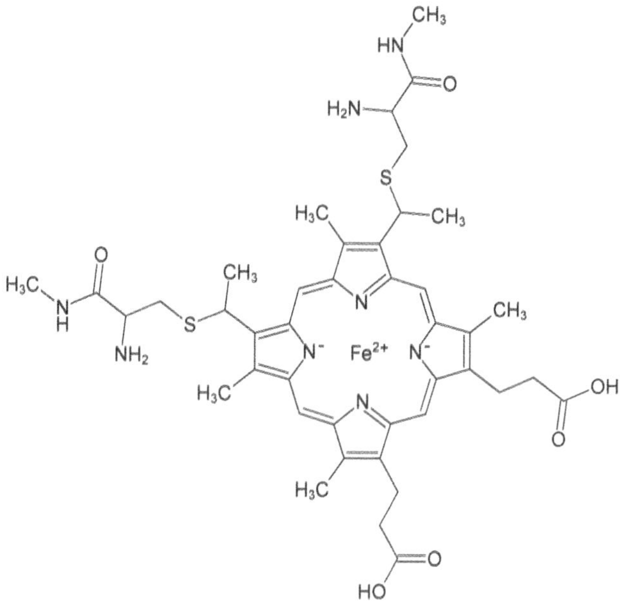

Figure 7.12 Cytochrome C.

7.2 Oil Spill Control

The separation of oil and water mixtures is of great importance across a wide range of technologies and industries. For example, oil and water separation problems gained national attention during the 2010 Gulf oil spill and subsequent cleanup efforts (34). The petroleum industry faces similar water and oil separation challenges as it attempts to extract oil from beneath the sea.

7.2.1 *Polyolefins*

The usage of natural and synthetic sorbents to control oil spills is of increasing importance due to environmental aspects. Polyolefins, including poly(propylene) (PP) and poly(ethylene) (PE), are the most commonly used oil sorbent materials because of their low cost. However, these materials suffer from their low oil absorption capacity (35).

7.2.1.1 *Poly(ethylene)*

Thin flexible PE sheets consisting of multiple materials with increased cost and complication in the production process have been shown to exhibit oil sorption capacities exceeding 200 $g\,g^{-1}$ (36, 37). The saturation time reported for these sheets is long, suggesting a slow uptake mechanism.

In a study, super oil sorbent polymer sheets consisting of porous ultrahigh-molecular-weight poly(ethylene) (UHMWPE) have been prepared. Such polymer films were prepared as follows (35):

Preparation 7–3: An UHMWPE powder was placed in a vacuum oven for 24 *h* and then dispersed in mineral oil and petrolatum at a ratio of 10:45:45 by weight. Irganox 1010 at a ratio of 0.5% of total weight of mixture was also added as antioxidant. The mixture was heated to 80°C with high-speed mixing until a homogeneity was observed.

Then, the mixture went through a twin-screw extruder model, where it was heated beyond the melting point of the polymer. Afterwards, it was hot-pressed at 200°C to achieve a flat sheet followed by solvent extraction in a hexane bath for 10 *min* under constant stirring at room temperature to remove around 50 to 60% of the solvent.

The sheets were stretched by a successive biaxial stretching technique at different stretching ratios using a crosshead speed of 400 $mm\,min^{-1}$ at

121°C. The thus stretched films were placed in boiling hexane to extract and remove the residual solvent and then dried to obtain PE porous films.

For each sample, the following parameters (37) were subsequently calculated (35):

1. The oil pick-up ratio mo/ms, which provides a dimensionless measure of the characteristic affinity for oil of the sorbent tested.
2. The pick-up density $g\,cm^{-3}$, the mass of oil taken up by a given volume of sorbent material as packaged in readiness for use.
3. Related oil thickness characterizing the particular sorbent is a measure of the maximum theoretical thickness of the film of sorbate which can be fully taken up by a sorbent sheet exposed to the sorbate film.

This sorbent has extremely high uptake and retention capacities along with a mechanically strong structure. The combination of these factors as well as the cost effectiveness of such a material makes these sheets viable candidates for widespread production and utilization (35).

7.2.1.2 Poly(propylene)

7.2.2 Porphyrin

A highly porous porphyrin-based organic polymer, i.e. a porous covalent porphyrin framework, was fabricated via the homocoupling reaction of a custom-designed porphyrin ligand, i.e., 5,10,15,20-tetrakis (4-bromophenyl)porphyrin, c.f. Figure 7.13.

The polymer has a large Brunauer-Emmett-Teller (BET) surface area of more than 1300 m^2g^{-1} (Langmuir surface area of over 2400 m^2g^{-1}) and exhibits strong hydrophobicity with a water contact angle of 135°; and these features afford it the highest adsorptive capacities for saturated hydrocarbons and gasoline among sorbent materials reported thus far, as well as render it the capability to remove oil from water (38).

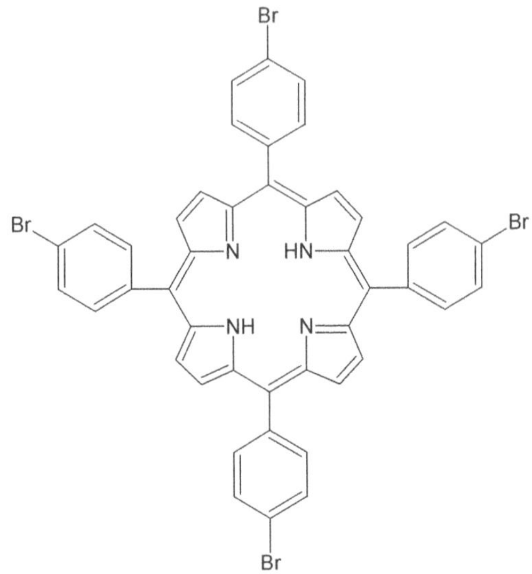

Figure 7.13 5,10,15,20-Tetrakis (4-bromophenyl)porphyrin.

7.2.3 *Poly(urethane) Sponge*

Conjugated microporous polymers based on tristyryl-*s*-triazine unit can be obtained through the coupling reaction of 2,4,6-tribromostyryl-*s*-triazine with 1,4-*p*-phenylenediacetylene monomers, c.f. Figure 7.14 (39). Due to the highly conjugated structure composed of C=C double bonds and C≡C triple bonds, tristyryl-*s*-triazine exhibits a superhydrophobicity, a rich porous structure, as well as an excellent chemical stability in harsh conditions.

Furthermore, the tristyryl-*s*-triazine can be coated on the skeleton of a poly(urethane) (PU) sponge, which endows the PU sponge with superhydrophobicity, high elasticity, rich macroporous structure, and good mechanical stability.

The coated PU sponge maintains its original superhydrophobicity even under extremely acidic/alkaline conditions at a pH ranging from 1 to 14.

The sponge can absorb a low-density grease to saturation within 10 *s*, and its adsorption saturation capacity can be up to 72 times of its own weight. In addition, the sponge can efficiently separate

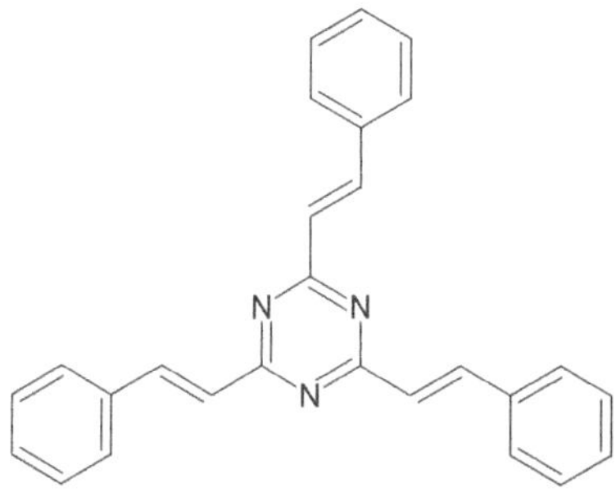

Tristyryl-s-triazine

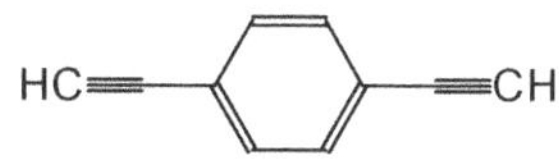

1,4-*p*-Phenylenediacetylene

Figure 7.14 Momomers.

a series of different oil/water mixtures, including floating oil and bottom oil, with oil content greater than 99.8%.

Also, the sponge can continuously absorb and discharge oil and organic solvents on the water surface by means of vacuum, and the separation efficiency remains unchanged after repeating the process 10 times, indicating good cycle stability.

Also, the sponge can demulsify a surfactant-stabilized oil-in-water emulsion, and the purity of the recovered oil in the filtrate is about 99.8% (39).

7.2.4 *Hierarchical Porous Membrane*

A hierarchical porous membrane, which is suitable for use in oil/water separation processes has been described (34).

The membrane may be used to separate trace amounts of water not greater than 50 *ppm* to 1000 *ppm* water from oil. By altering properties of the membrane, such as pore size, hydrophobicity, and layer length, the critical parameters for proper operation can be controlled, such as the breakthrough pressure and the permeability.

The membrane has a first layer with small pores and has a second layer, wherein the second layer is a support layer that is substantially thicker than the first layer and that has substantially larger pores than the first layer.

The polymer for the membrane is selected from poly(carbonate), poly(sulfone) (PSf), poly(acrylonitrile) (PAN), poly(ethersulfone), and any combination thereof. The first layer may contain a coating. The coating may be a silane coating. The silane coating may include at least one member selected from the group consisting of octadecyltrichlorosilane, methylsilane, phenylsilane, isobutylsilane, dimethylsilane, tetramethyldisilane, hexylsilane, octadecylsilane, and fluorosilane.

The membrane can be prepared by the following steps (34):

1. Combining a polymer, a pore former, and a solvent to make a polymer solution,
2. Casting the polymer solution on a plate (e.g., glass plate),
3. Immersing the plate in a non-solvent to allow the release of air bubbles, and
4. Optionally, cutting a layer of the membrane resulting from step 3 to undergo coagulation, thereby forming the hierarchical porous membrane.

The solvent preferably may be *N*,*N*-dimethyl acetamide and/or *N*-methyl-2-pyrrolidone. *N*,*N*-Dimethyl acetamide and *N*-methyl-2-pyrrolidone are shown in Figure 7.15. The non-solvent

N,*N*-Dimethyl acetamide *N*-Methyl-2-pyrrolidone

Figure 7.15 Solvents for membrane fabrication (34).

may be a mixture of water and ethanol.

To improve the permeability of the PSf or PAN hierarchical membranes, selective or plasma etching of the top layer (e.g., where the pore sizes are 30–300 *nm*) is performed. The plasma etching is preferably performed in an oxygen plasma chamber in a vacuum of 200 *mbar*, for a controlled etching time of 3 to 10 s. A scheme of a plasma etching process is shown in Figure 7.16.

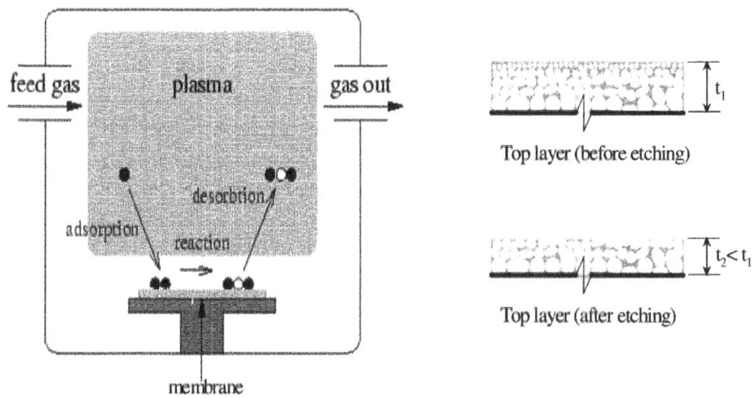

Figure 7.16 Plasma etching process (34).

The plasma etching removes a part of polymer material from the top layer surface, thereby decreasing an effective thickness of the top layer and opening up bigger pores (e.g., with size greater than 80 *nm*) beneath the original surface. The plasma etching process may be helpful to increase the overall permeability of the membrane (34).

7.2.5 *Waste Polymers*

Thermoplastic polymers, such as PP, PE, poly(ethylene terephthalate) (PET) and high-density poly(ethylene) (HDPE), constitute 5–15% of the municipal solid waste produced across the world (40).

A huge quantity of plastic waste is disposed of each year and is mostly either discarded in landfills or incinerated. On the other hand, the usage of synthetic polymers as oil sorbents, in particular, polyolefins, including PP and PE, are the most commonly used oil sorbent materials mainly due to their low cost.

The idea of using plastic wastes as the feed for the production of oil sorbents has gained interest. Various processes for usage have been reviewed (41).

7.2.5.1 *Polyolefins*

An innovative way has been reported to produce an oil-sorbent film with high practical oil uptake values from waste HDPE bottles for a rapid oil spill remedy (40). This fabrication can be done as follows:

Preparation 7–4: The HDPE bottle was cut into small pieces and then washed with a detergent. It was further washed with acetone and then dried in an oven for several hours to remove traces of the solvent. The blends were prepared via melt mixing in a static-mixer which was equipped with roller rotors. HDPE pellets were first charged in the chamber at 180°C and a mixing speed of 25 *rpm*. A mixture of dispersed ultra-high-molecular-weight poly(ethylene) (UHMWPE) in a solvent (mineral oil and petrolatum wax with equal amount per weight) was slowly added to the chamber. The mixture was allowed to mix for 30 *min* at a mixing speed of 25 *rpm* at 180°C. The partial gel obtained from the chamber was cut into fine pieces and was put back into the static-mixer for round 2 under the same operating conditions in order to achieve uniform mixing.

Then, it was hot-pressed at 180°C in order to achieve a flat sheet followed by solvent extraction in a hexane bath for 20 min at room temperature to remove around 50% to 60% of the solvent. The sheet was later stretched by using a successive biaxial stretching technique at a stretching ratio of 12 × 12 using a crosshead speed of 400 $mm\ min^{-1}$ at 121°C. The so stretched film was placed in boiling hexane to extract and remove the residual solvent and then dried to obtain the porous film. A similar procedure for material preparation was used as described previously (35, 42, 43).

The initial target of the study was to prepare sorbent films using only HDPE bottles (40). However, HDPE with relatively smaller molecular weight and molecular chains, could not be swollen significantly with the solvent and could not be converted into a pseudo-gel state. So, most of the solvent did not mix well with HDPE and remained unmixed, resulting in an inhomogeneous gel sheet.

So, the molecular weight of the polymer should be sufficiently higher with values ranging from 300 *kDa* to 6 *MDa*. Therefore, 25% of higher molecular weight polymer UHMWPE was added to help the blend undergo swelling by the solvent to get a pseudo gel state followed by extraction of the solvent to create pores (40).

7.2.5.2 Poly(ethylene terephthalate)

The fabrication of a composite oil sorbent from used PET bottles and waste carbonaceous residue from the gasification of oil refinery bottoms via an energy-efficient phase inversion method has been done (44).

The composite sorbent demonstrated a oil sorption capacity of ~8 $g\,g^{-1}$, which is comparable to some commercially available adsorbents. Also, an excellent retention capacity for mineral, paraffin and crude oils of different viscosities was found.

Furthermore, the composite is able to remove oil from mineral or paraffin oil-in-water emulsions by sorption, with an efficacy of >99.7%. So, this material can be a promising application for the treatment of industrial oily wastewater. Cost analysis showed that this waste-derived composite can be produced commercially at a competitive price (44).

7.3 Sorbents

7.3.1 *Purification of Ethylene*

Ethylene is the largest-volume product of the chemical industry (45). The purification involves energy-intensive processes, such as chemisorption (CO_2 removal), catalytic hydrogenation (C_2H_2 conversion), and cryogenic distillation (C_2H_6 separation).

Although advanced physisorbent or membrane separation could lower the energy input, one-step removal of multiple impurities, especially trace impurities, has not been feasible.

A synergistic sorbent separation method was described for the one-step production of polymer-grade C_2H_4 from ternary (C_2H_2/C_2H_6/C_2H_4) or quaternary (CO_2/C_2H_2/C_2H_6/C_2H_4) gas mixtures with a series of physisorbents in a packed-bed geometry (45).

Ultraselective microporous metal-organic materials were synthesized that could be readily regenerated (45).

For purification, the gas mixture is sent through a breakthrough column with three different sorbents. Each sorbent exhibits a strong selectivity for one of the impurity gases. All the sorbents were preactivated by heating under high vacuum before preparing the breakthrough column.

7.3.2 *Carbon Dioxide Capture*

Global investment in counteracting climate change has galvanized an increasing interest in carbon capture and sequestration as a versatile emissions mitigation technology. As decarbonization efforts accelerate, carbon capture and sequestration can target the emissions of large point-source emitters, such as coal- or natural gas-fired power plants, while also supporting the production of renewable or low-carbon fuels (46).

Furthermore, the carbon capture and sequestration can enable decarbonization of difficult-to-abate industrial processes and can support net CO_2 removal from the atmosphere through bioenergy coupled with carbon capture and sequestration or direct air capture.

The development of porous materials as next-generation sorbents for CO_2 capture applications has been reviewed (46).

In a chapter of a monograph, various functionalized organic porous polymers and their CO_2 uptake capacity, CO_2/N_2 selectivity, regenerability and also challenges and potential of this kind of material in gas separation are discussed (47).

Due to the uncontrolled release of greenhouse gases into the atmosphere through anthropogenic activities, the planet's temperature and natural ecosystem are being adversely affected globally. Fossil fuel power plants and transportation are the major sources for the release of CO_2 into the atmosphere. Carbon capture and sequestration is one of the promising alternatives for CO_2 mitigation.

To capture this CO_2, highly selective and high storage capacity adsorbent material is required (47). Also, the adsorbent should be chemically stable, highly porous, have large surface area, minimal energy input, easy to regenerate and low cost. Amine-based technology has long been used for CO_2 mitigation but this process is very much energy intensive.

Physical sorbents with high CO_2 selectivity are available in powder form and cannot be used for real-world applications. There is a need to transform it in some particulate form or one can form a porous framework that can be easily done using polymers, since and polymers are known to be mechanically, thermally and chemically very stable. Also, due to the presence of abundant functionalizable sites polymers can be functionalized to make the polymer surface rich in CO_2-philic moieties (47).

7.3.2.1 Poly(styrene)

A moisture-resistant CO_2 adsorbent was designed and prepared (48). The adsorbent used porous poly(styrene) as a skeleton, surface modified with amine groups. Due to the huge specific surface area covered by a large number of amine groups, this porous material could be applied for capturing carbon dioxide under atmospheric pressure.

Under a pressure of 100 *kPa*, 3.46 $mmol\,g^{-1}$ of CO_2 adsorption uptake was observed. The porous adsorbent also exhibited a high CO_2 adsorption rate, excellent moisture resistance, good selectivity for CO_2–N_2 separation, and easy CO_2 regeneration (48).

7.3.2.2 *Benzimidazole-linked Polymers*

A series of benzimidazole-linked porous polymers were obtained by the condensation reaction between the *o*-aminobenzol end groups of building blocks, i.e., 2,3,6,7,10,11-hexaaminotriphenylene, 3,3′-diaminobenzidine or 1,2,4,5-benzenetetraamine and the aldehyde groups of the building blocks, i.e., terephthalaldehyde, 4,4′-biphenyldicarboxaldehyde, 1,3,5-tris(4-acetylphenyl)benzene or 1,3,5-tris(4-formylbiphenyl)amine, in a one-pot synthesis without using a catalyst or a template (49). Some of the above-mentioned compounds are shown in Figure 7.17.

The polymers show Brunauer-Emmett-Teller specific surface area values of more than 600 m^2g^{-1}, they have hydrogen storage of up to 1.6%, at 77°K and $1 times 10^5$ *Pa*, and carbon dioxide capture up to 12.6%, at 273°K and $1 times 10^5$ *Pa* (49).

7.3.2.3 *2-Ethylhexyl Methacrylate Divinylbenzene Copolymer*

An effective adsorbent was developed for its use in CO_2 capture technologies. For this purpose, a porous polymer was prepared by high internal phase emulsion using 2-ethylhexyl methacrylate and divinylbenzene. This prepared porous polymer was then used as support for the wet impregnation of poly(ethylene imine), thus resulting in the final adsorbent.

At the optimal poly(ethylene imine) loading of 60% on the 2-ethylhexyl methacrylate and divinylbenzene polymer, the CO_2 sorption capacity reached 4 *mmol* CO_2 per *g*-sorbent using 10% by volume CO_2 and 3 vol% H_2O in N_2 at 70°C.

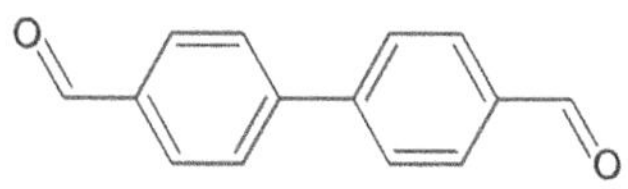

4,4′-Biphenyldicarboxaldehyde

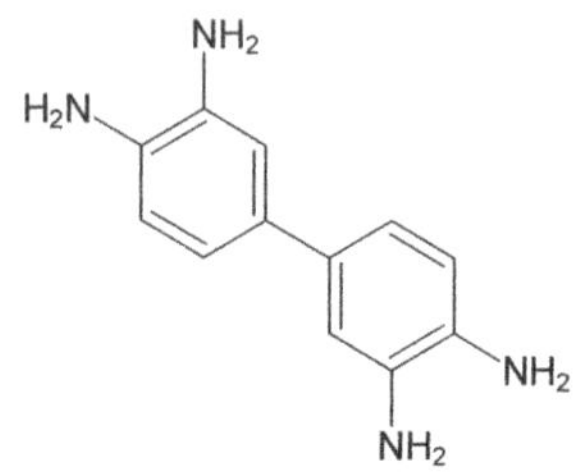

3,3′-Diaminobenzidine

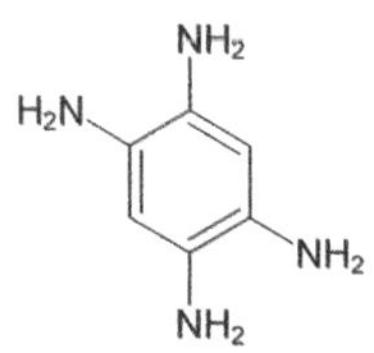

1,2,4,5-Benzenetetraamine

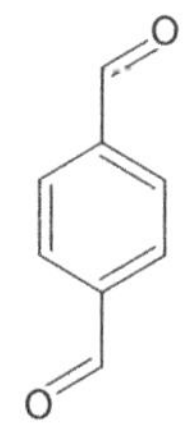

Terephthalaldehyde

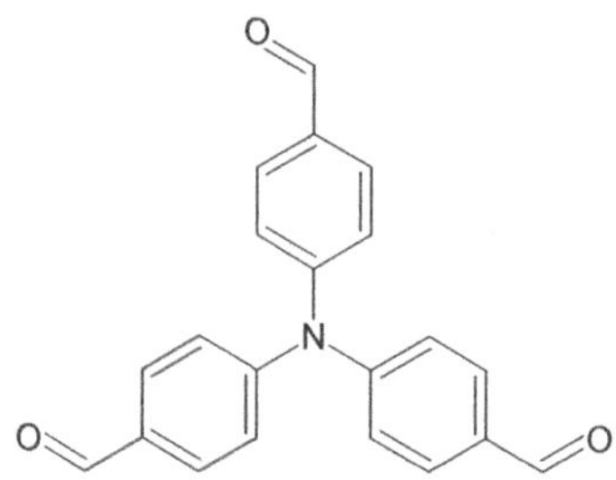

1,3,5-Tris(4-formylbiphenyl)amine

Figure 7.17 Monomers for benzimidazole-linked polymers.

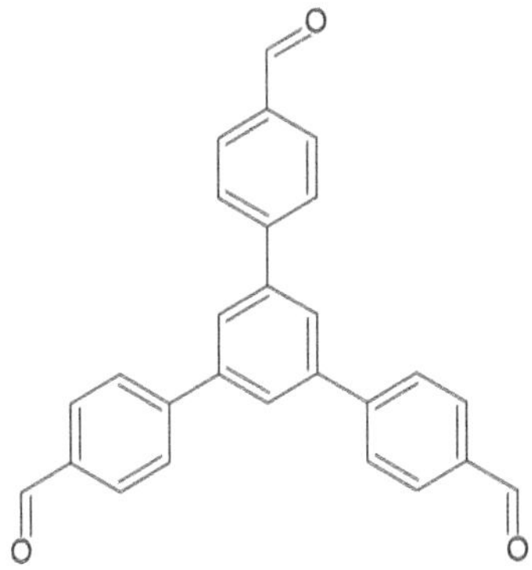

1,3,5-Tris(4-formylbiphenyl)benzene

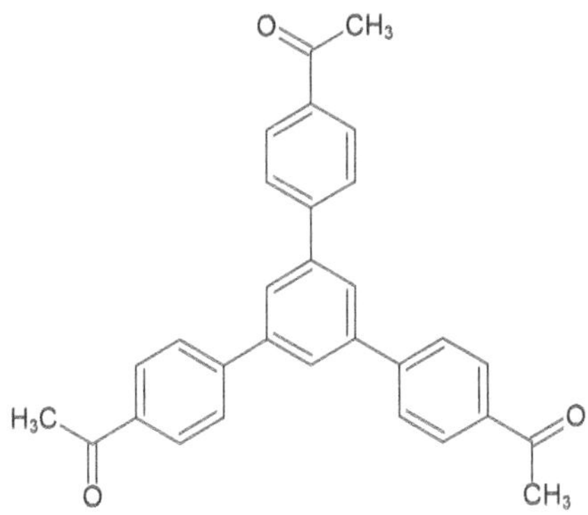

1,3,5-Tris(4-acetylphenyl)benzene

Figure 7.17 (cont) Monomers for benzimidazole-linked polymers.

The fabricated material has the potential to reduce the costs for CO_2 capture due to the low activation energy and the high CO_2 desorption rate (50).

7.3.2.4 *Torlon Polymer*

A facile approach for the fabrication of composite monoliths consisting of Torlon polymer and 13X and 5A zeolites via a 3D printing technique has been reported (51). A prominent distributor of poly(amide imide)s is Solvay Specialty Polymers, which uses the trademark Torlon (52).

This composite can be used for the removal of CO_2 from flue gas (51).

The physical and structural properties of 3D-printed composite monoliths were systematically evaluated and compared with their pristine powders.

The formation of monoliths was facilitated through a non-solvent induced phase separation technique. By the incorporation of ~31% of zeolite particles into the polymer matrix, the so obtained monoliths displayed CO_2 capture capacities proportional to the zeolite loading. The 3D-printed Torlon-zeolite monoliths exhibited high compressive strengths of approximately 210 *MPa*.

These results demonstrate the superiority of polymer-zeolite composite monoliths, formulated by 3D printing technique as robust structures with outstanding mechanical integrity and comparable adsorption capacity for various adsorption-based gas separation processes (51).

The here used printing strategy can be easily extended to other organic-inorganic composite materials with different polymers and inorganic particles (51).

7.4 Recovery of Organic Materials

7.4.1 *Adsorption of Acteoside*

Acteoside is a bioactive component, which has potential applications in the fields of food, healthcare, medicine, etc. The development of adsorbents that can efficiently and selectively adsorb acteoside is still challenging. Acteoside is shown in Figure 7.18.

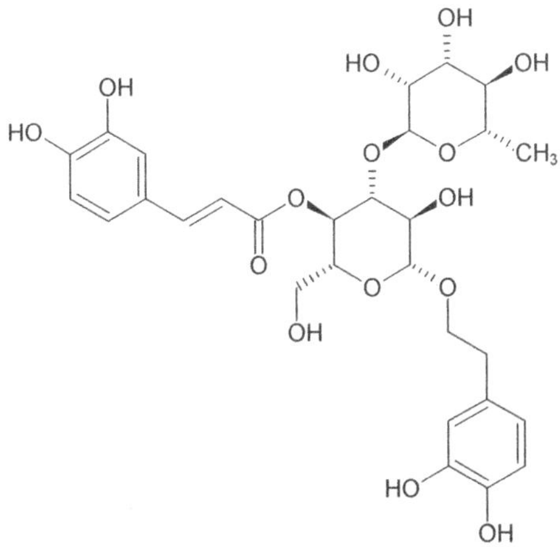

Figure 7.18 Acteoside.

Poly(ether imide) (PEI) modified heterostructure porous polymer microspheres were designed and synthesized via a double emulsion interfacial polymerization strategy for the efficient adsorption of acteoside (53).

These materials possessed a porous structure with a large pore volume of 1.58 m^3g^{-1}. The PEI moiety was identified on the interior pores and external surface of the microspheres. PEI played two significant roles in the adsorption performance of the material. First, the introduced PEI increased the pore volume and the specific surface area to provide more adsorption sites. Second, the PEI on interior pores and external surface allowed he pH to regulate the adsorption capacity and the selectivity for acteoside.

The adsorption mechanisms proposed that electrostatic attraction, hydrogen bonding, $\pi - \pi$ stacking, and van der Waals force were the main interaction forces between the adsorbent and acteoside (53).

In another study, a series of the functionalized mesoporous poly(styrene)-based microspheres with different functional comonomers, such as acrylamide, ethylene glycol dimethacrylate, hydroxyethyl methacrylate and ratios of styrene to divinylbenzene were designed and synthesized by a double emulsion interface polymerization method (54).

Among them, styrene and divinylbenzene existed in the oil phase, forming the skeleton structure of functionalized mesoporous poly(styrene)-based microspheres. Acrylamide, ethylene glycol dimethacrylate, or 2-hydroxyethyl methacrylate in the water phase formed functional layers on the inner and outer surfaces of functionalized mesoporous poly(styrene)-based microspheres.

The experimental results indicated that the optimal functional comonomers and the ratio of styrene to divinylbenzene were acrylamide (provided the hydrophilic $-CONH_2$ groups) and 1:1, respectively. Thus, A-functionalized mesoporous poly(styrene)-based microspheres-2 exhibited the highest adsorption capacity of 108.95 ± 8.13 $mg\,g^{-1}$ and the selectivity of 5.14 ± 0.17. These results were attributed to the hydrophilic $-CONH_2$ groups on A-functionalized mesoporous poly(styrene)-based microspheres-2, and these groups were beneficial to ACT molecules diffusion driven by concentration gradient, improving the adsorption performance.

Furthermore, the hydrophilic $-CONH_2$ groups on the inner and outer surfaces of A-functionalized mesoporous poly(styrene)-based microspheres-2 acted as hydrophilic sites that had a high-affinity interaction with ACT molecules, thus increasing the adsorption selectivity. In addition, A-functionalized mesoporous poly(styrene)-based microspheres-2 had the highest specific surface area and largest pore volume, resulting in the highest adsorption capacity and adsorption selectivity (54).

So, the development of adsorbents with adjustable pore structure and a large number of hydrophilic sites will provide a strategy for the selective separation of bioactive components from natural products (54).

7.4.2 Toxic Organic Materials

7.4.2.1 Toxic Chlorinated Compounds

Radiolabeled hexachlorocyclohexane, hexachlorobiphenyl, anthracene, and Parathion belonging to the chemical classes of chlorinated hydrocarbons, chlorinated aromatic compounds, polynuclear aromatic compounds, and organophosphate pesticides, were used to demonstrate the ability of supercritical carbon dioxide to desorb such environmentally significant compounds from the polymeric sorbent Tenax-GC, i.e., poly(2,6-diphenyl-*p*-phenylene oxide) (55). These materials are shown in Figure 7.19.

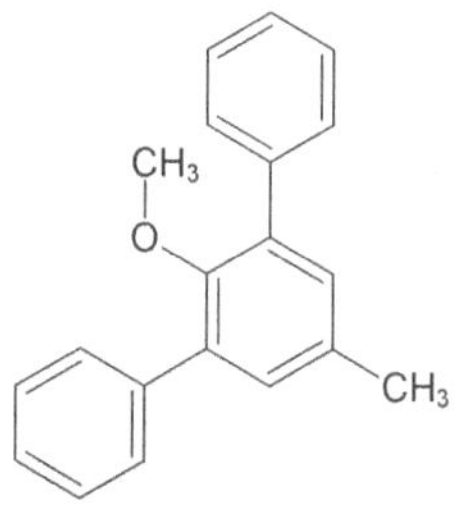

2,6-Diphenyl-*p*-phenylene oxide
Tenax monomer

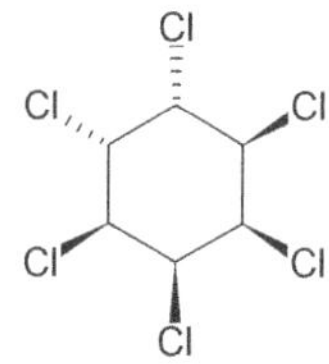

Hexachlorocyclohexane

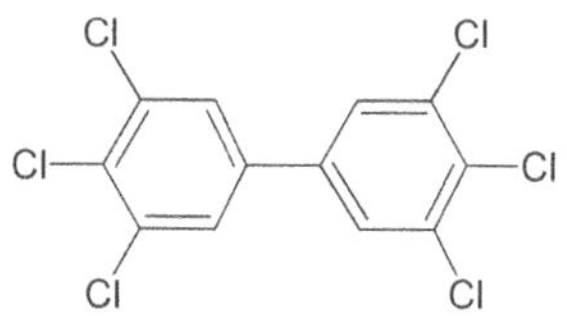

Hexachlorobiphenyl

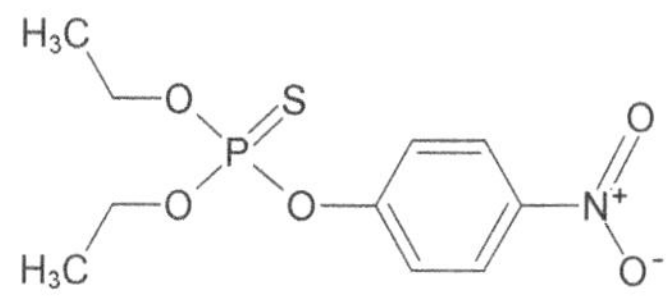

Parathion

Figure 7.19 Tenax and toxic organic materials.

All the compounds could be recovered in an excess of 90%. The recoveries of these compounds were examined as a function of the volume of carbon dioxide used for the desorption. The identities of the desorbed compounds were tested by the use of thin-layer chromatography and mass spectrometry.

When compared to the thermal desorption from Tenax for the same compounds, the hexachlorobiphenyl and Parathion were recovered in a ratio of only about 13% (55).

In another study, the sorption capacities of sorbents were studied measured using the radiation-chemical-modification and initial-matrix techniques (56).

It was shown that high values of the adsorption capacities of radiation-modified sorbents can be determined by the formed structural characteristics and the chemical nature of the surface.

Furthermore, radiation-modified sorbents with a high thermal stability allow the use of high temperatures in the desorption of the analyzed compounds and provide a higher speed and sensitivity for the analysis technique (56).

7.4.2.2 *Toxic Urea and Nitrophenol*

The urea oxidation reaction and nitrophenol reduction are safe and key limiting reactions for sustainable energy conversion and storage. Urea and nitrophenol are abundant in industrial and agricultural wastes, human wastewater, and in the environment (57).

A catalytic oxidative and reductive removal is the most effective process to remove urea and 4-nitrophenol from the environment. 2D carbon-supported, cobalt nanoparticle-based materials are emerging catalysts for nitrophenol reduction and an anode material for the urea oxidation reaction. A cobalt-modified porous organic polymer was synthesized and carbonized at 400°C and 600°C. In detail, the synthesis runs as (57, 58):

Preparation 7–5: First, 5.42 *mmol* ethylene diamine was dissolved in 30 *ml* of 1,4-dioxane at room temperature for 10 *min*. Subsequently, 10.8 *mol* potassium carbonate was added to the prepared solution, which was stirred further for 30 *min*. The solution was transferred to an ice bath maintained at 10°C to which 20 *ml* of 5.42 *mol* dissolved cyanuric chloride was added dropwise for 5 *h*. After this addition was complete, 1 *mmol* of cobalt nitrate $Co(NO_3)_2$times$6H_2O$ was added to the reaction mixture. This mixture was then transferred to a Teflon-lined autoclave and heated at 150°C for 10 *h*. After the completion of this reaction, the precipitate was washed several times with a solvent mixture of water and ethanol, and dried at 60°C overnight. This powder (CoPOP) was carbonized at 400°C and 600°C in a nitrogen atmosphere for 3 *h*, after which CoPOP-1 and CoPOP-2 samples, respectively, were finally obtained.

Some of the above-mentioned compounds are shown in Figure 7.21.

The formation of a cobalt-modified porous organic polymer was confirmed by Fourier transform infrared spectroscopy analysis. The catalytic activity for the 4-nitrophenol reduction was studied and the related electrocatalytic urea oxidation reaction could be done.

So, cobalt metal-doped porous organic polymers can be used as efficient catalysts to remove urea and nitrophenol from wastewater (57).

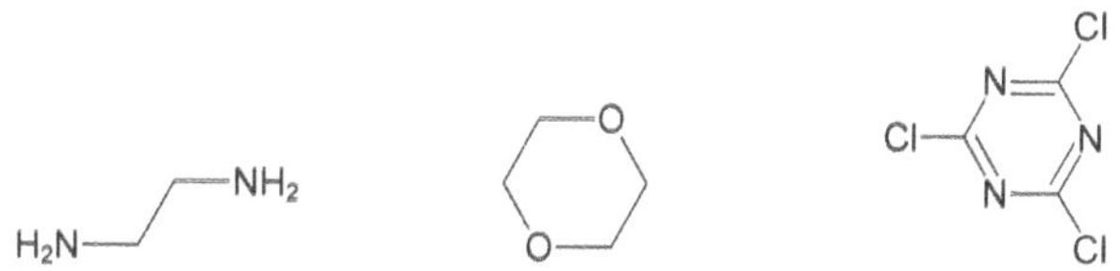

Ethylene diamine   1,4-Dioxane   Cyanuric chloride

Figure 7.20 Compounds for synthesis.

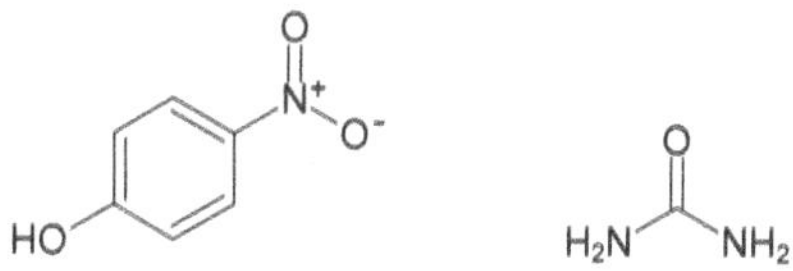

4-Nitrophenol   Urea

Figure 7.20 (cont) Toxic compounds.

7.4.3 *Removal of Organic Micropollutants*

The hydroxyl groups of macrocyclic β-cyclodextrin were crosslinked in the presence of tetrafluoroterephthalonitrile to obtain a mesoporous polymer network for the rapid removal of organic micropollutants (59).

Model compounds that were used are shown in Table 7.4 and in Figure 7.21.

Table 7.4 Model compounds (59).

Compound	Shortcut	Property
2,4-Dichlorophenol	2,4-DCP	Carcinogen
2-Naphthol	2-NO	Model for pollutants
1-Naphthyl amine	1-NA	Carcinogen
Bisphenol A	BPA	Endocrine disruptor
Bisphenol S	BPS	Endocrine disruptor
Metolachlor		Pesticide
Ethinylestradiol		Contraceptive
Propranolol hydrochloride		Beta blocker

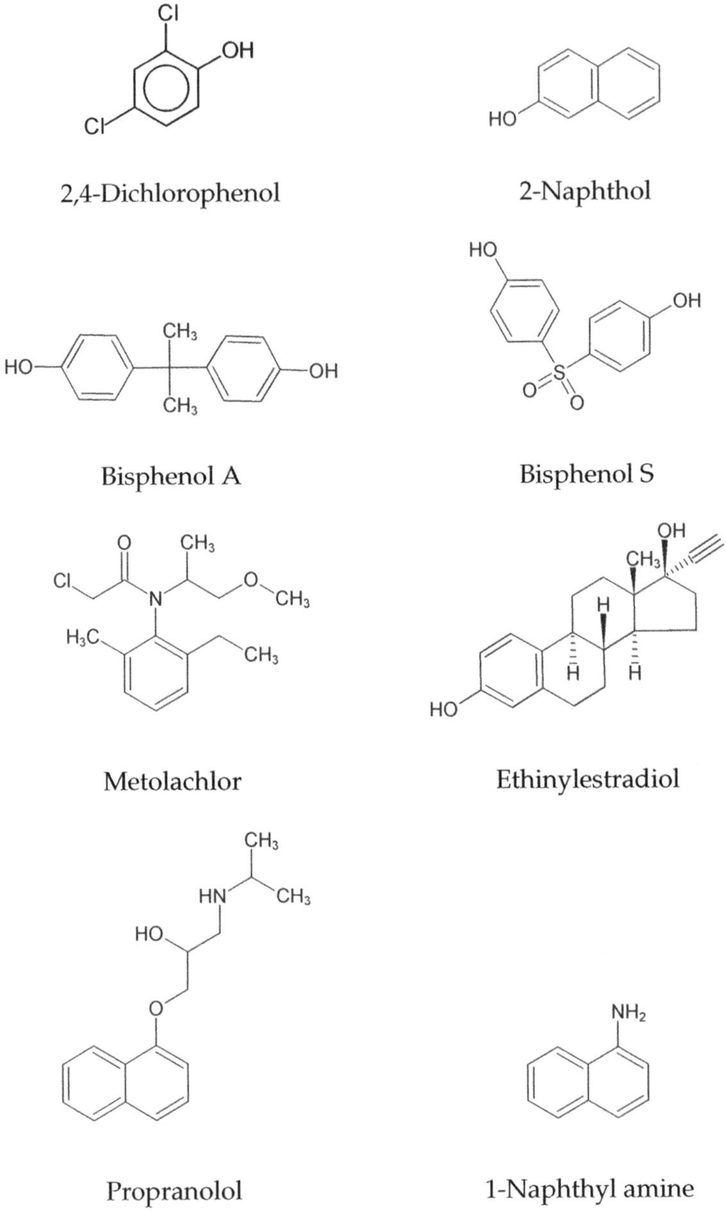

Figure 7.21 Model compounds.

7.4.3.1 *Polycalixarenes*

The contamination of water with petroleum products, organic solvents, micropollutants and textile dyes causes a wide array of health problems in animals, as well as a range of environmental issues. The importance of preserving the environment, spurs the growing need for novel multifunctional materials that effectively remove these contaminants from the environment (60). The synthesis and the characterization of a porous covalent calix[4]arene polymer (CalP) as well as its use for the removal of organic contaminants from water has been detailed (61).

CalP has a high surface, a large pore volume and a good sorption capacity for a range of organic solvents, oils, and toxic dyes. The polymer can selectively absorb up to seven times its weight of oil from oil/water mixtures. From aqueous solutions, it can adsorb both anionic and cationic dyes in under 15 *min*.

Its uptake capacity is significantly higher than that of the most adsorbent materials reported to date, including commercial activated carbon. Additionally, the polymer can be easily regenerated using mild washing procedures and reused several times with no loss of absorption efficiency (60).

A porous CalP can be synthesized by a Sonogashira-Hagihara cross coupling (60, 62). The Sonogashira-Hagira reaction is a convenient means for crosslinking calixarenes having reactive groups, e.g., bromine groups. This method has the additional benefits of alkyne functionality, which can endow materials with, for example, high surface area, high absorption capacity and provide superhydrophobicity. The synthesis is shown in Figure 7.22.

The reaction was carried out in dry tetrahydrofuran at 65°C for 60 *h*. After centrifugation and multiple washings of the reaction precipitate, the polymer was obtained as a brown powder that was completely insoluble in a lot of organic solvents.

A method of using a porous polymeric material can use the following steps (60):

1. Contacting the sample with a porous polymeric material composition,
2. Separating the porous polymeric material from the sample,
3. Heating the porous polymeric material device consisting of a porous polymeric material of the present disclosure, and

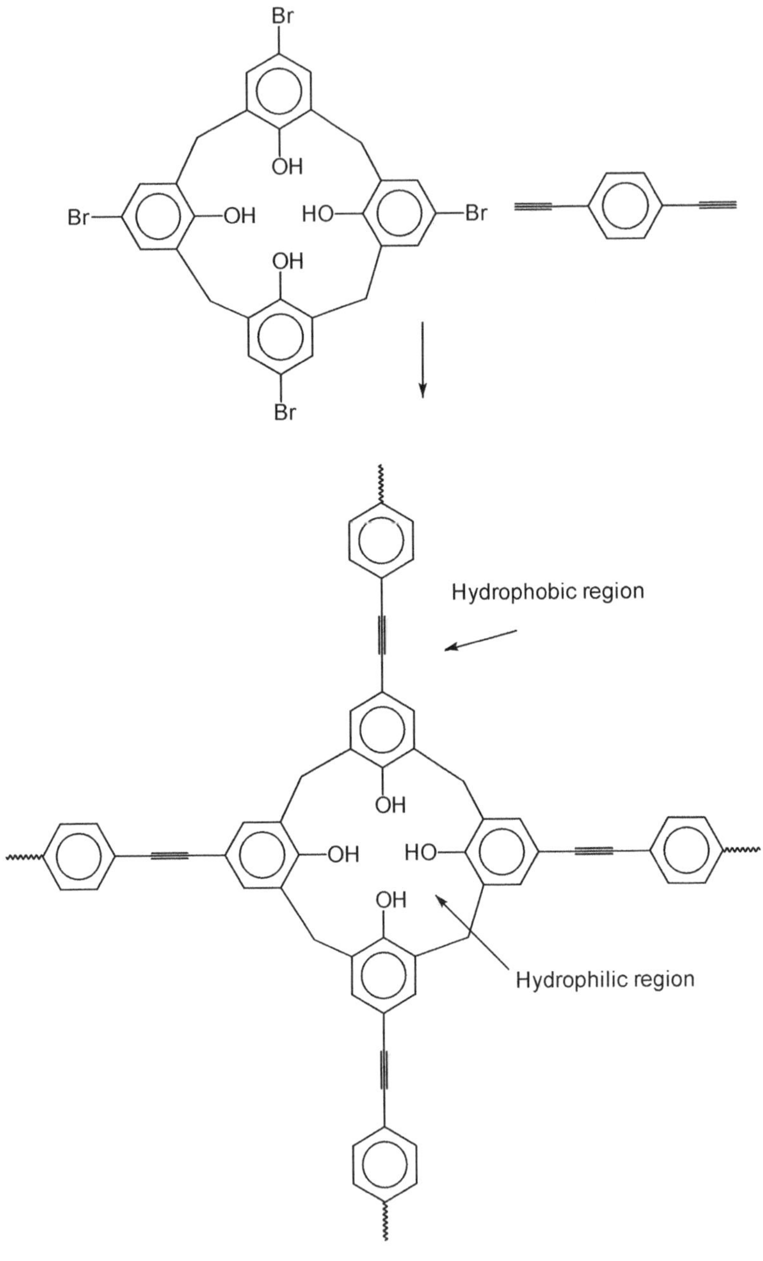

Figure 7.22 Synthesis of Polycalixarene (60).

4. Determining the presence or absence of any absorbed compounds.

7.4.3.2 *Cyclodextrin Compounds*

β-Cyclodextrin-tolylene-2,4-diisocyanate polymer. Adsorption processes can be employed to remove specific contaminants or contaminant classes from fluids like air and water (63). Insoluble β-cyclodextrin polymers have been formed by crosslinking with epichlorohydrin and other reactive compounds, and feature well-defined binding sites and high association constants.

Insoluble β-cyclodextrin polymers crosslinked with epichlorohydrin have been investigated as alternatives to activated carbon for water purification, but their low surface areas result in inferior sorbent performance relative to activated carbons. So, there is a need for an adsorbent that provides a rapid anionic micropollutant extraction, high total uptake, and facile regeneration and reuse procedures.

An adsorbent material for contaminants can be made from polymeric cyclodextrin materials produced from insoluble polymers of β-cyclodextrin, which are toroidal macrocycles with seven glucose units whose internal cavities are capable of binding organic compounds (63). β-Cyclodextrin is an inexpensive and sustainably produced monomer derived from cornstarch that is used extensively to formulate and stabilize pharmaceuticals, flavorants, and fragrances, as well as within chiral chromatography stationary phases. β-Cyclodextrin is shown in Figure 7.25.

The synthesis of a β-cyclodextrin-tolylene-2,4-diisocyanate polymer can be done as follows (63):

Preparation 7–6: β-Cyclodextrin (60.0 *g*, 0.0529 *mol*) was dissolved in 120 *ml* *N,N*-dimethylformamide in a 500 *ml* one-neck round-bottom flask at a magnetic stir rate of 400 *rpm* and the temperature was set to 80°C. An oil bath equipped with a thermocouple was used for heating. After completely dissolving the β-cyclodextrin, tolylene-2,4-diisocyanate (36.8 *g*, 0.2115 *mol*) was added subsequently to the flask at 80°C. Air bubbles were observed likely due to the presence of water in the reaction medium. After about 1 *min* when there was no bubble produced, the flask was capped with a rubber septum.

After 3 *h*, the reaction was stopped by adding 30 *ml* of methanol and turning off the heating. The resulting viscous clear solution was precipitated into 1.2 *l* methanol to obtain white powder product. After 1 *h* stirring, the crude product was filtered under vacuum using a Buchner funnel. The filtered polymer powder was transferred back to a 2*L* beaker and washed again two times with 1.5 *l* water and on time with 1.2 *l* methanol. During each cycle, the washing time was 1 *h*.

After final filtration, the wet solid product was transferred to an evaporating dish, which was placed into a vacuum oven at 80°C to yield 72.6 *g* dry polymer. It was observed that starting at a 6 equivalence of tolylene-2,4-diisocyanate and above, a hard gel was obtained which is difficult to work up. In contrast, tolylene-2,4-diisocyanate:cyclodextrin ratios in the range of 2:1-5:1 provide a powder material upon stopping the reaction with methanol.

The presence of amine groups into the polymerization reaction is believed to result in the formation of urea linkages in addition to the urethane linkages which result from the crosslinking of β-cyclodextrin and tolylene-2,4-diisocyanate under anhydrous conditions.

Rigid and Non-rigid Crosslinked Cyclodextrin. The usage of a non-rigid crosslinking agent and a rigid crosslinking agent simultaneously in the preparation of a cyclodextrin polymer has been tested (63).

The introduction of the non-rigid crosslinking agent can crosslink the cyclodextrin monomer and impart a certain swelling property to the polymer material. The rigid crosslinking agent can be used as a structure regulation agent to make the polymer material have certain microporosity and be capable of maintaining the stability of the polymer pores after swelling as a skeleton.

The non-rigid crosslinking agent is epichlorohydrin. The rigid crosslinking agent is preferably a material with a benzene ring structure, such as tetrafluoroterephthalonitrile, decafluorobiphenyl, decafluorobenzophenone, octafluoronaphthalene, or a combination thereof. Some of these compounds are shown in Figure 7.23.

The preparation of the polymer only needs to be carried out in an aqueous phase, and the synthesis process is very simple and easy for industrial production. The preparation can be done as follows (63):

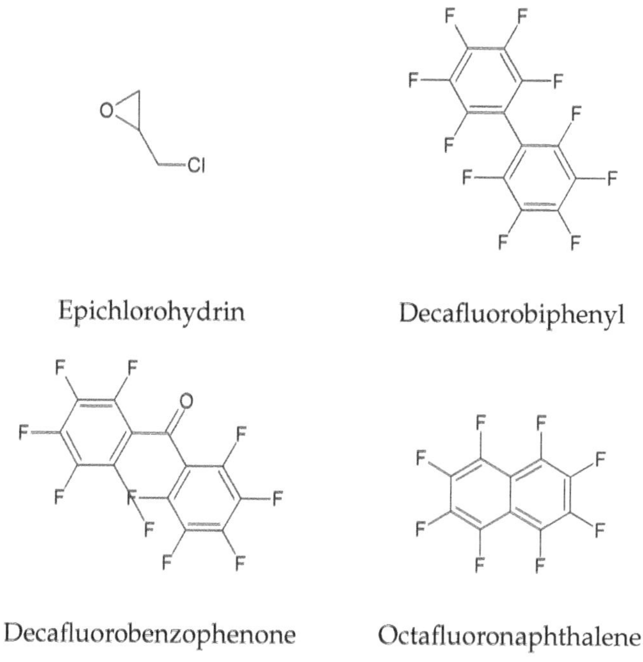

Figure 7.23 Rigid crosslinking agents (63).

Preparation 7–7: A certain amount of cyclodextrin is weighed into a round-bottom flask and dissolved completely by adding an alkaline aqueous solution. Then appropriate amounts of the rigid crosslinking agent and the flexible crosslinking agent are added.

The round-bottom flask is placed in an oil bath equipped with a magnetic stirring device, and a magnetic stirrer is put in. The oil bath temperature and the rotating speed are adjusted, and the reaction is allowed to continue for 3 *h* after reaching the set temperature. After the reaction is completed, the mixture is filtered, and the precipitate is washed successively with distilled water, an organic washing solvent (such as tetrahydrofuran, ethanol, etc., which can dissolve the unreacted crosslinking agent) several times. The obtained solid is further dried under vacuum for 12 *h* and the cyclodextrin polymer is obtained.

The cyclodextrin polymer combines the advantages of the rigid material and the flexible material and has a certain porosity and swellability. The material can maintain a certain pore structure after swelling. Therefore, the polymer exhibits a fast adsorption performance for a variety of organic pollutants.

The organic micropollutants that can be adsorbed are some organic pollutants that are harmful to the environment and are widely concerned. There are mainly endocrine disruptors, plastic components, and organic phenolic substances.

For example, the endocrine disruptors include ethinylestradiol, estriol, and estradiol. Plastic components include, but are not limited to, bisphenol A and bisphenol S, phenolic substances 2-naphthol, 3-phenylphenol, and 2,4,6-trichlorophenol. Some of these compounds are shown in Figure 7.24.

7.4.4 *Lysozyme Extraction*

Lysozyme is widely used in medical, food and industrial fields due to its bacteriolytic effect and thus it is significant to design and develop a specific adsorbent with high adsorption capacity and selectivity for lysozyme (64).

Polymers from tetrafluoroterephthalonitrile crosslinked porous β-cyclodextrin have a high uptake capacity and show a noncovalent interaction between lysozyme and the carboxyl groups. These compounds are shown in Figure 7.25.

So, such carboxyl-functionalized polymers were synthesized by base-catalyzed hydrolysis of the nitrile groups to carboxyl (64).

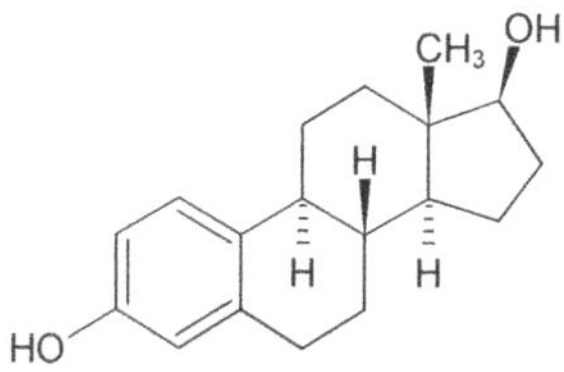

Estradiol

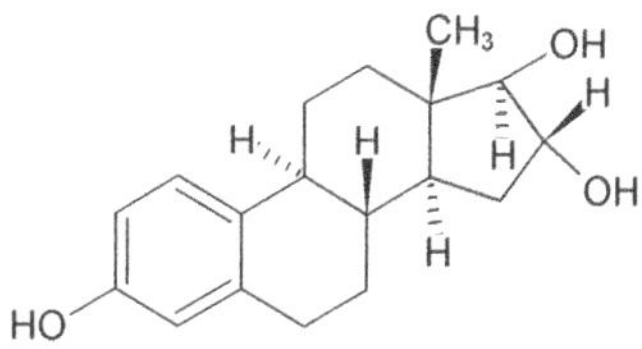

Estriol

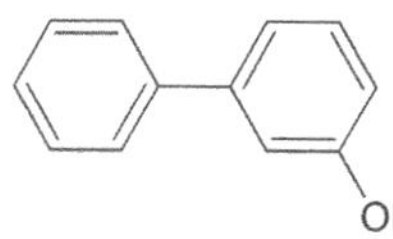

3-Phenylphenol

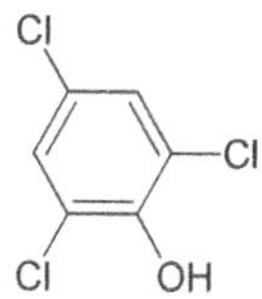

2,4,6-Trichlorophenol

Figure 7.24 Organic pollutants.

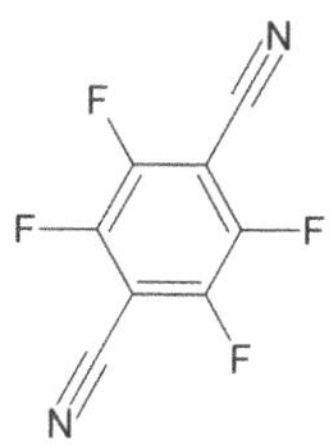

Tetrafluoroterephthalonitrile

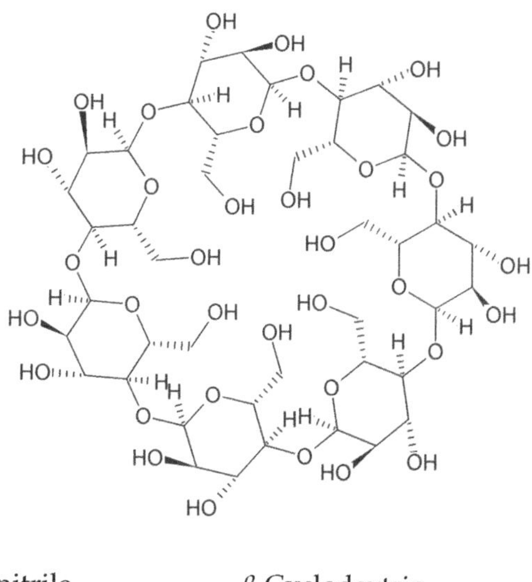

β-Cyclodextrin

Figure 7.25 Monomers for lysozyme extraction.

The porous structure, the large extent of carboxylic functional groups, and cyclodextrin's preventing aggregation show an outstanding adsorption capability for lysozyme. The maximum saturated adsorption capacity reaches 1520 $mg\,g^{-1}$, which is much better than the parent polymer and other known materials.

The here prepared material was successfully utilized for the selective extraction of lysozyme from egg white (64).

The chicken egg whites were treated by the method of Chen *et. al.* (65). Then, 5 *ml* egg whites solution and 30 *mg* of the crosslinked porous β-cyclodextrin-COO^- were mixed and stirred at 200 *rpm* for 2 *h* at room temperature. The extracted lysozyme on adsorbent was eluted with phosphate buffer (20 *mM*, pH 7.0) containing 1.0 *M* NaCl. The initial egg white solution, the supernatant after extraction, and the eluate were analyzed using HPLC (64).

7.5 Metal Recovery

7.5.1 Rice Straw in Poly(urethane) Foams

Not only marine species, but also human and natural environments have been seriously affected by the marine environment being polluted by oil spills, heavy metals and other pollutants (66).

A sorbent fabricated by adding rice straw into PU foams was investigated (66). Oil types such as crude oil, fuel oil, diesel oil, kerosene and a solution containing Cd^{2+}, Cu^{2+} metal ion were used to evaluate the absorption capacity of this sorbent.

After 150 *min* it was observed that the oil absorption capacity of the sorbent added 25% of rice straw mass and 500 μm of size was highest, correspondingly to 15.086 $g\,g^{-1}$, 13.964 $g\,g^{-1}$, 12.226 $g\,g^{-1}$, 10.746 $g\,g^{-1}$ of crude oil, fuel oil, diesel oil, kerosene, respectively. Moreover, the metal ion absorption capacity of Cd^{2+}, and Cu^{2+} was 212.9 $mg\,l^{-1}$ and 193.6 $mg\,l^{-1}$ (66).

The method based on ASTM F726-99 for determining the sorbent capacity of oil and water was used (67). Petroleum products, including crude oil, fuel oil, diesel oil, and kerosene with different densities, were used as study objects for evaluating the sorption capacities of rice straw PU material.

The reason for choosing the above oil was that fuel oil and diesel oil were fuels that are popularly used for ships or vessel fleets.

Crude oil was a product from Vietnamese oil rigs and kerosene was fuel provided to people living in the Mekong Delta. These fuels appeared regularly on the surface water of the river at a port in the Mekong Delta, hence it was considered as the study object. Properties of crude oil, fuel oil, diesel oil, and kerosene available in Vietnam are shown in Table 7.5.

Table 7.5 Properties of oil types.

Properties	Unit	Crude oil	Diesel oil	Fuel oil	Kerosene
Density	$kg\,m^{-3}$	937	890	850	810
Surface tension	$mN\,m^{-1}$	31.7	28.8	26	22.5

A 50 *ml* solution containing a Cd^{2+} and Cu^{2+} concentration of 100 $mg\,l^{-1}$ was poured into a beaker of 100 *ml* of water to prepare for an experiment about the evaluation of heavy metal ion absorption capacity. After that, absorbent material accurately weighed to the milligram was immersed in the above beaker. The metal adsorption capacity was evaluated after each 10, 20, 40 and 60 *min* by the inductively coupled plasma optical emission spectrophotometry (ICP-OES) method for determining the sorbent capacity of oil and water.

7.5.2 *Bonding of Metal-Containing Ions*

The removal and extraction, or the recovery of metal-containing ions, in particular heavy metal-containing ions, from industrial wastewater, e.g., in electroplating plants, from catalyst residues from the petrochemical or pharmaceutical industry, from mine drainage and seepage water, the renaturation of soil contaminated with heavy metals, etc., is a task of increasing importance because heavy metal-containing ions in particular have an environmentally damaging effect, and their recovery is of financial interest as well in view of rare or economically strategic metals (68).

Porous particles of a crosslinked hydroxy- or amino-group-containing polymer have a relatively low swelling factor. A composite material containing the porous particles dispersed in a continuous aqueous phase has been developed (68).

The porous particles, or the composite material, can be used for purifying organic molecules and for bonding metals from solutions.

A filter cartridge contains the porous particles of a crosslinked polymer or the composite material.

The sorbent should have a relatively high binding capacity for metals even under acidic conditions. Furthermore, the volume of the sorbent used for binding the target molecules is to be reduced as compared with metal binding sorbents and chromatographic sorbents. A method for producing such porous particles from a crosslinked polymer, can be done using the following steps (68):

1. Application of an organic polymer onto a porous inorganic carrier material in particulate form,
2. Crosslinking of the organic polymer in the pores of the inorganic carrier material, and
3. Dissolving out the inorganic carrier material to obtain the porous particles from a crosslinked organic polymer.

The porous inorganic carrier material in particulate form is preferably a mesoporous or microporous carrier material. The mean pore size of the porous carrier material is most preferably in the range from 10 *nm* to 150 *nm*. In addition, the porous carrier material should have a pore volume in the range from 60% to 70% by volume, with reference to the total volume of the porous carrier material in each case. The mean pore size and the pore volume of the porous carrier material can be determined by means of the pore filling method with mercury according to the ASTM D4404-18 standard (69).

The polymer used in the first step is preferably a polymer which contains hydroxyl or amino groups, such as PVA, poly(vinylamine), poly(ethylene imine) and poly(lysine).

After the first step of the application of the polymer and after drying of the polymer applied onto the porous inorganic carrier material, crosslinking is performed. The degree of crosslinking can be adjusted by using the desired quantity of the crosslinking agent.

The here used crosslinking agent has preferably two, three or more functional groups. The crosslinking agent is preferably selected from the group comprising dicarboxylic acids, tricarboxylic acids, urea, bis-epoxides or tris-epoxides, diisocyanates or triisocyanates, dihalogen alkylenes or trihalogen alkyls and halogen epoxides. Dicarboxylic acids, bis-epoxides and halogen epoxides are preferred.

These materials are collected in Table 7.6 and some of these compounds are shown in Figure 7.26.

Table 7.6 Crosslinking agents.

Compound
Terephthalic acid
Biphenyldicarboxylic acid
Ethylene glycol diglycidyl ether
Chlorendic acid
Epichlorohydrin

Furthermore, the crosslinked polymer can be derivatized in its side groups. Here, an organic group is preferably bound to the polymer. This group may be any conceivable group, such as an aliphatic or aromatic group, which may also contain heteroatoms. These groups may also be substituted with anionic or cationic groups and groups which can be protonated or deprotonated, respectively.

Organic groups with the characteristics of a Lewis base are in particular groups which form a complex bond with the metal to be bound. Organic groups which have a Lewis base are ones which have heteroatoms with free electron pairs such as N, O, P, S or As (68).

Preferred organic groups for the derivatization of the polymer are ligands that are collected in Table 7.7. Some of these materials are shown in Figure 7.27.

The here described materials can be used for the examples collected in Table 7.8 among others.

7.5.3 *Porous Porphyrin Polymer*

Recycling precious metals from electronic waste not only protects the environment, but is also favorable for alleviating resource shortages. Ionic porous organic polymers, as one type of burgeoning material, are regarded as excellent adsorbents due to their high ion density, but their application in precious metal recovery is still limited (70).

A porous porphyrin polymer and a method of recovering precious metal elements using the same were described (71). The porous porphyrin polymer shows high selectivity for precious metal

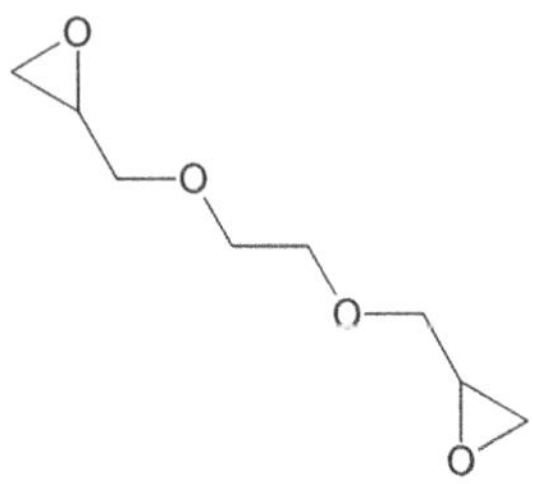

Ethylene glycol diglycidyl ether

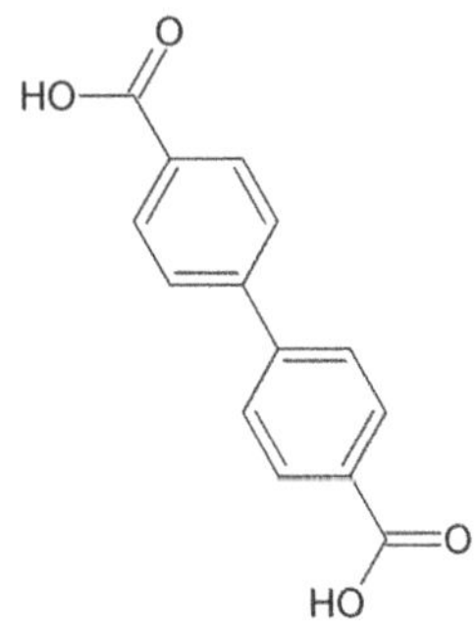

Biphenyldicarboxylic acid

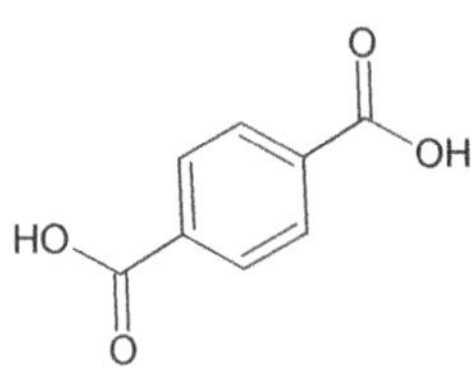

Terephthalic acid

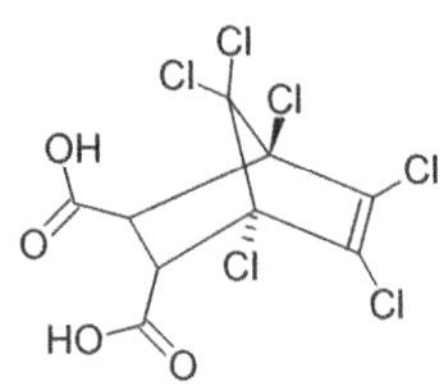

Chlorendic acid

Figure 7.26 Crosslinking agents.

Table 7.7 Organic groups for polymer derivatization (68).

Ligand group
6-Amino-nicotinic acid groups
Arginine groups
Succinic acid-*N*-methyl-piperazine
4-[(4-Aminopiperazine-1-yl)amino]- 4-oxobutyric acid groups
Succinic acid groups
Creatine groups
Diaminobicyclooctane carboxylic acid
Diethylene triamine
Diglycolic acid groups
Ethylene diamine tetraacetic acid groups
Ethylphosphonylcarbonyl group
N-Ethanethiol group
N,N-Diacetic acid groups
4-Aminobutyric acid groups
Glutaric acid groups
4-Piperidine carboxylic acid groups
4-Imidazolylacetyl groups
4-Imidazole acrylic acid groups
Isonicotinic acid groups
Lysine acid groups
Methylthiourea groups
Nitrilotriacetic acid
Phosphoric acid group
Proline
Purine-6-carboxylic acid groups
Pyrazine-2-carboxyclic acid groups
Thymine-*N*-acetic acid groups
Theophyllin-7-acetic acid groups
Citric acid groups

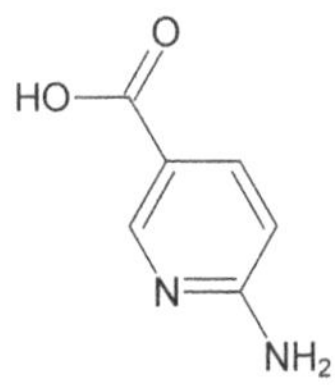

6-Amino-nicotinic acid

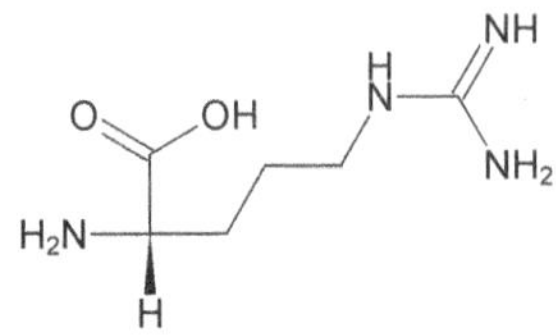

Arginine

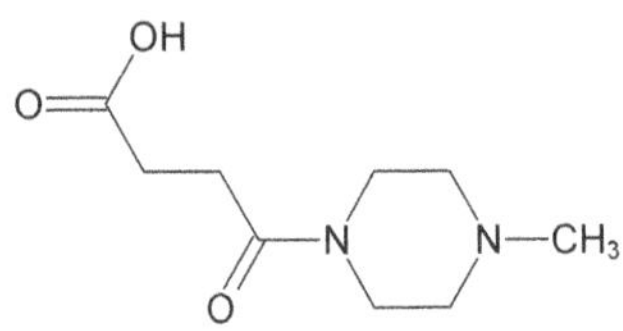

4-[(4-Aminopiperazine-1-yl)amino]-
4-Oxobutyric acid

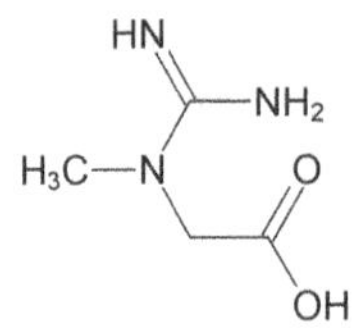

Creatine

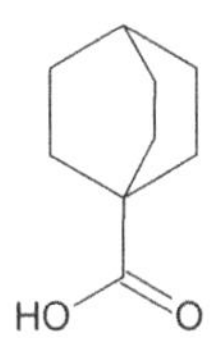

Diaminobicyclooctane carboxylic acid

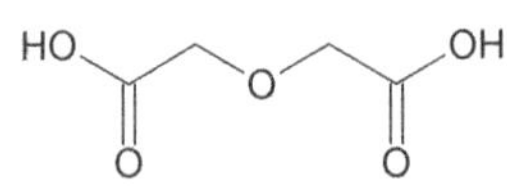

Diglycolic acid

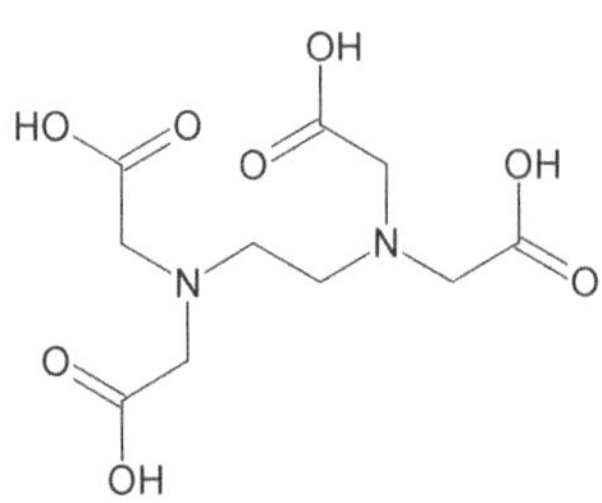

Ethylene diamine tetraacetic acid

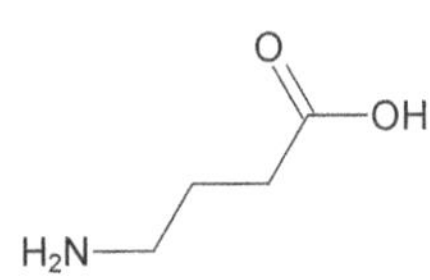

4-Aminobutyric acid

Figure 7.27 Compounds for derivatization (68).

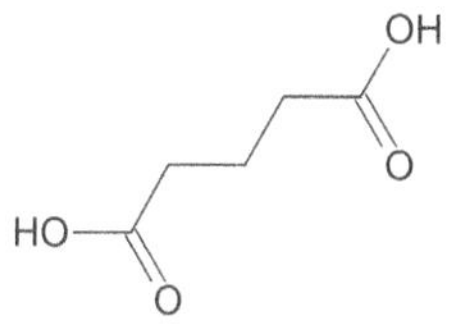

Glutaric acid

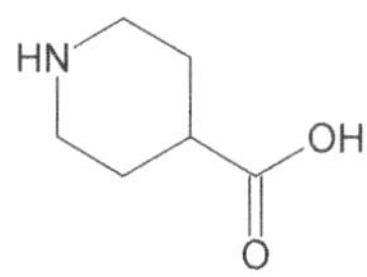

4-Piperidine carboxylic acid

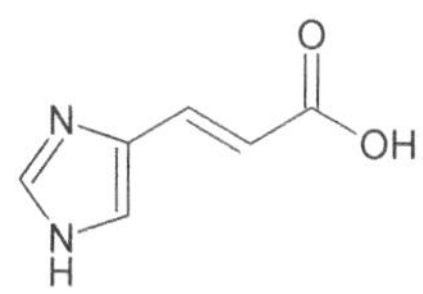

4-Imidazole acrylic acid

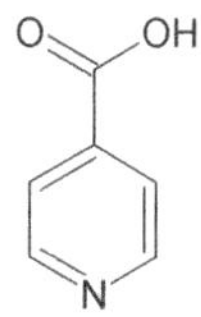

Isonicotinic acid

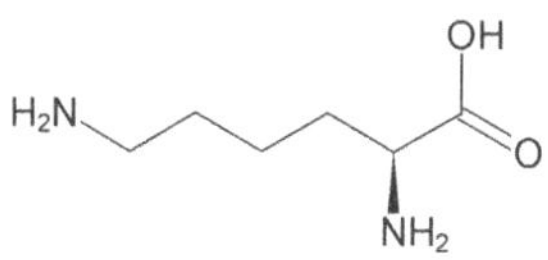

Lysine acid

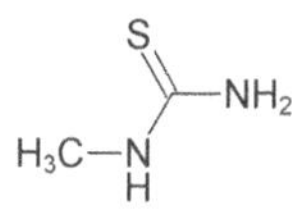

Methylthiourea

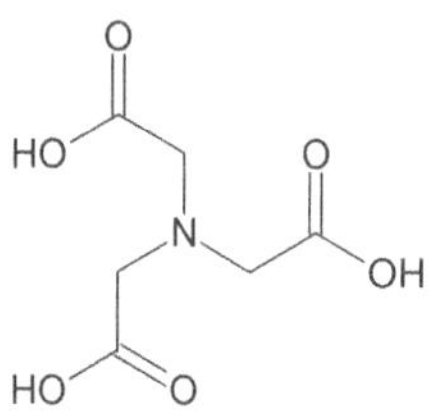

Nitrilotriacetic acid

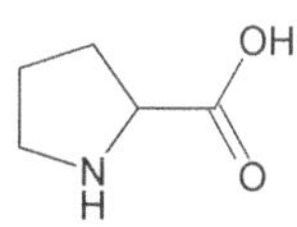

Proline

Figure 7.27 (cont) Compounds for derivatization (68).

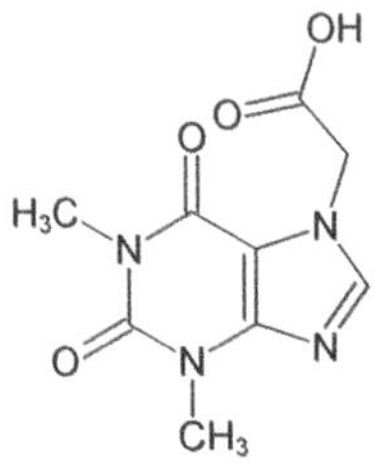

Theophyllin-7-acetic acid

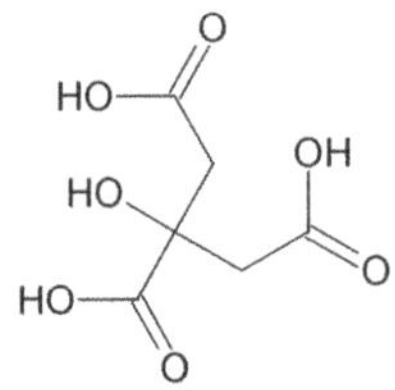

Citric acid

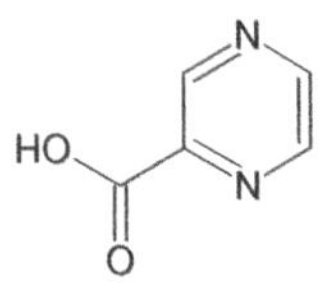

Pyrazine-2-carboxyclic acid

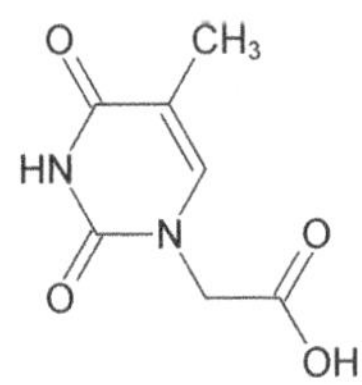

Thymine-*N*-acetic acid

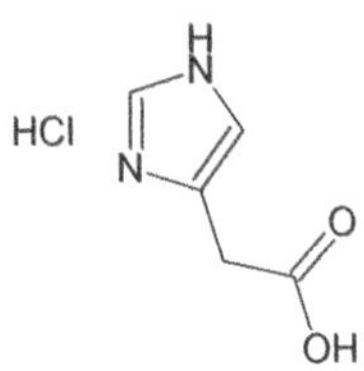

4-Imidazolyl acetic acid

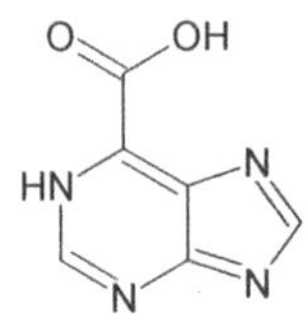

Purine-6-carboxylic acid

Figure 7.27 (cont) Compounds for derivatization (68).

Table 7.8 Fields of use for the described materials (68).

Method	Description
Analytic Methods	Determination of the concentration of the amino groups of a sorbent with breakthrough measurement with 4-Methylbenzenesulfonic acid (titration analysis)
Separation of Metals from Solutions	Reduction of the copper content in an aqueous solution
Drinking Water	Copper as well as nickel with an initial concentration of 0.5 *ppm* each in tap water are considerably reduced
Dynamic binding of Nickel	
Dynamic binding of platinum	The very strong binding of the platinum (at pH 1.1) is largely reversible. In that way, the phase can be reused again and again. A recovery of the platinum is possible without burning. Due to the very strong binding, solutions with very low concentration can also be processed effectively and with a high yield (here 100%); 77% of the bound platinum was recovered by means of elution with 50 *ml* 1*M* HNO_3

elements and a high ability to adsorb precious metal elements. So, it can be applied to the recovery of precious metal elements either from metal leachates of waste electronic products or from river water or seawater. The porphyrin monomer is shown in Figure 7.28.

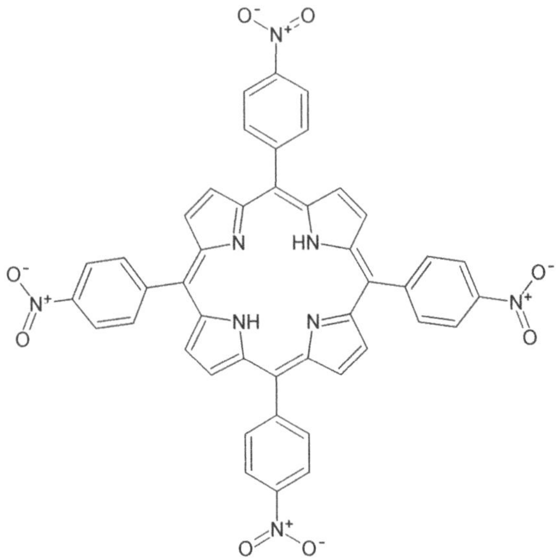

Figure 7.28 5,10,15,20-Tetrakis (4-nitrophenyl)-21*H*,23*H*-porphyrin.

This porphyrin monomer may be obtained by dissolving 4-nitrobenzaldehyde in propionic acid to obtain a solution, and then adding and reacting acetic anhydride and pyrrole with the solution. The porphyrin polymer may be represented by the structure shown in Figure 7.29.

It has been found that the porphyrin polymer has a high selectivity for gold or platinum metal ions in a mixture solution containing various metal ions, and it has high adsorption efficiency at almost all pHs. Thus, when the porphyrin polymer is applied to metal leachates from waste electronic products or to seawater, it can adsorb and recover gold or platinum metal ions with high selectivity in comparison to other metal ions (71).

When the precious metal is gold (Au), the solution preferably has a pH of 4 or less, and when the precious metal is platinum (Pt), the solution preferably has a pH of 2-9.

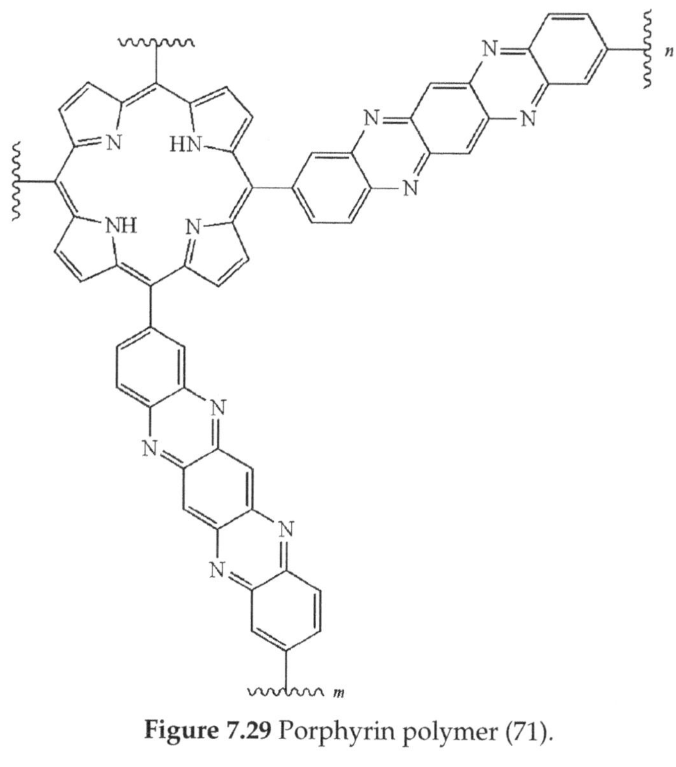

Figure 7.29 Porphyrin polymer (71).

The thermal durability of the porous porphyrin polymer was measured by thermogravimetry (TG) under air and nitrogen atmospheres. A first decrease in weight of about 3 to 4% at about 100°C or lower was attributable to the loss of water absorbed into the polymer.

In addition, the weight started to decrease at around 150°C, and a second decrease to 93% in air and 90% in nitrogen appeared, and this decrease was attributable to *N,N*-dimethylformamide used as the solvent in each polymer synthesis and Soxhlet extraction. It was confirmed that the polymer was stable up to 330°C under the air and nitrogen atmospheres.

The recovery was tested under several specific conditions, e.g., gold recovery from waste electronic products, the increase in the gold ion adsorption ability by light irradiation, and platinum recovery from seawater.

In summary, it was found that porous porphyrin polymer has an excellent ability to adsorb precious metal elements, such as gold and platinum, and thus may be applied to the recovery of precious metal elements either from metal leachates of waste electronic products or from natural river water or seawater (71).

A porphyrin and viologen-linked cationic porous organic polymer was prepared via the Zincke reaction (70).

By linking porphyrins with viologens, this material obtained the characteristics of a hierarchical porous structure, a high ion density skeleton, and a rich nitrogen content, which gave it an ultrahigh adsorption capacity ($Q_{max} = 792.22\ mg\,g^{-1}$) and rapid adsorption rate for Au(III).

The material also had an effective Au(III) recovery capability from SIM cards. The investigation of the mechanism confirmed that the adsorption performance was attributed to the interplay of ion exchange, redox reactions and coordination.

Also, the material had an excellent recyclability and could maintain an ultrahigh adsorption efficiency of 81% after eight consecutive adsorption-desorption experiments (70).

7.5.4 *Iminodiacetic Acid-Functionalized Polymer*

The synthesis of an iminodiacetic acid-functionalized hyper-crosslinked polymer was reported (72). The hyper-crosslinked porous polymer was prepared via a $FeCl_3$ catalyzed Friedel-Crafts reaction.

This material was used for purification of water polluted by toxic metal ions via coordination of carboxylate and amino active sites on the surface of the porous polymer.

The obtained porous polymer was stable under harsh conditions and the structural features on the polymer work together to help the removal of Pb^{2+} with 1138 $mg\ g^{-1}$ uptake capacity.

In addition, the iminodiacetic acid-functionalized hyper--crosslinked polymer reveals a reuseability and very promising capture efficiency not only for Pb^{2+}, but also for Hg^{2+} and Cd^{2+}, from a mixture of Pb^{2+}, Hg^{2+}, Cd^{2+}, Co^{2+}, Fe^{3+}, Zn^{2+}, Mg^{2+}, and Na^{+} metal ions. So, this polymer material can solve the pollution problem caused by various metal ions (72).

7.5.5 *Removal of Toxic Elements*

Water contamination by toxic heavy elements is becoming an urgent problem in environmental science and separation technologies (73). However, the design of sophisticated absorbents with high stability and outstanding removal efficacy for ion coadsorption is still a technical challenge.

Inspired by a biological Hg/Se antagonism detoxification, a porous polymer network (PPN) was designed for the concurrent removal of Hg/Se species in aqueous solutions (73).

The MoS_4^{2-} functionalized PPN-150-MoS_4 exhibits a rapid and highly efficient simultaneous removal of toxic anions SeO_4^{2-} and SeO_3^{2-} and metals, e.g., Hg^{2+}. The high thiophilicity of Hg^{2+} leads to a 99.9% removal within minutes. Selenite and selenate, which are known for being difficult to remove from aqueous environments, can be removed by PPN-150-MoS_4, exhibiting a removal greater than 99% within minutes when in the presence of Hg^{2+}. At the same time, the removal efficiency for Se(IV) and Se(VI) oxoanions in the absence of Hg^{2+} is very low, reaching only 14% removal.

Overall, PPN-150-MoS_4 exhibits one of the highest adsorption capacities toward SeO_3^{2-} (124 mg/g), making it a promising and cheap sorbent material for water remediation applications (73).

7.5.6 *Polyfunctional Sorbent Materials*

Clean drinking water is a health concern for the entire world population, and as such, is a focal point for the World Health Organization (WHO) (74). Sorbent technologies, like functionalized silica substrates, are prime candidates for water remediation and analytical separations due to their ability to selectively remove harmful contaminants at trace levels.

These materials also hold promise for uranium mining from seawater, collection of desired metals from mining streams, and trace collection of specific contaminants in industrial aqueous waste streams. The growth in research in mesoporous silicas and the breadth of application of these materials have expanded greatly over the last twenty years.

Functionalized silicas are also applicable in fields such as biomedical engineering, supported catalysis, and molecular motors. The high specific surface area, absorption capacity, robust nature, and tunability of the surface chemistry make functionalized silica substrates particularly well suited as sorbents. Many of these properties are governed by the substrate, but the affinity of the material for a target analyte is dependent on the ligand used for functionalization.

7.5.6.1 *Thioalkylene Functionalization*

A material has been described with a functionalized solid support surface, wherein the functionalization is a thioalkylene linker bound to the support surface and the thioalkylene linker is coupled to a moiety derived from a ligand, wherein the ligand includes a terminal alkenyl and at least one first functional group configured to bind to at least one predetermined target species (74).

Thiol-based sorbent materials have been shown to be excellent sorbents for the capture of soft heavy metals such as mercury, silver, and lead. The functionalization of silica materials with 3-mercaptopropyl ligands has been optimized, reaching near theoretical limits for surface coverage with thiol groups.

A substrate functionalized with 3-mercaptopropyltrimethoxysilane has been prepared as follows (74):

Preparation 7–8: In a 500 *ml* long-neck flat-bottom flask, 1.5 *g* of CAB-O-SIL® EH5 was stirred for 2 *h* with 100 *ml* of toluene and 0.14

ml of deionized water to hydrate the silica surface. Then, 3 *ml* of 3-mercaptopropyltrimethoxysilane, c.f. Figure 7.30, was added, and the flask was heated to reflux for 84 *h*. After the reflux, the excess water and the methanol were distilled off to ensure complete surface coverage. The flask was slowly cooled to room temperature and filtered through a medium fitted glass funnel. The material was washed with 3 × 50 *ml* isopropyl alcohol followed by 50 *ml* of methanol and dried under vacuum overnight.

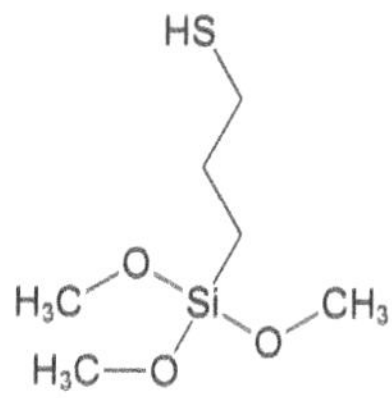

Figure 7.30 3-Mercaptopropyltrimethoxysilane.

For the thiol-ene click reaction surface, thiols are reacted with a terminal alkene on the chelating ligand. The resulting materials exhibit affinities for soft metals such as gold, due to the thiol, and harder metals such as uranium, due to the specific chelating ligand.

The reaction conditions needed for creation of such materials and the impact of the sorbent support material were investigated, and the performance of these materials were evaluated. A variety of mercaptopropyl-functionalized silica substrates were prepared.

The thiol coverage loadings were varied on all substrates, ranging from the theoretical maximum around 3 thiols per nm^2 down to 0.5 thiols per nm^2. TG was used to determine the organic content of the material and calculate the ligands per surface area. The loading ranged from 0.5 to 3.1 mercaptopropyl groups per nm^2.

These thiol-functionalized materials were then used as a base for further thiol-ene click functionalization, which resulted in a polyvalent material. The thiol-ene reaction is a UV-promoted click reaction: a rapid method for coupling a thiol to an olefin (74).

The reaction for the thiol-ene surface functionalization is shown in Figure 7.31.

Other compounds for surface functionalization are shown in Figure 7.32.

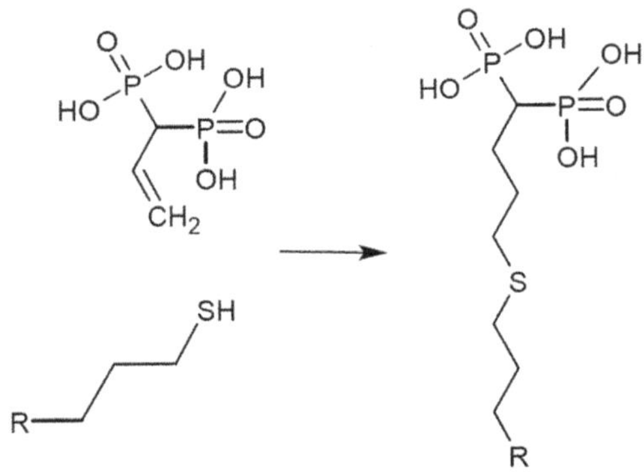

Figure 7.31 Thiol-ene surface functionalization with tetraethyl but-3-ene-1,1-diylbis(phosphonate).

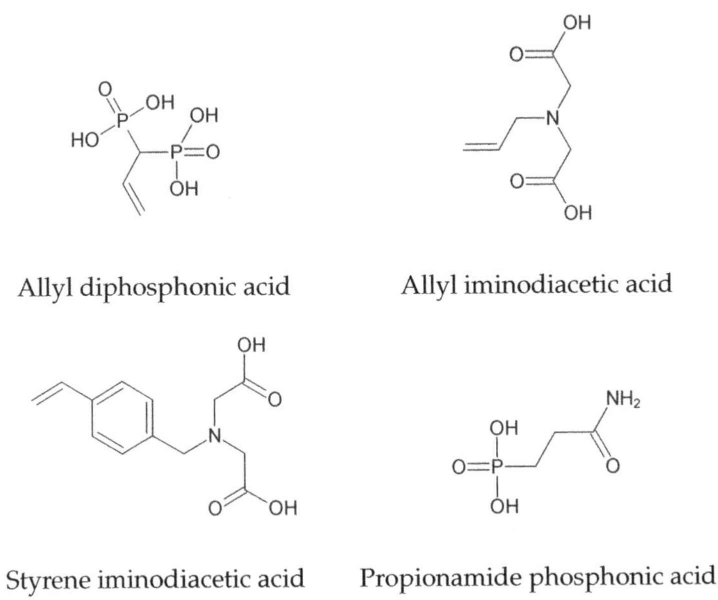

Figure 7.32 Compounds for surface functionalization (74).

But-3-ene-1,1-diylbis(phosphonate) (DiPhos), propionamide phosphonic acid (Prop-Phos), and styrene iminodiacetic acid were selected for testing, due to promising initial results. Phosphonic acids have been shown to be excellent chelators for rare earth elements, and iminodiacetic acid groups, a close relative of ethylene diamine tetraacetic acid (EDTA), have been extensively used for chelation.

Two base silica materials, either Cab-O-Sil (NF silica) or MCM-41 (NP silica), were functionalized with the above-mentioned compounds. In order to determine the increase in performance due to the addition of the ligands, the silica base materials were also tested.

These materials were also compared with commercially available sorbents, Chelex100 and activated carbon. Chelex 100 is a commercially available resin that is functionalized with EDTA. Activated carbon is a very common and cheap sorbent but it lacks affinity and selectivity (74).

The amount of sorption of these materials was tested in natural waters. All organic functionalized materials outperformed commercial resins in river water and seawater. Both DiPhos-NF silica and Prop-Phos had outstanding performance, with the base silica controls having almost no affinity for any metals tested. The DiPhos-NF silica removed almost all metals tested from seawater, and maintained high affinity for the tested metals even in acidic solutions.

Other materials tested lost almost all affinities for rare earth elements in acidic conditions. Activated carbon showed a slight uptake with seawater and removed almost half of most metals in river water, significantly underperforming when compared to the functionalized sorbents. Chelex 100 performed poorly in all the solutions (74).

7.5.6.2 *Lanthanides*

The increased demand for the lanthanides in commercial products resulted in increased production of lanthanide-containing ores, which increases public exposure to the lanthanides, both from various commercial products and from production wastes and effluents (75).

The binding properties for lanthanides on self-assembled monolayers on mesoporous silica supports (SAMMS™) were investigated (75). The supports were functionalized with diphosphonic acid (DiPhos), acetamide phosphonic acid (AcPhos), propionamide phosphonic acid (Prop-Phos), and 1-hydroxy-2-pyridinone (1,2-HOPO), c.f. Figure 7.33. As materials for elution, natural waters (river, ground and sea waters), acid solutions (to mimic certain industrial process streams), and dialysate were used.

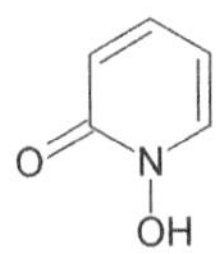

Figure 7.33 1-Hydroxy-2-pyridinone.

The affinity, capacity, and kinetics of the lanthanide sorption, as well as regenerability of the SAMMS materials, were investigated (75).

Going from the acid side over to the alkaline side, the AcPhos- and DiPhos-SAMMS maintain their outstanding affinity for lanthanides, which enable the use of the materials in systems where the pH may fluctuate. In acid solutions, Prop-Phos- and 1,2-HOPO-SAMMS have differing affinity along the lanthanide series, suggesting their use in chromatographic lanthanide separation. Over 95% of 100 $g\,l^{-1}$ of Gd in dialysate was removed by the Prop-Phos-SAMMS after 1 *min* and 99% over 10 *min*.

SAMMS can be regenerated with an acid wash (0.5 *M* HCl) without losing their binding properties. Thus, they have great potential for use as in large-scale treatment of lanthanides, lanthanide separation prior to analytical instruments, and in sorbent dialyzers for treatment of acute lanthanide poisoning (75).

7.5.6.3 Mercury

Mercury is uniquely dangerous due to the natural processes that convert the organic and inorganic forms of mercury into its highly volatile elemental form (76). This creates a constant cycle of mercury entering and reentering the atmosphere, soil, and water, ultimately accumulating in the food supply as methylmercury, a potent neurotoxin.

Certain methods like ion exchange resins were explored but were found to be expensive and ineffective (77). Further research employed the strong soft-soft interactions between the thiol group and mercury (78) and much work has been done on thiol-functionalized adsorbent materials, including clays (79), activated carbon (80, 81), mesoporous silica (82–86), crosslinked poly(ethylene imine) (87), and others (88).

However, these materials display low surface areas, small pore sizes, and a low density of their functional groups. Metal-organic framework (MOF) and covalent organic frameworks (COFs) offer an amenable design strategy to create a high density of chelating sites with promising uptake capacities, but suffer from stability issues which inhibit their use in a wide range of conditions (78, 89–95).

Compositions for binding mercury have been developed that can overcome the above-mentioned deficiencies (76). These compositions include a porous organic polymer having repeat units having a structure according to the formula $--CH_2-A_1-CH2--$. Here, A_1 is a conjugated core having one or more heavy metal chelator moieties covalently attached thereto. For example, A_1 can be a substituted phenyl, a substituted biphenyl, a substituted terphenyl, a substituted naphthalene, or substituted anthracene. In addition, a plurality of pores have a hierarchical pore size distribution over a range of pore sizes.

A thiol-functionalized porous organic polymer (POP-SH) was synthesized using simple free radical polymerization techniques. In detail, a synthesis can run as follows (76):

Preparation 7–9: To a solvothermal autoclave, 1 *g* 3,5-divinylbenzyl chloride was added with 25 *mg* azobisisobutyronitrile and 10 *ml* dimethylformamide. The autoclave was placed in an oven and heated to 100°C for 24 *h*. The resulting product was collected by filtration, washed with acetone, and dried under vacuum to produce POP-Cl. Then, 250 *mg* POP-Cl and 1.3 *g* sodium hydrosulfide were placed in a 100 *ml* Schlenk tube, with 50 *ml* ethanol being introduced under N_2 atmosphere and stirred at 75°C for 3 *d*.

The resulting product was collected by filtration, washed extensively with deionized water and ethanol, and dried under vacuum, thus producing POP-SH as a white powder.

Some of these compounds are shown in Figure 7.34.

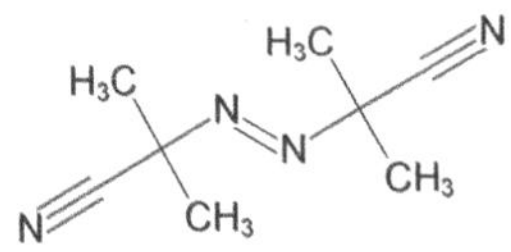

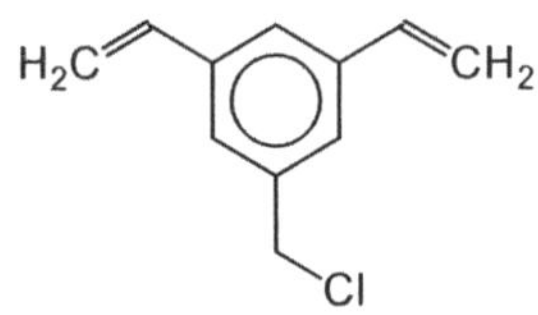

Azobisisobutyronitrile   3,5-Divinylbenzyl chloride

Figure 7.34 Material for polymerization.

With the so synthesized product, several tests were performed, i.e., sorption experiments, mercury vapor adsorption tests, recyclability of the POP-SH (76).

The maximum uptake capacity of POP-SH to remove Hg^{2+} from an aqueous solution was investigated. Equilibrium data were obtained for aqueous solutions with initial Hg^{2+} concentrations ranging from 25 *ppm* to 800 *ppm* after treatment with POP-SH.

The uptake capacities were calculated and the maximum was found to be 1216 $mg\,g^{-1}$, with an equilibrium concentration of 200 *ppm*. With only a simple washing procedure, the activity of POP-SH could be fully regenerated. Based on the sulfur content of POP-SH, the theoretical uptake capacity was calculated to be 1018 $mg\,g^{-1}$.

This increased capacity can be attributed to the conjugated nature of POP-SH, resulting in cation-*p* interactions between the benzene rings of POP-SH and the mercury cations (76).

References

1. N. Fontanals, R.M. Marcé, and F. Borrull, Porous polymer sorbents in C.F. Poole, ed., *Solid-Phase Extraction*, Handbooks in Separation Science, chapter 3, pp. 55–82. Elsevier, 2020.
2. P.J.-J. Lee and J.E. O'Gara, Ion exchange porous resins for solid phase extraction and chromatography, US Patent 8 574 433, assigned to Waters Technologies Corp., November 05, 2013.
3. H. Ritchie, A. Ahmed, P. Myers, and H. Zhang, Porous particles for liquid chromatography and processes for the preparation thereof, US Patent 10 493 428, assigned to Thermo Electron Manufacturing Ltd, December 03, 2019.

4. D.W. Brousmiche and K.D. Wyndham, Porous materials for solid phase extraction and chromatography and processes for preparation and use thereof, US Patent 10 258 979, assigned to Waters Technologies Corp., April 16, 2019.
5. J.J. Kirkland, F.A. Truszkowski, C.H. Dilks, and G.S. Engel, *Journal of Chromatography A*, Vol. 890, p. 3, 2000. 19th Int. Symp. on the Separation and Analysis of Proteins.
6. K.D. Wyndham and B.W. Muriithi, Superficially porous materials comprising a substantially nonporous hybrid core having narrow particle size distribution; process for the preparation thereof; and use thereof for chromatographic separations, US Patent 10 092 893, assigned to Waters Technologies Corp., October 09, 2018.
7. Y. Okamoto, N. Enomoto, R. Ohishi, Y. Ogasawara, H. Akano, and Y. Kawamura, Separation of chiral compounds on polysaccharide supports, US Patent 5 679 572, assigned to Daicel Chemical Industries Ltd., October 21, 1997.
8. Y. Okamoto, C. Yamamoto, and T. Kubota, Separating agent for enantiomeric isomers, US Patent 7 223 334, assigned to Daicel Chemical Industries Ltd, May 29, 2007.
9. P. Sun, C. Wang, Z.S. Breitbach, Y. Zhang, and D.W. Armstrong, *Analytical Chemistry*, Vol. 81, p. 10215, 2009.
10. P. Sun and D.W. Armstrong, *Journal of Chromatography A*, Vol. 1217, p. 4904, July 2010.
11. D.W. Armstrong and Z.S. Breitbach, High efficiency, ultra-stable, bonded hydrophilic interaction chromatography (HILIC) matrix on superficially porous particles (SPPS), US Patent 10 265 643, assigned to Azyp LLC, April 23, 2019.
12. D.W. Armstrong and Z.S. Breitbach, Superficially porous particle (spp) chiral phases for liquid chromatography, US Patent 10 272 364, assigned to Azyp LLC, April 30, 2019.
13. M. Lämmerhofer and W. Lindner, *Journal of Chromatography A*, Vol. 829, p. 115, 1998.
14. X. Wang, H. Li, K. Quan, L. Zhao, Z. Li, and H. Qiu, *Journal of Chromatography A*, Vol. 1651, p. 462338, 2021.
15. Y. Lv, Z. Lin, T. Tan, and F. Svec, *Journal of Chromatography A*, Vol. 1316, p. 154, 2013.
16. S.R. Jim and M.J. Brett, Ultrathin and nanostructured stationary phases in C.F. Poole, ed., *Instrumental Thin-Layer Chromatography*, chapter 3, pp. 53–72. Elsevier, Boston, 2015.
17. A.T. James and A.J.P. Martin, *The Biochemical Journal*, Vol. 50, p. 679, March 1952.
18. K.D. Bartle and P. Myers, *TrAC Trends in Analytical Chemistry*, Vol. 21, p. 547, 2002.

19. C.J. Cowper, Gas analysis: Gas chromatography in I.D. Wilson, M. Cooke, and C.F. Poole, eds., *Encyclopedia of Separation Science*, pp. 2925–2932. Academic Press, Oxford, 2000.
20. Z. Yumin, *Chinese Journal of Analytical Chemistry*, Vol. 8, 1993.
21. M.V. Bermeshev and E.S. Finkelshtein, *INEOS Open*, Vol. 1, p. 39, 2018.
22. E. Yakubenko, A. Korolev, P. Chapala, M. Bermeshev, A. Kanateva, and A. Kurganov, *Analytica Chimica Acta*, Vol. 986, p. 153, 2017.
23. L. Starannikova, M. Pilipenko, N. Belov, Y. Yampolskii, M. Gringolts, and E. Finkelshtein, *Journal of Membrane Science*, Vol. 323, p. 134, 2008.
24. V.G. Berezkin, A.A. Korolev, I.V. Malyukova, T.P. Popova, V.E. Shiryaeva, and V.S. Khotimskii, *Journal of Chromatography A*, Vol. 960, p. 151, 2002.
25. S.-M. Xie, J.-H. Zhang, and L.-M. Yuan, Enantioseparations by gas chromatography using porous organic cages as stationary phase in G.K.E. Scriba, ed., *Chiral Separations: Methods and Protocols*, Vol. 1985 of *Methods in Molecular Biology*, pp. 45–55. Springer New York, New York, NY, 2019.
26. H.-X. Li, T.-P. Xie, S.-M. Xie, B.-J. Wang, J.-H. Zhang, and L.-M. Yuan, *Chromatographia*, Vol. 83, p. 703, 2020.
27. C. Lu, S. Liu, J. Xu, Y. Ding, and G. Ouyang, *Analytica Chimica Acta*, Vol. 902, p. 205, 2016.
28. M. Gewehr, K. Nakamura, N. Ise, and H. Kitano, *Die Makromolekulare Chemie*, Vol. 193, p. 249, 1992.
29. Q. Meng, T. Su, and L. Deng, *Science in China Series B-Chemistry, Biological, Agricultural, Medical & Earth Sciences*, Vol. 28, p. 906, 1985.
30. R.V. Vivilecchia, B.G. Lightbody, N.Z. Thimot, and H.M. Quinn, *Journal of Chromatographic Science*, Vol. 15, p. 424, 09 1977.
31. J. Hradil, D. Horák, Z. Pelzbauer, E. Votavová, F. Švec, and J. Kálal, *Journal of Chromatography A*, Vol. 259, p. 269, 1983.
32. M. Slater, M. Snauko, F. Svec, and J.M.J. Fréchet, *Analytical Chemistry*, Vol. 78, p. 4969, 2006.
33. X. Sun, D. Lin, X. He, L. Chen, and Y. Zhang, *Talanta*, Vol. 82, p. 404, 2010.
34. B.R. Solomon, K.K. Varanasi, and N. Hyder, Hierarchical porous membrane for emulsion separation, US Patent 10 384 961, assigned to Massachusetts Institute of Technology, August 20, 2019.
35. J. Saleem, A. Bazargan, J. Barford, and G. McKay, *Chemical Engineering & Technology*, Vol. 38, p. 482, 2015.
36. W. Herkenberg, Thin flexible sheet sorption material for the removal of oil from oil spills, WO Patent 1 991 008 347, assigned to Wolf Herkenberg, June 13, 1991.
37. W. Herkenberg, Method for the removal of oil from oil spills, US Patent 5 451 325, assigned to Wolf Herkenberg, September 19, 1995.

38. X.-S. Wang, J. Liu, J.M. Bonefont, D.-Q. Yuan, P.K. Thallapally, and S. Ma, *Chem. Commun.*, Vol. 49, p. 1533, 2013.
39. M. Zheng, H. Huang, Z. Jiang, Y. Zhao, Y. Sun, and C. Zhong, *Journal of Porous Materials*, p. in press, 2022.
40. J. Saleem, C. Ning, J. Barford, and G. McKay, *Waste Management*, Vol. 44, p. 34, 2015.
41. J. Saleem, M. Adil Riaz, and M. Gordon, *Journal of Hazardous Materials*, Vol. 341, p. 424, 2018.
42. J. Saleem, P. Gao, J. Barford, and G. McKay, *J. Mater. Chem. A*, Vol. 1, p. 14335, 2013.
43. J. Saleem, A. Bazargan, J. Barford, and G. McKay, *Polymers for Advanced Technologies*, Vol. 25, p. 1181, 2014.
44. D. Mitra, M.H. Tai, E.B. Abdullah, C.-H. Wang, and K.G. Neoh, *Separation and Purification Technology*, Vol. 279, p. 119664, 2021.
45. K.-J. Chen, D.G. Madden, S. Mukherjee, T. Pham, K.A. Forrest, A. Kumar, B. Space, J. Kong, Q.-Y. Zhang, and M.J. Zaworotko, *Science*, Vol. 366, p. 241, 2019.
46. R.L. Siegelman, E.J. Kim, and J.R. Long, *Nature Materials*, Vol. 20, p. 1060, 2021.
47. R. Vaghasia, B. Saini, and A. Dey, Advanced functional polymer-based porous composites for CO_2 capture in N.K. Subramani, S.K. Nataraj, C. Patel, and S. Shivanna, eds., *Polymer-Based Advanced Functional Materials for Energy and Environmental Applications*, pp. 147–175. Springer Singapore, Singapore, 2022.
48. Z. Liu, Z. Du, W. Zou, J. Mi, H. Li, Y. Wang, and C. Zhang, *RSC Adv.*, Vol. 3, p. 18849, 2013.
49. Y. Cui, Y. Zhao, T. Wang, and B. Han, *Chinese Journal of Chemistry*, Vol. 33, p. 131, 2015.
50. M. Irani, A.T. Jacobson, K.A. Gasem, and M. Fan, *Energy*, Vol. 157, p. 1, 2018.
51. H. Thakkar, S. Lawson, A.A. Rownaghi, and F. Rezaei, *Chemical Engineering Journal*, Vol. 348, p. 109, 2018.
52. Wikipedia contributors, Polyamide-imide — Wikipedia, the free encyclopedia, https://en.wikipedia.org/w/index.php?title=Polyamide-imide&oldid=1023149266, 2021. [Online; accessed 24-September-2021].
53. H. Xu, X. Li, Y. Hao, X. Xu, Y. Zhang, and J. Zhang, *Journal of Molecular Liquids*, Vol. 347, p. 118253, 2022.
54. Y. Hao, H. Xu, X. Li, Z. Wei, and Y. Zhang, *Journal of Chromatography A*, Vol. 1662, p. 462720, 2022.
55. J.H. Raymer and E.D. Pellizzari, *Analytical Chemistry*, Vol. 59, p. 1043, 1987.
56. P.V. Zibarev, O.A. Zubkova, T.S. Shepelenko, and O.I. Nedavnii, *Russian Journal of Nondestructive Testing*, Vol. 42, p. 418, 2006.

57. S. Gopi, A.G. Ramu, S. Sakthivel, G. Maia, C.-H. Jang, D. Choi, and K. Yun, *Chemosphere*, Vol. 265, p. 129052, 2021.
58. S. Gopi, K. Giribabu, M. Kathiresan, and K. Yun, *Sustainable Energy Fuels*, Vol. 4, p. 3797, 2020.
59. A. Alsbaiee, B.J. Smith, L. Xiao, Y. Ling, D.E. Helbling, and W.R. Dichtel, *Nature*, Vol. 529, p. 190, 2016.
60. A. Trabolsi and D. Shetty, Polycalixarene materials, methods of making same, and uses thereof, US Patent 11 027 257, assigned to New York University in Abu Dhabi Corp., June 08, 2021.
61. D. Shetty, I. Jahovic, J. Raya, F. Ravaux, M. Jouiad, J.-C. Olsen, and A. Trabolsi, *Journal of Materials Chemistry A*, Vol. 5, p. 62, 2017.
62. Wikipedia contributors, Sonogashira coupling — Wikipedia, the free encyclopedia, https://en.wikipedia.org/w/index.php?title=Sonogashira_coupling&oldid=1046506138, 2021. [Online; accessed 30-September-2021].
63. G. Barin, J.M. Spruell, M. Brown, and S. Li, Charge-bearing cyclodextrin polymeric materials and methods of making and using same, US Patent 11 001 645, assigned to Cyclopure Inc., May 11, 2021.
64. H.-L. Duan, Q.-L. Niu, J. Wang, S.-Y. Ma, J. Zhang, and Z.-Q. Zhang, *Journal of Chromatography A*, Vol. 1600, p. 80, 2019.
65. J. Chen, Y. Lin, and L. Jia, *Journal of Chromatography A*, Vol. 1388, p. 43, 2015.
66. A.T. Hoang, X.L. Bui, and X.D. Pham, *Energy Sources, Part A: Recovery, Utilization, and Environmental Effects*, Vol. 40, p. 929, 2018.
67. Standard test method for sorbent performance of adsorbents for use on crude oil and related spills, ASTM Standard F726-17, ASTM International, West Conshohocken, PA, 2017. https://www.astm.org/Standards/F726.htm.
68. M. Welter, C. Meyer, K. Lungfiel, and T. Schwarz, Porous polymer material for bonding metal-containing ions or for purifying organic molecules, US Patent 11 014 071, assigned to instrAction GmbH, May 25, 2021.
69. Standard test method for determination of pore volume and pore volume distribution of soil and rock by mercury intrusion porosimetry, ASTM Standard D4404-18, ASTM International, West Conshohocken, PA, 2018. https://www.astm.org/Standards/D4404.htm.
70. Y. Chen, Z. Li, R. Ding, T. Liu, H. Zhao, and X. Zhang, *Journal of Hazardous Materials*, Vol. 426, p. 128073, 2022.
71. C.T. Yavuz, Y. Hong, D. Thirion, and S. Subramanian, Porous porphyrin polymer and method of recovering precious metal elements using the same, US Patent 10 961 343, assigned to Korea Advanced Institute of Science and Technology KAIST, March 30, 2021.
72. D.A. Anito, T.-X. Wang, Z.-W. Liu, X. Ding, and B.-H. Han, *Journal of Hazardous Materials*, Vol. 400, p. 123188, 2020.

73. L. Feng, W.-M. Chen, J.-L. Li, G. Day, H. Drake, E. Joseph, and H.-C. Zhou, *ACS Applied Materials & Interfaces*, Vol. 11, p. 14383, 2019.
74. K. Nell, D.W. Johnson, J. Pittman, W. Chouyyok, R.S. Addleman, and M.G. Warner, Polyfunctional sorbent materials, US Patent 10 315 185, assigned to University of Oregon, Battelle Memorial Institute Inc., June 11, 2019.
75. W. Yantasee, G.E. Fryxell, R.S. Addleman, R.J. Wiacek, V. Koonsiripaiboon, K. Pattamakomsan, V. Sukwarotwat, J. Xu, and K.N. Raymond, *Journal of Hazardous Materials*, Vol. 168, p. 1233, 2009.
76. S. Ma, Q. Sun, and B. Aguila, Mercury capture using functionalized porous organic polymer with hierarchical porosity, US Patent 11 065 598, assigned to University of South Florida, July 20, 2021.
77. M. Dujardin, C. Cazé, and I. Vroman, *Reactive and Functional Polymers*, Vol. 43, p. 123, 2000.
78. S.-Y. Ding, M. Dong, Y.-W. Wang, Y.-T. Chen, H.-Z. Wang, C.-Y. Su, and W. Wang, *Journal of the American Chemical Society*, Vol. 138, p. 3031, 2016.
79. A.J. Tchinda, E. Ngameni, I.T. Kenfack, and A. Walcarius, *Chemistry of Materials*, Vol. 21, p. 4111, 2009.
80. Y. Shin, G.E. Fryxell, W. Um, K. Parker, S.V. Mattigod, and R. Skaggs, *Advanced Functional Materials*, Vol. 17, p. 2897, 2007.
81. W. Kiciński, M. Szala, and M. Bystrzejewski, *Carbon*, Vol. 68, p. 1, 2014.
82. A. Bibby and L. Mercier, *Chemistry of Materials*, Vol. 14, p. 1591, 2002.
83. S.J.L. Billinge, E.J. McKimmy, M. Shatnawi, H. Kim, V. Petkov, D. Wermeille, and T.J. Pinnavaia, *Journal of the American Chemical Society*, Vol. 127, p. 8492, 2005.
84. P.I. Girginova, A.L.D. da Silva, C.B. Lopes, P. Figueira, M. Otero, V.S. Amaral, E. Pereira, and T. Trindade, *Journal of Colloid and Interface Science*, Vol. 345, p. 234, 2010.
85. A. Walcarius and L. Mercier, *J. Mater. Chem.*, Vol. 20, p. 4478, 2010.
86. O. Hakami, Y. Zhang, and C.J. Banks, *Water Research*, Vol. 46, p. 3913, 2012.
87. D.M. Saad, E.M. Cukrowska, and H. Tutu, *Applied Water Science*, Vol. 3, p. 527, 2013.
88. H. Wang, B. Khezri, and M. Pumera, *Chem*, Vol. 1, p. 473, 2016.
89. X.-P. Zhou, Z. Xu, M. Zeller, and A.D. Hunter, *Chem. Commun.*, Vol. 36, p. 5439, 2009.
90. Q.-R. Fang, D.-Q. Yuan, J. Sculley, J.-R. Li, Z.-B. Han, and H.-C. Zhou, *Inorganic Chemistry*, Vol. 49, p. 11637, 2010.
91. K.-K. Yee, N. Reimer, J. Liu, S.-Y. Cheng, S.-M. Yiu, J. Weber, N. Stock, and Z. Xu, *Journal of the American Chemical Society*, Vol. 135, p. 7795, 2013.
92. F. Luo, J.L. Chen, L.L. Dang, W.N. Zhou, H.L. Lin, J.Q. Li, S.J. Liu, and M.B. Luo, *J. Mater. Chem. A*, Vol. 3, p. 9616, 2015.

93. M. Mon, F. Lloret, J. Ferrando-Soria, C. Martí-Gastaldo, D. Armentano, and E. Pardo, *Angewandte Chemie International Edition*, Vol. 55, p. 11167, 2016.
94. Q. Sun, B. Aguila, J. Perman, L.D. Earl, C.W. Abney, Y. Cheng, H. Wei, N. Nguyen, L. Wojtas, and S. Ma, *Journal of the American Chemical Society*, Vol. 139, p. 2786, 2017.
95. N. Huang, L. Zhai, H. Xu, and D. Jiang, *Journal of the American Chemical Society*, Vol. 139, p. 2428, 2017.

8

Other Fields of Use

8.1 Ceramic Articles

The microstructure-property relation in poly(methyl methacrylate) (PMMA)-based porous mold materials that are used for high-pressure casting of ceramic articles was investigated (1). Here, porous plastic materials were produced by the polymerization of water-in-oil emulsions with various compositions of emulsion constituents and particle sizes of the filler PMMA beads. A water-soluble nonionic surfactant of an ethylene oxide-propylene oxide block copolymer was used.

The pore morphology, porosity and water permeability of the materials were measured. The compressive stress-strain behavior, collapse stress and elastic modulus values of the macroporous materials were determined by a performing compressive mechanical testing method. Also, the fracture toughness values of the materials were measured using the single-edge notched bending method.

The variation of the porosity and permeability as a function of total weight fraction of surfactant and water is shown in Figure 8.1. As can be seen in this figure, the percentage of the porosity increased from 34% to 47% when the amount of water-surfactant increased from 29% to 40%, respectively.

The results of the study showed that the concentration of emulsion constituents and PMMA bead sizes has significant effects on the pore morphology, porosity, water permeability and mechanical properties of the porous plastics (1).

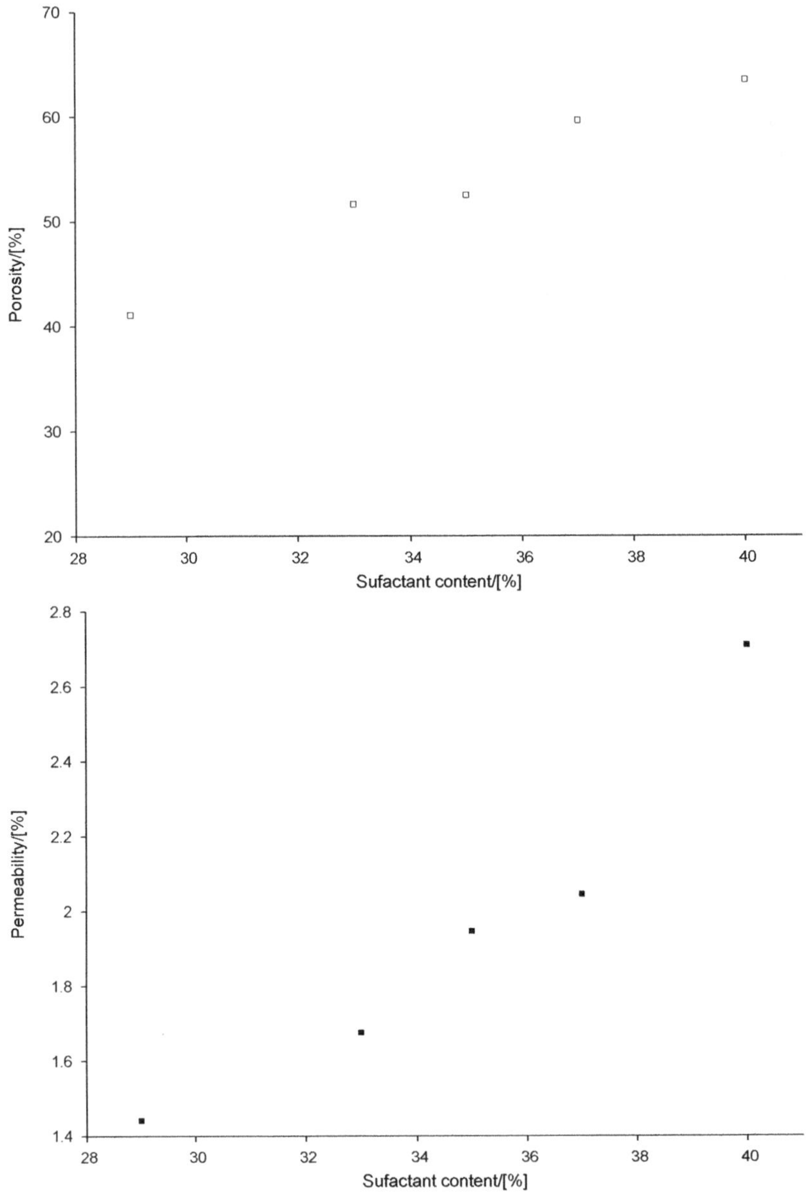

Figure 8.1 Porosity and permeability (1).

8.2 Polymer-Modified Porous Cement

The impact of the different grain sizes of sand on the ultimate stress of a hand-mixed cement-grouted sand was investigated. The material was modified with poly(carboxylate) ether-based polymer using two different test standards.

Laboratory tests for different types of cement-based grout mixtures prepared with the five grains of sands were carried out as recommended by ASTM C938 (2).

The flow rate of the cement-based grout was quantified using the flow cone method (ASTM C939 (3)).

The fresh and hardened properties of cement-grouted sands modified with polymer up to 0.16% of the weight of cement were tested and quantified. Five types of sand with different grain sizes were used in this study. Adding polymer decreased the water/cement ratio (w/c) by 21.9–54.1%, and kept the flow time of the cement-based grout in the range of 18–23 s.

The addition of the polymer creates an amorphous gel that fills the pores between the cement particles, which causes a reduction in the voids, porosity and enhances the dry density of the cement.

So, the compression strength of the cement-grouted sands increased significantly. Linear and nonlinear approaches were employed to estimate the compressive strength of cement-grouted sand with a different grain size of sand, w/c, amount of polymer, and curing age. The compressive strength of the cement-grouted sands following the BS standard was 71% larger than the compression strength of the same mix using the ASTM standard (4).

8.3 Flame Retardant Foams

8.3.1 Poly(urethane) Foam

A composition for poly(urethane) (PU) foaming has been described (5). The foam has a flame retardant property, that is to say, it can be used for construction insulation, and can also be used for packaging a buffer material, has a good thermal insulation, adhesion and dimensional stability, and has a low odor during the process of construction. The foaming agents all use water, and do not contain chlorofluorocarbon substances that destroy the ozone layer and climate.

Previously, HCFC-141B was used for the foaming of PU, which is a substance that destroys the ozone layer that has been banned. HCFC stands for hydrochlorofluorocarbons and HCFC-141B is the abbreviation for 1,1-dichloro-1-fluoroethane.

The foaming agents that can replace HCFC-141B are HFC-245fa, 365mfc, and cyclopentane, which are all water foaming. These materials are expensive and are flammable, thus they cannot be used for *in-situ* spraying, and the flame retardance of the foam is poor. Foaming agents that are made with only water is a technical route that is low cost, environmentally friendly and easy to achieve flame retardance. This foaming method is especially suitable for buildings with thick interlayers; for example, inside the roof of a single-family house, in buildings with curtain wall structure, in ceilings with cavities, in internal walls of wood houses, and in brick wall structures with cavities.

The PU resin is obtained by the reaction of the above composition with polymethylene polyphenyl poly(isocyanate), PAPI ™. Here, two polyether polyol types are used. The first polyether polyol is produced by the polymerization of propylene oxide, ethylene oxide with 2,4-diamino-phenol and ethylene diamine as initiators. The second polyether polyol is produced by the polymerization of propylene oxide, ethylene oxide and the mixture of one or more of dipropylene glycol, diethylene glycol or triethanolamine and glycerol.

The catalyst is a mixture of dimethylamino ethoxyethanol and trimethylethoxylethylene diamine in any proportion, and may also contain one or more compounds from *N*,*N*,*N*′-trimethyl-*N*′-hydroxyethyl diamino ethyl ether, trimethyl hydroxyethyl propanediamine, and dimethyl ethanolamine. Some of the above compounds are shown in Figure 8.2. Most preferably, the catalyst is the composition of dimethylamino ethoxyethanol, trimethylethoxylethylene diamine and *N*,*N*,*N*′-trimethyl-*N*′-hydroxyethyl diamino ethyl ether in a mass ratio of 3:1:1.

The here used flame retardant is a mixture of one or two compounds from tris(2-chloropropyl) phosphate, tris(2-chloroethyl) phosphate and triethyl phosphate, c.f. Figure 8.3.

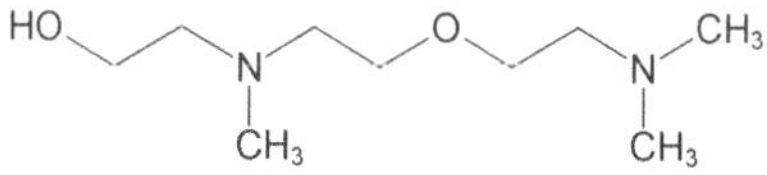

N,N,N'-Trimethyl-
N'-hydroxyethyl diamino ethyl ether

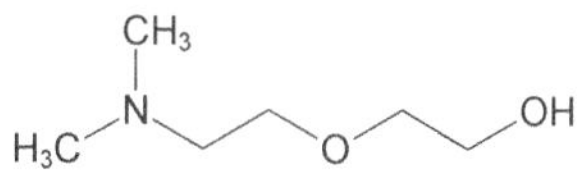

Dimethylamino
ethoxyethanol

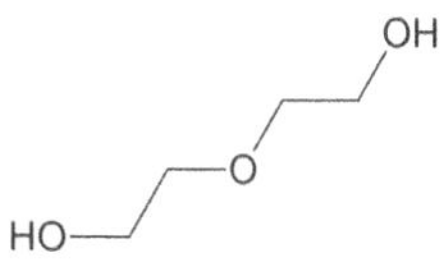

Diethylene glycol

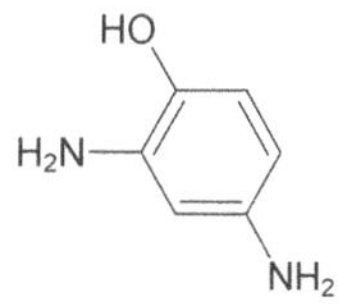

2,4-Diamino-phenol

Figure 8.2 Catalysts.

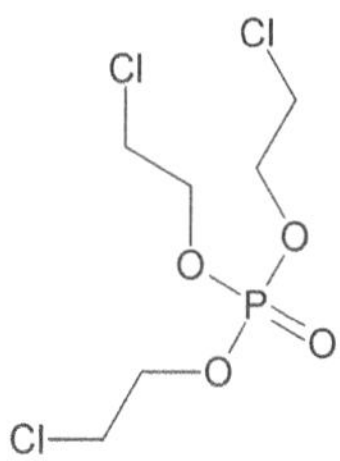

Tris(2-chloroethyl) phosphate

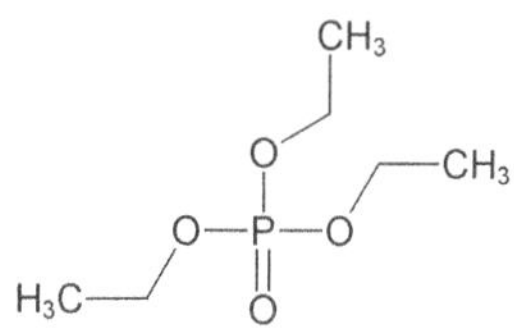

Triethyl phosphate

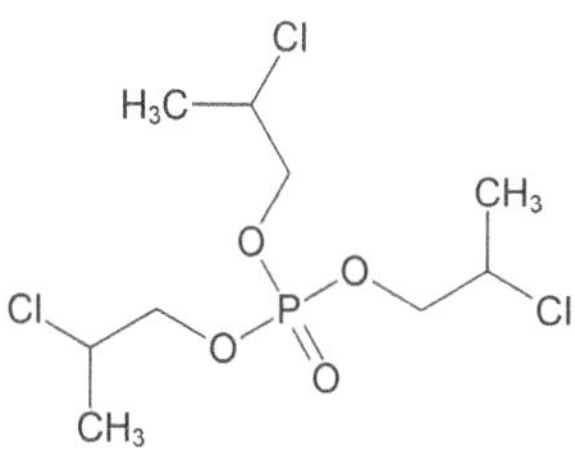

Tris(2-chloropropyl) phosphate

Figure 8.3 Flame retardants.

8.4 Clay-Containing Composites

8.4.1 *Tissue Engineering*

Porous biopolymer structures have attracted a lot of attention in recent years because of their potential applications in tissue engineering (6).

Porous structures of poly(lactic-*co*-glycolic acid) (PLGA) reinforced with organically modified montmorillonite (Cl15A) were fabricated. A ternary, co-continuous blend consisting of PLGA, poly(styrene) (PS), and Cl15A was prepared by using melt extrusion.

Then, a porous PLGA composite was created by the sacrificial extraction of the PS phase. The morphological characterization revealed the creation of a well-formed 3D porous network consisting of Cl15A-reinforced PLGA.

Results obtained from scanning electron microscopy micrographs of the fabricated porous structures showed that small variations in the clay loading affect the geometrical characteristics (% porosity and pores average diameter) of these porous structures. The results suggest that these porous PLGA/clay structures may be promising candidates for mechanically strong scaffolds in tissue engineering applications (6).

8.4.2 *Poly(methyl methacrylate) Composites*

The preparation of PMMA-based porous nanocomposites that contain a clay, such as montmorillonite platelets, were prepared as reinforcements within the cell walls of a porous structure (7). To render the clay layers organophilic, montmorillonite was surface treated by an ion exchange reaction between interlayer cations of the clay and ammonium ions of a surfactant.

The clay/PMMA-based porous nanocomposites were prepared by polymerization of water-in-oil emulsions with and without clay addition. The microstructure and compressive mechanical behavior of the nanocomposites were investigated. The results of mechanical tests showed that the porous systems with the addition of 1% of organoclay exhibited a 90% and 50% increase of collapse stress and elastic modulus values, respectively, in comparison to neat porous PMMA (7).

8.4.3 *Hectorites*

Synthetic hectorites were hydrothermally crystallized with direct incorporation of a cationic polymer, i.e., poly(dimethyl diallyl ammonium chloride), and two neutral cellulosic polymers, hydroxypropyl methyl cellulose (HPMC) and hydroxyethyl cellulose (HEC). The monomer, dimethyl diallyl ammonium chloride is shown in Figure 8.4.

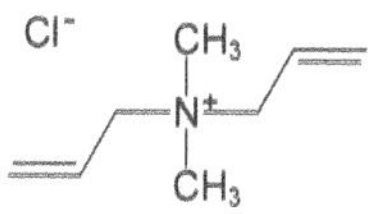

Figure 8.4 Dimethyl diallyl ammonium chloride.

Synthetic poly(dimethyl diallyl ammonium chloride)-hectorite displays the lowest d-spacing at 15.8 Angstrom (Å) along with less polymer incorporation (7.8% organic) than the neutral polymers (18% to 22% organic).

Thermal analysis and small angle neutron scattering were used to further examine the polymer-clay systems. Clay platelets of the largest size and best stacking order occur when cationic poly(dimethyl diallyl ammonium chloride) polymer is used. Poly(dimethyl diallyl ammonium chloride) also enhances these properties over the crystallites prepared for a control mineral, where no polymer is used.

The HEC acts to aggregate the silica, leaving less to react to form clay. The clay platelets which result from HEC are small, not stacked to a large degree, and oriented randomly. Neutral HPMC acts more like cationic poly(dimethyl diallyl ammonium chloride) in that larger clay platelets are allowed to form. The extended microstructure of the clay network remains undisturbed after polymer is removed by calcination. When no polymer is used, the synthetic hectorite has a N_2 Brunauer-Emmet-Teller (BET) surface area of 200 m^2g^{-1}, even after calcination. This increases by 20–50% for the synthetic polymer-hectorites after the polymer is removed by calcination (8).

8.4.4 Catalyst Supports

8.4.4.1 Mesoporous Clays

Mesoporous synthetic clays are obtained when polymer-containing silicate gels are hydrothermally crystallized to form layered magnesium silicate hectorite clays containing polymers that are incorporated *in-situ* (9).

Poly(*N*-vinyl-2-pyrrolidone) (PVP) in gel loadings varying from 5% to 30 wt% were used. The organic polymer template molecules were removed from synthetic polymer-clay complexes via calcination.

The pore radii, surface areas, and pore volumes of the resulting porous inorganic networks were then measured. For the most, part there is a direct correlation between both PVP molecular weight and polymer loading on the diameter of the average pore. In addition to conventional techniques, the polymer-clay materials were also characterized by small-angle X-ray scattering to ascertain the disposition of the polymeric matrix (9).

After calcination, the mesoporous synthetic clay materials were examined as potential supports for hydrodesulfurization. They were loaded with a bimetallic Co/Mo catalyst system for comparison with a commercial Co/Mo alumina catalyst.

Dibenzothiophene diluted with hexadecane (0.75% S) wa s utilized as a liquid feed for the hydrodesulfurization tests. Dibenzothiophene is shown in Figure 8.5. This feed was chosen as

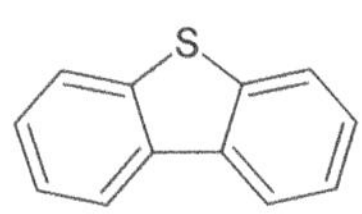

Figure 8.5 Dibenzothiophene.

a deep hydrodesulfurization test of a heavy model oil. The pore diameters of the mesoporous synthetic clay catalysts were found to have a strong effect on both the hydrodesulfurization activity and selectivity (9).

8.4.4.2 Alcohol Alkylation

The formation of carbon-carbon bonds is fundamental in the field of organic synthesis (10, 11).

Solid supports functionalized with molecular metal catalysts combine many of the advantages of heterogeneous and homogeneous catalysis. A (NNN)Ru-incorporated porous organic polymer exhibits a high catalytic efficiency and a broad functional group tolerance in the C–C cross coupling of secondary and primary alcohols to give β-alkylated secondary alcohols (12).

The synthesis of a porous organic polymer with a Ru catalyst can be done as follows (12):

Preparation 8–1: Under a N_2 atmosphere, 2,6-bis(benzimidazo-1-yl) pyridine (c.f. Figure 8.6) (0.311 *g*, 1 *mmol*) and biphenyl (0.154 *g*, 1 *mmol*) were dissolved in anhydrous chloroform (20 *ml*), followed by the addition of anhydrous aluminum chloride (10 *mmol*, 1.33 *g*) with vigorous stirring at 58°C. After 24 *h*, the reaction mixture was cooled to ambient temperature, filtered and the solid product was washed with methanol, $HCl–H_2O$ (v/v = 1:1), H_2O, CH_2Cl_2 and EtOH, followed by washing with methanol in a Soxhlet thimble for 72 *h*. The polymer was then dried in a vacuum oven at 100°C for 24 *h* to give a brown solid. Under a N_2 atmosphere, a mixture of porous organic polymer (0.2 *g*), $RuCl_3 \times H_2O$ (0.017 *g*), and anhydrous ethanol (25 *ml*) was introduced into a round-bottom flask (100 *ml*). The resulting mixture was refluxed for 12 *h*). After this period, the brown solid product was separated by centrifugation, washed sequentially with ethanol and ethyl ether, and then dried under vacuum.

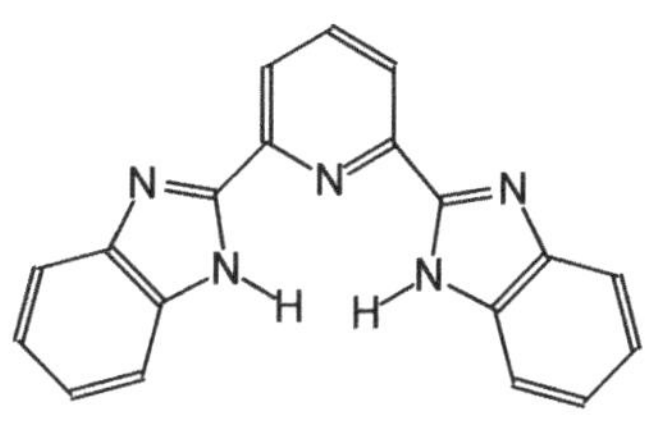

Figure 8.6 2,6-Bis(Benzimidazo-1-yl) pyridine.

The synthesis of the Ru-incorporated porous organic polymer is shown in Figure 8.7.

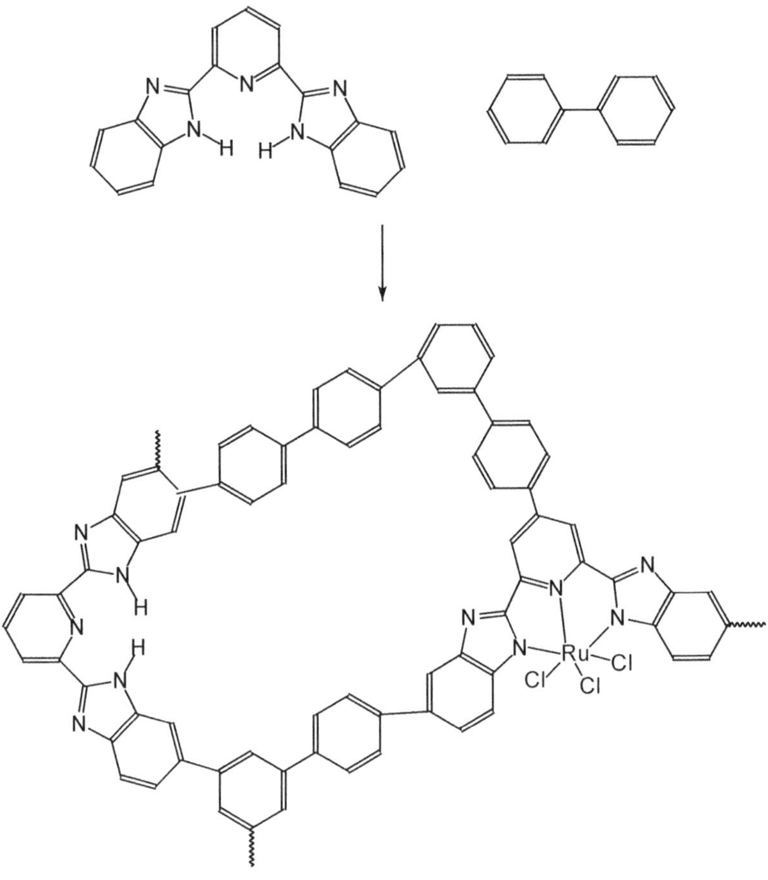

Figure 8.7 Synthesis of a Ru-incorporated porous organic polymer.

This catalyst demonstrated an excellent durability during successive recycling without Ru leaching. This is ascribed to the strong binding of the pincer ligands to the metal ions (12).

8.4.4.3 Manganese Catalysts for Alcohol Oxidation

The catalytic oxidation of alcohols to carbonyl compounds is an important reaction type both in organic synthesis and industry (13–15). The development of an efficient catalytic system for this process under mild conditions is of interest. Several metalloenzymes are capable of the efficient and selective oxidation of organic substrates.

Porous organic polymers have high surface areas, tunable components, and designable hierarchical pores, thus showing versatile applications, including catalysis, separation, or gas storage. However, their hydrophobic property may be unfavorable for the catalytic reaction that involves hydrophilic reactants.

By using the hydrophilic principle of the amide bond, R–NH–CO–R can be constructed in the porous frameworks to regulate the wettability. The construction of an amphiphilic and hierarchical porous $Mn{-}N_4$ catalyst via the solvothermal copolymerization has been reported (15), in which *N,N′*-methylenebisacrylamide is used as a hydrophilic monomer. *N,N′*-Methylenebisacrylamide is shown in Figure 8.8.

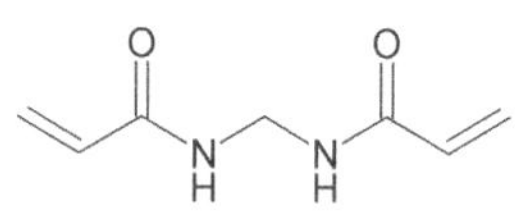

Figure 8.8 *N,N′*-methylenebisacrylamide.

This manganese catalyst can effectively promote the oxidation of a secondary alcohol to produce the ketone using aqueous hydrogen peroxide as the oxidant under mild conditions.

Such an amphiphilic catalyst displays a catalytic activity as its homogeneous counterpart in the selective oxidation of alcohols (15).

8.5 Lubricant Additives

Zwitterionic polymers functionalized with *N*-enriched porous carbon nanosheets were prepared via pyrolysis and etching of pyridine-coordinated polymer for porous coordination networks (16).

Subsequently, the porous coordination network was modified with poly(dopamine) to grow the zwitterionic polymer, thereby acquiring poly(ethylene imine) functionalized network. The exceptional tribological performance of this material was attributed to the synergistic lubrication effects along with the zwitterionic polymer, which facilitates the formation of a stable protective carbon film over the worn surface via complicated tribochemical reaction.

The presence of a hydration layer encapsulated by poly(ethylene imine) as the boundary lubricating film proves highly conducive for friction reduction. The average coefficient of friction showed a dramatic decrease to 0.17 from 0.63, and the wear depth of stainless steel dropped to 64% (16).

8.6 Cosmetic Compositions

Scoria is known as a kind of natural ceramic, which is a pyroclastic deposit made by burning clay at high temperatures due to volcanic activity (17). When a volcano erupts, the scoria is formed by the hardening of the first erupted lava. At this time, the lava mixed with the gas is ejected first, and the scoria with a porous structure containing numerous fine holes is created in this process.

A composite powder, scoria sphere, has been described, where the surface and the inside of a porous polymer are uniformly impregnated with scoria powder particles prepared by spraying in one-step a dispersion in which scoria powder is dispersed in a solution in which a polymer is dissolved (17).

When the scoria impregnated powder, in which the porous polymer is uniformly impregnated with the scoria powder particles, is formulated as a cosmetic composition, the porous polymer supplements the sebum absorption performance of the scoria miniaturized in a micro size, and thus the sebum absorption power is greatly improved.

Poly(methyl methacrylate) (PMMA) was applied as material for the porous polymer, but it may be selected from other materials, such as PVP, poly(ε-caprolactone), poly(l-lactic acid), poly(ethylene oxide) (PEO), poly-(propylene oxide), poly(dimethyl siloxane), PLGA, PS and copolymers from these compounds (17).

8.7 Packaging Materials

8.7.1 *Breathable Films*

Various packaging technologies have been developed to meet the continuously increasing consumer demand for comfortable and safe use, convenience of use, and fresh quality of packaged products (18).

Porous and breathable films provide various advantages such as the minimization of anaerobiosis originating from the accumulation of CO_2 in packaging and the automatic release of steam in a microwave system.

Nonporous breathable films are among the alternative candidates for packaging applications and they can respond to desired and undesired temperature fluctuations by reversibly changing the gas permeability depending on the temperature change. They can also prevent problems related to packaging integrity such as the intrusion of insects and microorganisms through perforations, undesired permeation of moisture, and loss of flavor in agricultural products and food.

The issues of porous and nonporous breathable films that are responsive to stimuli like temperature changes have been detailed in a review (18). Furthermore, the importance and possible packaging application areas of nonporous breathable films with temperature-dependent gas permeabilities were briefly accessed.

8.8 Char Layer

A char layer is an important barrier that resists ablation. During chemical vapor deposition (CVD), hydrocarbons from a pyrolysis gas densify the porosity of the char layer, which can enhance the char layer by improving its mechanical properties and its ability to resist mechanical erosion by gases and particulate erosion (19).

The conditions for depositing gaseous hydrocarbons in the porous structure of the char layer differ from those under circumfluence and large-scale deposition in the pores. Therefore, investigating the deposition characteristics of pyrolysis gases in the char layer is of great significance for establishing a strength model of the char layer and an ablation model of the polymer composites.

In a study, the effects of the deposition temperature and the gas partial pressure on the CVD characteristics of propylene at 1000°C to 1400°C were studied by simulating the deposition mode in the porous structure of the char layer based on a high-temperature reaction platform (19). The experimental results of the study showed that the deposition rate in the char layer is much higher than that under circumfluence and large-scale deposition in the pores. Moreover, in contrast with the exponential relationship between the deposition rate and reaction temperature under circumfluence and large-scale pore deposition, there is a well-established linear relationship between these variables during deposition of the char layer. Chemical vapor deposition and deposition temperature is shown in Figure 8.9.

Based on these results, it can be inferred that the micro-nano and honeycomb porous structure of the char layer increased the reaction area, affected the flow state, and changed the kinetics of the depositional reaction (19).

8.9 Batteries

8.9.1 Electrodes

8.9.1.1 Metal Oxides

Metal oxides are of high interest for their applications as catalysts or electrochemical devices. The tailored design of metal oxides featuring a high specific surface area and additional functionalities is of the utmost importance for the performance of the resulting materials (20).

A new method has been reported for preparing freestanding films consisting of hierarchically porous metal oxides (titanium and niobium based) by combining emulsion polymerization and shear-in-

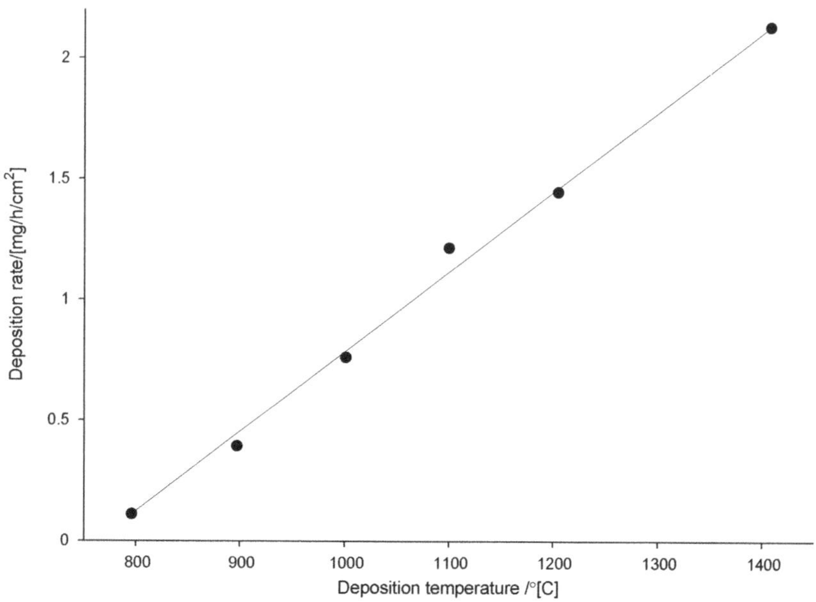

Figure 8.9 Chemical vapor deposition and deposition temperature (19).

duced monodisperse particle self-assembly in the presence of sol-gel precursors (20).

After a thermal treatment, the resulting porous materials can be used as electrodes in Li-ion batteries. The titanium and niobium sol-gel precursors were partially immobilized on the surface of organic core-interlayer particles featuring hydroxyl groups to obtain hybrid organic-inorganic particles through the melt-shear organization process.

Freestanding particle-based films, in analogy to elastomeric opal films and colloidal crystals, can be prepared in a convenient one-step preparation process. After thermal treatment, ordered pores are obtained, while the pristine metal oxide precursor shell can be converted to the (mixed) metal oxide matrix. Heat treatment under CO_2 leads to mixed-TiNb oxide/carbon hybrid materials. The highly porous derivative structure enhances electrolyte permeation. When tested as Li-ion battery electrodes, it shows a specific capacity of 335 $mA\,h\,g^{-1}$ at a rate of 10 $mA\,g^{-1}$. After 1000 cycles at 250 $mA\,g^{-1}$, the electrodes still provided a specific capacity of 191 $mA\,g^{-1}$ (20).

8.9.1.2 *Electrodes for Lithium-Air Batteries*

Li-air cells have aroused intense interest because of their ultrahigh specific energies, but their practical application is still hindered by many problems, such as the slurry reaction kinetics and serious parasitic reactions at the cathode, obvious liquid electrolyte decomposition/leakage, and passivation/pulverization issues at the anode (21).

Au-nanoparticle-coated Ni nanosheets grown on Ni foam (Au/NNS-NF) were first fabricated and then reacted with molten Li to form Li-based Au/NNS-NF (LiAu/NNS-NF), which was applied as an electrode in quasi-symmetrical Li-air cells in which a poly(methyl methacrylate)-based electrolyte is used.

Due to the porous/stable framework, good conductivity, and uniform Au coating layer, the Au/NNS-NF cathodes possess excellent oxygen reduction reaction properties and oxygen evolution reaction properties and inhibit parasitic reactions in Li-air batteries, while the LiAu/NNS-NF anodes exhibit superior lithiophilicity, reducing dendritic Li formation and electrode dimension changes.

A LiAu/NNS-NF-based symmetric battery exhibits an excellent lifespan beyond 1100 cycles at 1 $mA\,cm^{-2}$ and small overpotentials (20 mV), while the Au/NNS-NF cathode also shows better performance than a carbon electrode. Additionally, the polymer electrolyte possesses good stability and is leakproof, offering superior anode protection.

Therefore, a quasi-symmetrical Au/NNS-NF-based polymer Li-air battery shows a large capacity of 13 764.2 $mA\,h\,g^{-1}$ at 500 $mA\,g^{-1}$ and long cycling life (150 cycles) with low overpotentials of less than 2 V; and corresponding soft-package cells also present excellent properties under bending and can even drive many devices (21).

8.9.1.3 *Electrodes for All-Solid-State Lithium Batteries*

Composite polymer electrolytes can replace the traditional liquid electrolytes for next-generation all-solid-state batteries with enhanced safety and uncompromising electrochemical properties (22).

However, short circuits due to lithium dendrite growth in the composite polymer electrolytes largely hinders its practical applications.

A thin composite polymer electrolyte was described. This material is composed of hollow ZIF-8 encapsulating Li-ionic liquids (ZIF-8/IL) nanofillers and a PEO matrix. It shows an ionic conductivity of 1.41×10^{-4} $S\,cm^{-1}$ at 25°C with an improved surface roughness, thermal stability, and electrochemical stability.

The preparation of ZIF-8 can be done as follows (22):

Preparation 8–2: First, 3.36 *g* $Zn(NO_3)_2 \cdot 6H_2O$ was dissolved in 40 *mL* CH_3OH to form solution A, and 8 *g* dimethylimidazole was dissolved in another 120 *ml* CH_3OH to form solution B under stirring for 30 *min*. Subsequently, the solution A was poured into solution B at a gallop under stirring. The mixed solution was stirred continuously for 1 *h* and then left to stand for 24 *h*. Thereafter, the precipitate was gathered by centrifugation and washed using CH_3OH for 3 times. Finally, ZIF-8 was collected after drying at 60°C under vacuum for 12 *h*.

The specific core-shell structure of fillers benefits from its insulation shell for inhibiting lithium dendrite growth and the inner cavity for storing sufficient Li–IL with an enhanced Li diffusivity. The composite polymer electrolyte with a content of 30% filler (30 H-Z composite polymer electrolyte) exhibits an excellent cycle stability of 770 *h* in the Li symmetric cell and an outstanding discharge capacity of 182.5 $mA\,h\,g^{-1}$ and capacity retention of 73% after 100 cycles at 0.1 C and 60°C in the Li/30 H-Z composite polymer electrolyte/NCM811 asymmetric cell (22).

8.9.2 *Rechargeable Batteries*

8.9.2.1 *Poly(vinylidene fluoride-co-hexafluoropropylene)*

A foaming process was used to prepare porous polymer membranes based on a poly(vinylidene difluoride-*co*-hexafluoropropylene) copolymer. The material can be used for lithium-ion batteries (23). Here, urea as the foaming agent, was decomposed into gases and was removed at an elevated temperature to get the porous structure within the polymer matrix.

When the weight ratio of urea to poly(vinylidene difluoride-*co*-hexafluoropropylene) is 5:6, the porous polymer membranes have the highest porosity of 70.2% (23).

The so prepared gelled polymer electrolyte shows an ionic conductivity up to 1.43×10^{-3} $S\,cm^{-1}$ at room temperature. This method is an easy way to prepare gelled polymer electrolytes for their applications in rechargeable lithium batteries (23).

8.9.2.2 *Metal-Organic Frameworks*

Metal-organic framework (MOF) nanoparticles with an inherently ordered porous structure are a promising component for improving the safety of rechargeable batteries (24). However, a separator combining the poly(benzimidazole) (PBI) matrix and an MOF filler has not been developed to optimize lithium-ion batteries.

An effective strategy has been designed to prepare a dual transport pathway separator containing poly(2,2′-(*p*-oxydiphenylene)-5,5′-bibenzimidazole) (OPBI) and an UiO-66 (zirconium-based MOF) filler using a phase separation method.

The composite membranes show excellent properties, such as superior wettability, thermostability, flame resistance and interfacial compatibility, resulting from the interconnected porous network structure formed by the synergistic regulation of the PBI porous structure and nano-MOF channels. Compared with the pristine OPBI separator, the composite separator that contains 40% UiO-66 delivers a higher ionic conductivity, a significant improvement in $LiFePO_4$/Li cell performance (151.6 $mA\,h\,g^{-1}$), and an excellent rate capacity.

Furthermore, this composite separator exhibits a good uniformity of lithium electrodeposition and an effective inhibition of the lithium dendrite issues. So, the dual-channel strategy provides broader prospects for the application of the PBI separator aided by the MOF crystals towards advanced-safety and high-performance lithium-ion batteries (24).

8.9.2.3 *Oxygen Electrocatalysts*

The controllable design of low-cost oxygen electrocatalysts with outstanding activity and durability is a top priority for rechargeable metal-air batteries (25). A facile approach has been proposed for obtaining hierarchically porous cobalt-nitrogen codoped carbon for reversible oxygen reduction and evolution reactions via the pyrolysis of Co/Zn bimetallic coordinated polymers with gradient Co:Zn ratio.

The evaporation of the Zn species during the pyrolysis process combined with the Kirkendall effect of Co species resulted in the formation of hierarchical porosity, which further creates highly accessible defective sites and improves multiscale mass transfer.

The C–7Co93Zn catalyst with optimized defective Co–N_x and small-sized Co_3O_4 nanoparticles delivers an optimal bifunctional activity with a potential gap of 0.79 V for reversible oxygen reduction reaction (E1/2) and oxygen evolution reaction (Ej=10). Moreover, it displays a superior power density of 122 $mW\,cm^{-2}$ and long-term durability in a rechargeable Zn-air battery. Density functional theory calculations indicate that the defects adjacent to Co–N_x display improved electrocatalytic oxygen reduction with a smaller energy barrier. This work provides a facile method for optimizing the Co : Zn ratio towards the synthesis of bifunctional oxygen electrocatalyst derived from a class of bimetallic coordinated polymers, resulting in hierarchically porous structure and defective cobalt-based active sites (25).

8.10 Light Emission

8.10.1 Porous Conjugated Polymer

Separating silver ions from lead ions is one of the many merits of a special here reported porous polymer framework (26). The porous polymer framework is shown in Figure 8.10.

The selective metal binding stems from the well-defined chelating unit of *N*-heterocycles, which consists of a triazine (C_3N_3) ring bonded to three 3,5-dimethylpyrazole moieties, c.f. Figure 8.11. Such a rigid and open triad also serves as the distinct building unit in a fully conjugated 3D polymer scaffold.

Because of its strong fluorescence and porosity, e.g., a BET surface area of 355 $m^2\,g^{-1}$, and because of the various types of metal species that can be readily taken up, this versatile framework is especially suitable for functionalization.

For example, when $AgNO_3$ is loaded, the framework solid exhibits a brown color in response to water solutions of H_2S, even at the dilution of 5.0 μM (0.17 ppm).

On the other hand, cysteine and other biologically relevant thiols do not effect a notable color change (26).

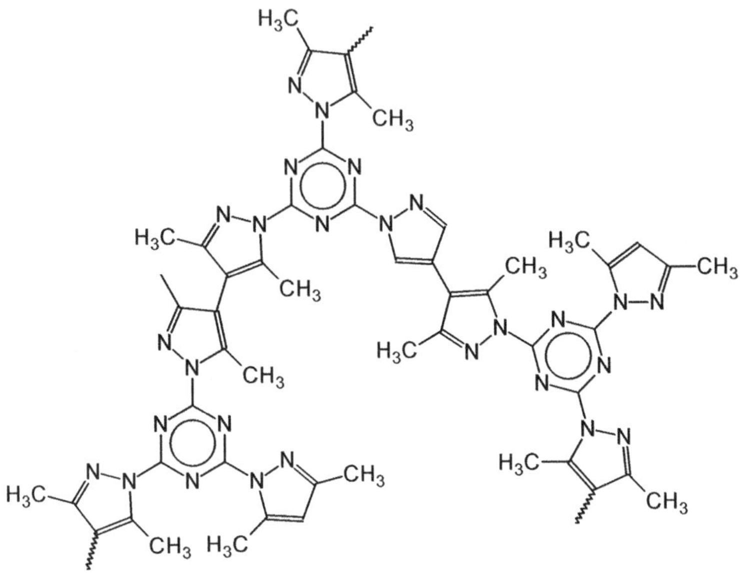

Figure 8.10 Porous polymer framework (26).

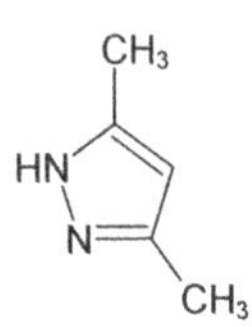

Figure 8.11 3,5-Dimethylpyrazole.

In another example, tunable white-light emission was produced when an Ir(III) complex was doped onto the framework, e.g., about 0.02% of the polymer weight. The bound Ir(III) centers become highly emissive in the orange-red region, complementing the broad, bluish emission from the polymer host to result in the overall white-light quality.

So, the color attributes of the emission are easily tunable with the concentration of the Ir(III) dopant (26).

8.10.2 Oxacalixarene Macrocycle

A calixarene is a macrocycle or a cyclic oligomer based on a hydroxyalkylation product of a phenol and an aldehyde (27). Calixarenes are used in commercial applications as sodium selective electrodes for the measurement of sodium levels in blood. Calixarenes can also form complexes with cadmium, lead, lanthanides and actinides.

A tetraphenylethylene-based oxacalixarene macrocycle, c.f. Figure 8.12, with a high fluorescence quantum yield (70%) was used as monomer to construct a fluorescent macrocycle-based porous organic polymer (28). With its high surface area, specific

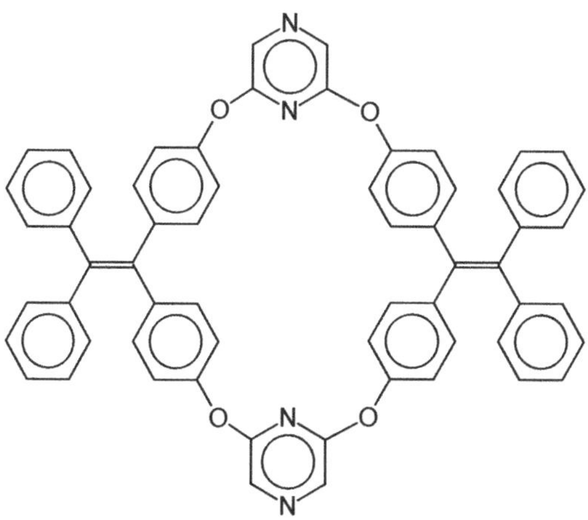

Figure 8.12 Tetraphenylethylene-based oxacalixarene (29).

pore size, and emissive properties, the blue-greenish fluorescent polymer can be developed into white-light-emission materials by adsorbing guests of tris(bipyridine)ruthenium with complementary emission color by the Förster resonance energy transfer effect (28).

8.10.3 *Tetraphenylcyclopentadiene*

A tetraphenylcyclopentadiene-based multifunctional solution processable, fluorescent, ultramicroporous polymer was fabricated (30). Tetraphenylcyclopentadiene is shown in Figure 8.13.

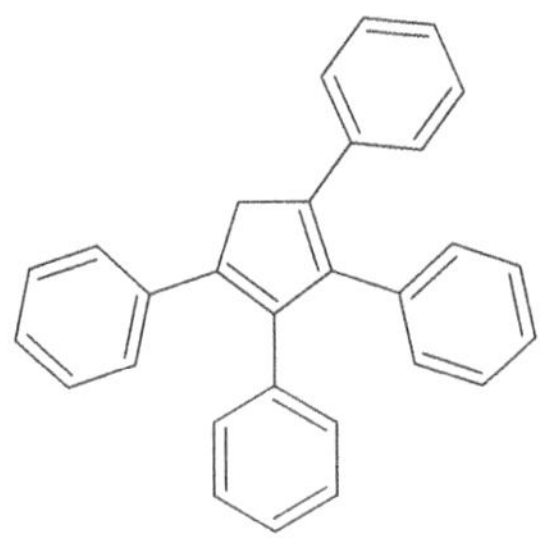

Figure 8.13 Tetraphenylcyclopentadiene.

This polymer shows a high hydrogen uptake. It was employed for the encapsulation of dyes to obtain enhanced white-light emission in solution, nanoparticles, gel and transparent thin film. Hybrid nanoparticles showed a quantum yield of 35% with a high color rendering index (30).

8.10.4 *Porous Silicone*

Porous silicon was mainly prepared from *p*-type substrates (31, 32). After preparation, the porous silicon was rinsed with ethanol, dried with nitrogen and evacuated in a Schlenk flask (33).

A ionically conducting polymer (polyethylene oxide/polyethylene glycol/tetrabutylammonium tetrafluoroborate) was spun cast onto the porous silicon (34).

The spun-cast polymer did produce visible light emission from porous silicon, although under relatively high bias. The first observation of light emission occurred under a bias of 20 *V* (34).

8.10.5 Light-Emitting Diodes

Light-emitting diodes (LEDs) may be used as white light sources in various applications such as flash sources for cellular telephone cameras and filament lamps (35).

White LEDs may appear to emit white light from the perspective of the viewer when the LEDs are in an on state. However, they may actually be made up of light-emitting semiconductor structures that emit non-white light, as well as wavelength converting structures that make the non-white light appear white to the viewer. For example, a white LED may be formed from a blue light-emitting semiconductor structure covered by a yellow-emitting phosphor layer. Photons of blue light emitted by the light-emitting semiconductor structure may either pass through the yellow-emitting phosphor layer as blue photons or may be converted into yellow photons by the yellow-emitting phosphor layer. The blue and yellow photons that are ultimately emitted out of the LED combine to make the light emitted from the LED appear white to the viewer.

A system and a method for tuning light scatter in an optically functional porous layer of an LED has been described (35).

The layer comprises a non-light-absorbing material structure having a plurality of sub-micron pores and a polymer matrix. The non-light-absorbing material forms either a plurality of micron-sized porous particles dispersed throughout the layer or a mesh slab, wherein a plurality of sub-micron pores is located within each micron-sized porous particle or forms an interconnected network of sub-micron pores within the mesh slab, respectively.

A polymer matrix, such as a high refractive index silicone, fills the plurality of sub-micron pores, creating an interface between the materials. Refractive index differences between the materials allow for light scatter to occur at the interface of the materials. Light scatter can also be decreased as a function of temperature, creating a system for tuning light scatter in both an off state and on state of an LED (35).

8.11 Sorbents

8.11.1 *Porous Hyper-Crosslinked Polymers*

Adsorbents with high surface area, thermal stability and microwave absorption ability are highly desired for cyclic adsorption and microwave regeneration processes (36).

Most polymeric adsorbents are transparent to microwaves. Herein, porous hyper-crosslinked polymers of (4,4′-bis((chloromethyl)-1,1′-biphenyl-benzyl chloride)) with different carbon black contents were synthesized via the Friedel-Crafts reaction. Carbon black was selected as the filler due to its low cost and high dielectric loss and was embedded inside the polymer structure during polymerization. Carbon black-containing composites showed enhanced thermal stability at elevated temperatures, and more than a 90-times increase in the dielectric loss factor, which is favorable for microwave regeneration.

Nitrogen physisorption analysis by the Brunauer-Emmett-Teller isotherms demonstrated that the presence of carbon black in the polymer structure nonlinearly decreases the surface area and total pore volume (by 38% and 26%, respectively, at the highest carbon black load). Based on the characterization testing, 4% by weight of carbon black was found to be an optimum filler content, having the highest microwave absorption and minimal effect on the adsorbent porosity. Hyper-crosslinked polymers with 4% carbon black allowed a substantial increase in the desorption temperature and yielded more than a 450% enhancement in the desorption efficiency compared to hyper-crosslinked polymers without carbon black (36).

References

1. Y. Ergűn, C. Dirier, and M. Tanoğlu, *Materials Science and Engineering: A*, Vol. 385, p. 279, 2004.
2. Standard practice for proportioning grout mixtures for preplaced-aggregate concrete, ASTM Standard C938, ASTM International, West Conshohocken, PA, 2019. https://www.astm.org/Standards/C938.htm.
3. Standard test method for flow of grout for preplaced-aggregate concrete (flow cone method), ASTM Standard C939, ASTM International, West Conshohocken, PA, 2016. https://www.astm.org/Standards/C939.

4. W. Emad, A. Salih, and R. Kurda, *Case Studies in Construction Materials*, Vol. 15, p. e00600, 2021.
5. J. Wang, X. Zhu, and J. Chen, Composition for polyurethane foaming, polyurethane foam and use thereof, US Patent 10 450 403, assigned to Wanhua Chemical Ningbo Rongwei Polyurethane Co. Ltd., October 22, 2019.
6. A. Baklavaridis, I. Zuburtikudis, and C. Panayiotou, *Polymer Engineering & Science*, Vol. 55, p. 1856, 2015.
7. M. Tanoğlu and Y. Ergün, *Composites Part A: Applied Science and Manufacturing*, Vol. 38, p. 318, 2007.
8. K.A. Carrado, P. Thiyagarajan, and D.L. Elder, Porous networks derived from synthetic polymer-clay complexes, Technical report, Argonne National Laboratory, IL (United States), 1995.
9. K.A. Carrado, L. Xu, C.L. Marshall, D. Wei, S. Seifert, and C.A.A. Bloomquist, Mesoporous synthetic clays: Synthesis, characterization, and use as HDS catalyst supports in A. Sayari and M. Jaroniec, eds., *Nanoporous Materials II*, Vol. 129 of *Studies in Surface Science and Catalysis*, pp. 417–424. Elsevier, 2000.
10. T. Irrgang and R. Kempe, *Chem. Rev.*, Vol. 119, p. 2524, 2019.
11. S.L. Shevick, C.V. Wilson, S. Kotesova, D. Kim, P.L. Holland, and R.A. Shenvi, *Chem. Sci.*, Vol. 11, p. 12401, 2020.
12. Y. Cui, J. Wang, L. Yu, Y. Xu, D.J. Young, H. Li, and H. Li, *Polymers*, Vol. 14, p. 1, 2022.
13. Y. Wang, L. Yan, C. Li, M. Jiang, Z. Zhao, G. Hou, and Y. Ding, *Journal of Catalysis*, Vol. 368, p. 197, 2018.
14. Y. Zhang, Y. Lyu, Y. Wang, C. Li, M. Jiang, and Y. Ding, *Chinese Journal of Catalysis*, Vol. 40, p. 147, 2019.
15. B. Wang, J. Lin, C. Xia, and W. Sun, *Journal of Catalysis*, p. in press, 2022.
16. Q. Lu, T. Zhang, B. He, F. Xu, S. Liu, Q. Ye, and F. Zhou, *Tribology International*, Vol. 167, p. 107421, 2022.
17. H.S. Lee, D.-H. Paik, J.S. Shin, M.-S. Yang, S.-W. Choi, Y.J. Kim, and J.H. Lee, Composite powder in which porous polymer is impregnated with scoria powder particles, cosmetic composition containing same, and process for producing same, US Patent 10 472 485, assigned to Amorepacific Corp., November 12, 2019.
18. D. Kim and J. Seo, *Trends in Food Science & Technology*, Vol. 76, p. 15, 2018.
19. H. Kun, J. Li, K. Li, N. Yan, C. bian, Y. Guan, Y. Yang, and H. Li, *Polymer Degradation and Stability*, Vol. 196, p. 109816, 2022.
20. A.K. Boehm, S. Husmann, M. Besch, O. Janka, V. Presser, and M. Gallei, *ACS Applied Materials & Interfaces*, Vol. 13, p. 61166, 2021.
21. H. Zhao, X. Liu, Q. Zhang, H. Yin, Z. Zhan, Z. Chi, P. Yang, Z. Li, Z. Guo, and L. Wang, *Mater. Chem. Front.*, p. in press, 2022.

22. Y. Hu, T. Feng, L. Xu, L. Zhang, and L. Luo, *Batteries & Supercaps*, Vol. n/a, p. e202100303, 2022.
23. Z.H. Li, C. Cheng, X.Y. Zhan, Y.P. Wu, and X.D. Zhou, *Electrochimica Acta*, Vol. 54, p. 4403, 2009.
24. X. Liu, Y. Wu, F. Yang, S. Wang, B. Zhang, and L. Wang, *Journal of Membrane Science*, Vol. 626, p. 119190, 2021.
25. Z. Meng, G. Zhu, J. Wu, R. Wang, T. Tian, H. Tang, R. Luo, D. Ye, R. Zhang, F. Kwofie, Y. Cheng, and H. Tang, *Materials Today Energy*, p. 100935, 2022.
26. J. Liu, K.-K. Yee, K.K.-W. Lo, K.Y. Zhang, W.-P. To, C.-M. Che, and Z. Xu, *Journal of the American Chemical Society*, Vol. 136, p. 2818, 2014.
27. Wikipedia contributors, Calixarene — Wikipedia, the free encyclopedia, https://en.wikipedia.org/w/index.php?title=Calixarene&oldid=1040345531, 2021. [Online; accessed 30-September-2021].
28. Z. Wang, S. Yan, H.-C. Cui, G. Cheng, H. Ma, Q.-M. Zhang, Q.-P. Zhang, J.-M. Liu, B. Tan, and C. Zhang, *Macromolecules*, Vol. 51, p. 7863, 2018.
29. C. Zhang, Z. Wang, S. Song, X. Meng, Y.-S. Zheng, X.-L. Yang, and H.-B. Xu, *The Journal of Organic Chemistry*, Vol. 79, p. 2729, 2014.
30. P. Pallavi, S. Bandyopadhyay, J. Louis, A. Deshmukh, and A. Patra, *Chem. Commun.*, Vol. 53, p. 1257, 2017.
31. J.M. Lauerhaas, G.M. Credo, J.L. Heinrich, and M.J. Sailor, *Journal of the American Chemical Society*, Vol. 114, p. 1911, 1992.
32. V.V. Doan, R.M. Penner, and M.J. Sailor, *The Journal of Physical Chemistry*, Vol. 97, p. 4505, 1993.
33. D.F. Shriver and M.A. Drezdzon, *The Manipulation of Air-Sensitive Compounds*, A Wiley-Interscience publication, Wiley, 1986.
34. J.M. Lauerhaas, *The Electrochemical Society Interface*, Vol. 4, p. 59, March 1995.
35. M.R. Bohmer, J.J.F.G. Heuts, D. Estrada, and K. Shimizu, Light emitting semiconductor device having polymer-filled sub-micron pores in porous structure to tune light scattering, US Patent 10761246, assigned to Lumileds LLC, September 01, 2020.
36. A. Peyravi, Z. Hashisho, D. Crompton, and J.E. Anderson, *Journal of Colloid and Interface Science*, Vol. 612, p. 181, 2022.

Index

Acronyms

Chemicals

Boldface numbers refer to Figures

General Index

Also of Interest

Check out these other books by the author published by Scrivener Publishing

Porous Plastics
By Johannes Karl Fink
Published 2022. ISBN 978-1-119-89638-8

Contact Lenses
Materials, Chemicals, Methods and Applications
By Johannes Karl Fink
Published 2022. ISBN 978-1-119-85735-8

Polymers and Additives in Extreme Environments
Application, Properties, and Fabrication
By Johannes Karl Fink
Published 2021. ISBN 978-1-119-85137-0

Plastics Process Analysis, Instrumentation, and Control
By Johannes Karl Fink
Published 2021. ISBN 978-1-119-79573-5

Flame Retardants
By Johannes Karl Fink
Published 2020. ISBN 978-1-119-75061-1

The Chemistry of Environmental Engineering
By Johannes Karl Fink
Published 2020. ISBN 978-1-119-70774-5

The Chemistry of Bio-based Polymers, second edition
By Johannes Karl Fink
Published 2019. ISBN 978-1-119-68131-1

Liquid Silicone Rubber
By Johannes Karl Fink
Published 2019. ISBN 978-1-119-63133-0

Functional Synthetic Polymers
By Johannes Karl Fink
Published 2019. ISBN 978-1-119-59202-0

3D Industrial Printing with Polymers
By Johannes Karl Fink
Published 2018. ISBN 978-1-119-55526-1

Polymer Waste Management
By Johannes Karl Fink
Published 2018. ISBN 978-1-119-53608-6

Materials, Chemicals and Methods for Dental Applications
By Johannes Karl Fink
Published 2018. ISBN 978-1-119-51031-4

Fuel Cells, Solar Panels, and Storage Devices
Materials and Methods
By Johannes Karl Fink
Published 2018. ISBN 978-1-119-48010-5

Chemicals and Methods for Conservation and Restoration
Paintings, Textiles, Fossils, Wood, Stones, Metals, and Glass
By Johannes Karl Fink
Published 2017. ISBN 978-1-119-41824-5

Additives for High Performance Applications
Chemistry and Applications
By Johannes Karl Fink
Published 2017. ISBN 978-1-119-36361-3

Metallized and Magnetic Polymers
By Johannes Karl Fink
Published 2016. ISBN: 978-1-119-24232-1

Marine, Waterborne, and Water-Resistant Polymers
Chemistry and Applications
By Johannes Karl Fink
Published 2015. ISBN 978-1-119-018486-7

The Chemistry of Printing Inks and Their Electronics and Medical Applications
By Johannes Karl Fink
Published 2015. ISBN 978-1-119-04130-6

The Chemistry of Bio-based Polymers
By Johannes Karl Fink
Published 2014. ISBN 978-1-118-83725-2

Polymeric Sensors and Actuators
By Johannes Karl Fink
Published 2012. ISBN 978-1-118-41408-8

Handbook of Engineering and Specialty Thermoplastics
Part 1: Polyolefins and Styrenics
By Johannes Karl Fink
Published 2010. ISBN 978-0-470-62483-5

Handbook of Engineering and Specialty Thermoplastics
Part 2: Water Soluble Polymers
By Johannes Karl Fink
Published 2011. ISBN 978-1-118-06275-3

A Concise Introduction to Additives for Thermoplastic Polymers
by Johannes Karl Fink.
Published 2010. ISBN 978-0-470-60955-2

Printed and bound by CPI Group (UK) Ltd, Croydon, CR0 4YY
17/09/2026
14964275-0005